CO_2 STORAGE IN CARBONIFEROUS FORMATIONS AND ABANDONED COAL MINES

PROCEEDINGS OF THE INTERNATIONAL WORKSHOP ON CO_2 STORAGE IN CARBONIFEROUS FORMATIONS AND ABANDONED COAL MINES, BEIJING, CHINA, 8–9 JANUARY 2011

CO_2 Storage in Carboniferous Formations and Abandoned Coal Mines

Editors

Manchao He

State Key Laboratory for GeoMechanics and Deep Underground Engineering
China University of Mining and Technology, Beijing, China

Luis Ribeiro e Sousa

State Key Laboratory for GeoMechanics and Deep Underground Engineering, China
Department of Civil Engineering, University of Porto, Portugal

Derek Elsworth

EMS Energy Institute and Energy and Mineral Engineering
Pennsylvania State University, USA

Eurípedes Vargas Jr.

Civil Engineering Department, Catholic University, Rio de Janeiro, Brazil

CRC Press
Taylor & Francis Group
Boca Raton London New York Leiden

CRC Press is an imprint of the
Taylor & Francis Group, an **informa** business
A BALKEMA BOOK

Published by:
CRC Press/Balkema
P.O. Box 447, 2300 AK Leiden, The Netherlands
e-mail: Pub.NL@taylorandfrancis.com
www.crcpress.com – www.taylorandfrancis.com

Typeset by Vikatan Publishing Solutions (P) Ltd., Chennai, India

ISBN 13: 978-0-415-62079-6 (hbk)
ISBN 13: 978-0-203-13422-1 (ebk)

Visit the Taylor & Francis Web site at
http://www.taylorandfrancis.com

and the CRC Press Web site at
http://www.crcpress.com

CO_2 Storage in Carboniferous Formations and Abandoned Coal Mines –
He, Ribeiro E Sousa, Elsworth & Vargas Jr. (eds)
 ISBN 978-0-415-62079-6

Table of contents

CO_2 Storage in Carboniferous Formations and Abandoned Coal Mines –
He, Ribeiro E Sousa, Elsworth & Vargas Jr. (eds)
 ISBN 978-0-415-62079-6

Preface

Underground geological storage of carbon dioxide (CO_2) has a considerable potential for mitigating climate change. CO_2 can be safely injected and stored at well characterized and properly managed sites. Injecting carbon dioxide in deep geological formations can store it underground for long periods of time. Depleted oil and gas reservoirs, saline aquifers and carboniferous formations can all be used for storage of CO_2, as well as in abandoned coal mines. At depths below about 800-1000m, CO_2 has a liquid-like density that permits the potential for an efficient use of the underground reservoirs in porous and sedimentary rocks.

An International Workshop was held by the State Key Laboratory for GeoMechanics and Deep Underground Engineering of the China University of Mining and Technology on *CO_2 Storage in Carboniferous Formations and Abandoned Coal Mines*. The purpose of this workshop was to promote research activities in China and to strengthen international cooperation among scientists and engineers from different countries.

The Workshop provided an excellent opportunity to hold high level discussions and to define novel approaches for the solution of the problems involved with the injection and storage in coal formations and in particular within abandoned coal mines, especially in China. During the event several topics were addressed, namely the present situation of Carbon Capture and Storage (CCS) projects in the world and particularly in China, the complexity of the coupling between physical processes in coal seams, considerations regarding numerical simulation of injection and storage in coal formations, considerations related to CO_2 storage in abandoned coal mines and finally methodologies for risk assessment of CO_2 storage in carboniferous formations.

Chinese and foreigner researchers from Australia, Belgium, Brazil, Portugal and the USA, participated actively in the event. The book contains ten Chapters summarizing research activities sourced in five countries. The contents will be of use to researchers and engineers involved in CCS.

Manchao He
L. Ribeiro e Sousa
Derek Elsworth
Eurípedes Vargas Jr.

CO_2 *Storage in Carboniferous Formations and Abandoned Coal Mines –*
He, Ribeiro E Sousa, Elsworth & Vargas Jr. (eds)
© 2012 Taylor & Francis Group, London, ISBN 978-0-415-62079-6

Organization

ORGANIZERS

State Key Laboratory for GeoMechanics and Deep Underground

Engineering of China University of Mining & Technology (Beijing)

Soft Rock Engineering & Dee Disaster Control Sub-society of Chinese Society for Rock Mechanics and Engineering

ORGANIZING COMMITTEE

Chair

Lin Xueyu (China)

Co-Chairs

Luis Sousa (Portugal)
He Manchao (China)

Members

Euripedes Vargas Jr. (Brazil)
Derek Elsworth (USA)
Liu Jishan (Australia)
Wang Jianguo (Australia)
Kris Piessens (Belgium)
Rita Sousa (USA)
Zhang Na (China)

 ISBN 978-0-415-62079-6

Sponsors

Chinese Foreign Experts Bureau

National Natural Science Foundation of China

Chinese Society for Rock Mechanics and Engineering

CO_2 Storage in Carboniferous Formations and Abandoned Coal Mines –
He, Ribeiro E Sousa, Elsworth & Vargas Jr. (eds)
 ISBN 978-0-415-62079-6

Contributers

André Muller, PhD, Catholic University of Rio de Janeiro, Brazil
Dr. Muller is a research engineer at TECGRAF, a software development division of Catholic University in Rio de Janeiro. He is currently involved in the development of simulation software for various divisions of Petrobras including simulation of rock mechanics problems both in well and reservoir scale.

Derek Elsworth, EMS Energy Institute and G3 Center, The Pennsylvania State University, USA
Prof. Elsworth is currently professor at Penn State University and he is now one of the outstanding leaders for the underground laboratory DUSEL, South Dakota, USA. Prof. Elsworth' research interest includes physicochemical properties of geomaterials, poromechanics, continuum and particulate modeling of fluid infiltrated media.

Eurípedes Vargas Jr., Professor, Catholic University of Rio de Janeiro, Brazil
Prof. Vargas is currently professor at the Catholic University of Rio de Janeiro, Brazil. Also he works for the Rock Mechanics Laboratory from PETROBRAS, and is participating in a project on Risk Analysis and Assessment of CO_2 Injection Processes in Carboniferous Formations of the State Key Laboratory for GeoMechanics and Deep Underground Engineering, Beijing.

Ghazal Izadi, EMS Energy Institute and G3 Center, The Pennsylvania State University, USA
Ghazal Izadi is currently a Ph.D. candidate at the Penn State University. Specified research interests are: Thermal, hydrologic, mechanical, and chemical process coupling in numerical representation of stimulated natural fractures and porous media, with application to geothermal and petroleum reservoirs.

Hongyan Qu, PhD Candidate, the University of Western Australia, Australia
Hongyan Qu is currently a PhD student at the University of Western Australia. She received B.S. and M.S. in Applied Mathematics from China University of Petroleum. Her research interests include CO_2-enhanced coalbed methane recovery, and CO_2 storage in coal seams. The main area she works on is the thermal effect on CO_2 injection into coal seams.

He Manchao, Professor, China University of Mining & Technology (Beijing), China
Prof. He is an expert in emergency prevention and control of geological disasters of Ministry of Land and Resources. Prof. He has made exceptional technical and scientific contributions on a wide variety of rock mechanics, rock engineering and engineering hazard control in deep mining, related to soft rock characterization, nonlinear mechanics design methods of caverns and tunnels, underground excavation, rock support, rockburst mechanism and control, in deep mines, slope stability, remote monitoring technique for geology hazards, etc. Prof. He published 7 books and more than 140 papers. He got three second Prizes of National Invention Award and twelve National Invent patents. At present, he is Dean of School of Mechanics and Civil Engineering, China University of Mining & Technology and he is the Director of the State Key Laboratory for GeoMechanics and Deep Underground Engineering.

Hemant Kumar, EMS Energy Institute and G3 Center, The Pennsylvania State University, USA
Hemant Kumar is a PhD student in the department of Energy and Mineral Engineering at Penn State University. His research interests include CO_2 enhanced gas recovery from coalbed and shale gas reservoirs.

Kris Piessens, PhD, Geological Survey of Belgium, Belgium
Dr. Piessens is working at the Geological Survey of Belgium, Department of the Museum for Natural Sciences and is specialized in geoenergy with focus on carbon capture and storage. He is coordinator of the national project Policy Support System for Carbon Capture and Storage, and has developed the PSS simulator that combines techno-economic and GIS simulation techniques, as well as prospection forecasts. He has authored several publications on the concept and opportunities of CO_2 storage in coal mines, as well as on other aspects of carbon capture and storage.

Liu Jishan, Professor, University of Western Australia
Prof. Jishan is currently a professor and program leader of petroleum engineering at the University of Western Australia. His current research interests include CO_2-enhanced oil and gas recovery, and CO_2 storage in geological formations.

Dae-Sung Lee, Korea Institute for Geosciences and Mineral Resources, Daejeon, South Korea
Dr. Lee is a senior researcher with KIGAM with interests in coalbed methane, tight gas shales and reservoir geomechanics related to these resources.

Luke Connell, Principal Scientist, CSIRO
Luke Connell leads the Carbon Dioxide Storage and Transport area within CSIRO. He has extensive experience in the management of coal seam gas for coal mining and for gas production. He works on a range of reservoir engineering topics including characterization, gas drainage optimization and carbon dioxide storage.

Luiz Eloy Vaz, Professor, Fluminense Federal University, Niteroi, Brazil
Prof. Vaz is currently a full professor at the Civil Engineering Department at Fluminense Federal University in Niteroi, Brazil and he currently participates as researcher in research projects with Petrobras. Prof. Vaz is a specialist in numerical analysis, finite elements in particular and has advised a considerable number of both MSc and PhD thesis in this subject.

Luís Ribeiro e Sousa, Professor, University of Porto, Portugal
Prof. Sousa has an extensive international experience on a range of projects. He is an expert in the domain of dams, foundations, slopes, numerical modeling, Rock Mechanics tests, underground construction, mining, Risk Analysis and Assessment, and monitoring. Prof. Sousa is multilingual and has authored or co-authored over 20 books and hundreds of journal articles, presentations and technical reports. He is very active in a number of professional societies and has served as President of the Portuguese Geotechnical Society and as Vice-President for the International Society for Rock Mechanics (ISRM). He was also Chairman of the 2007 ISRM Congress and he is now President of SKEC Engineering Consultants and Consultant of the State Key Laboratory for GeoMechanics and Deep Underground Engineering, Beijing.

Jonathan P. Mathews, Assistant Professor, The Pennsylvania State University, USA
Dr. Mathews is an assistant professor in the John and Willie Leone Family Department of Energy and Mineral Engineering. He is a coal scientist with interests in both experimental and molecular modeling approaches for coal and coal-related areas, including the CO_2 sequestration within coal.

Denis Pone, ConocoPhillips, Bartlesville, Oklahoma, USA
Dr. Pone is currently Sr. Reservoir Engineer within ConocoPhillips Upstream Technology. His research interests include traditional and novel enhanced oil recovery methods applied to both conventional and unconventional resources.

Raquel Velloso, PhD, Catholic University of Rio de Janeiro, Brazil
Dr. Velloso is currently a research engineer, participating in several research projects with PETROBRAS related to simulation of coupled fluid mechanical problems at the micro

(pore) scale). She recently finished her PhD program with a thesis related to development and implementation of discrete element coupled with lattice-Boltzmann procedures for simulation of the problems mentioned above.

Rita Leal e Sousa, PhD, Massachusetts Institute of Technology, USA
Dr. Sousa is a researcher at Massachusetts Institute of Technology and PhD at the same Institute and she got the MSc degree at Ecole Central de Paris, France. She worked at several companies as a consulting engineer, namely at Ferconsult (Portugal), Consulmar (Portugal), Fressinet (France), Sarooj (Oman) and SKEC (Portugal). Dr. Sousa's major interest is related to Risk Analysis and Risk Management. She is also participating in a project on Risk Analysis and Assessment of CO_2 Injection Processes in Carboniferous Formations of the State Key Laboratory for GeoMechanics and Deep Underground Engineering, Beijing.

Wang J.G., Assistant Professor, The University of Western Australia, Australia
Dr. Wang (Jianguo) is currently an assistant professor in the School of Mechanical and Chemical Engineering, The University of Western Australia. His contributions include RBF-based meshless method, homogenization method of geomaterials, and constitutive laws of geomaterials. His current research interest includes enhancement of methane recovery in tight sandstone reservoirs and coalbed seam, CO_2 geological sequestration, poromechanics, and protective technology of civil structures.

Shugang Wang, EMS Energy Institute and G3 Center, The Pennsylvania State University, USA
Shugang Wang is a Ph.D. Candidate at the Penn State University, USA. His research interests are flow and mechanical characteristics of coal, with applications to instability of coal, carbon sequestration, and coalbed methane.

W.R. Ribeiro, PhD, Catholic University of Rio de Janeiro, Brazil
Dr. Ribeiro is currently a research engineer, working in research projects connected to simulation of geological processes. He recently completed his PhD program with a thesis related to the simulation of coupled fluid-mechanical problems in rock masses incorporating two phase flow problems.

Zhang Na, Lecturer, China University of Mining & Technology (Beijing), China
Dr. Zhang is a staff in the State Key Laboratory for GeoMechanics and Deep Underground Engineering, Beijing (SKLGDUE) of China University of Mining and Technology (Beijing). She was graduated from Pekin University and majored in Environmental Geochemistry. Based on her research experience, Dr. Zhang is now working in a program of Water Adsorption on Soft Rock in SKLGDUE. She is also participating in a project on Risk Analysis and Assessment of CO_2 Injection Processes in Carboniferous Formations. Up to now, she has published and submitted a number of papers in international peer-reviewed scientific journals.

Zhejun Pan, Senior Research Scientist, CSIRO Earth Science and Resource Engineering, Australia
Dr. Pan is a senior research scientist with CSIRO Earth Science and Resource Engineering. His main research interest is primary and enhanced coalbed methane recovery and CO_2 storage in coal. He is practically interested in the mechanisms of gas storage in coal and sorption-induced coal swelling/shrinkage. He holds a Ph.D in Chemical Engineering from Oklahoma State University, USA. He received B.S. and M.S. from Zhejiang University, China, both in Chemical Engineering.

Zhongwei Chen, PhD candidate, The University of Western Australia
Zhongwei Chen is a PhD candidate at the University of Western Australia. His current study is focused on the influence of coal-gas interaction on CBM production as well as CO_2 injection enhanced CBM recovery (CO_2-ECBM). These include the development of coal permeability model and the prediction of the wellbore stability during gas production.

CO_2 Storage in Carboniferous Formations and Abandoned Coal Mines –
He, Ribeiro E Sousa, Elsworth & Vargas Jr. (eds)

Present day conditions in the world of Carbon Capture and Storage (CCS) projects

L. Ribeiro e Sousa
State Key Laboratory for GeoMechanics and Deep Underground Engineering, Beijing, China
University of Porto, Portugal

ABSTRACT: The special report on Carbon dioxide Capture and Storage (CCS) under the auspices of the United Nations—Framework Convention on Climate Changes constitutes a milestone in the world, which needs to urgently develop strategies that take into account the control and reduce emissions of greenhouse gases (GHG). Among the various GHG, carbon dioxide (CO_2) is the most abundant component, whose underground storage has been proven in low risk. This publication describe the present day conditions of CCS projects with particular emphasis in the description of the processes associate to the capture and transportation of CO_2; the different types of solutions of underground storage of carbon; and the existing projects. Finally the storage in carboniferous formations is examined separately, since China has the major resources in coal in the world, as well as a large number of abandoned coal mines.

1 INTRODUCTION

There are several ways of mitigating GHG emissions to the atmosphere, however, the storage of large quantities of carbon in geological formations is one of the most effective and visible results today. Carbon dioxide Capture and Storage (CCS) is a process consisting in the separation of CO_2 from industrial and energy-related sources. Figure 1 gives a schematic diagram of possible CCS systems showing the sources, for which CCS might be relevant, and the transport and storage of CO_2 options.

The CCS systems are considered as one of the options for reducing atmospheric emissions of CO_2 from human activities and consequently permit the stabilization of greenhouse gas concentrations in the atmosphere at a level that prevents dangerous interference with climate system. Other technological options include:

- reducing energy demand
- decarbonizing energy supplies
- increasing of renewable energies and (or) use nuclear energy
- sequestering CO_2 by biological fixation reducing non-CO_2 greenhouse gases

Figure 2 illustrates the options for storing CO_2 in deep underground geological formations. Geological storage of CO_2 can be undertaken in a variety of possibilities in sedimentary basins. Sedimentary basins occur throughout the world, both onshore and on the continental shelves (Figure 3). Other geological options which may serve as storage sites include caverns in basalt, organic-rich shales and in rock salt.

In 2007 China became the largest emitter of CO_2 (He, 2011). In 2006 reached a peak of 1.6 GtC/year (Zeng et al., 2008). In spite of this, China releases much less GHG per capita than other countries (Figure 4).

Currently CO_2 capture is a costly and energy consuming. The costs obviously depend on the dimension of the industrial unit and of the type of fuel used.

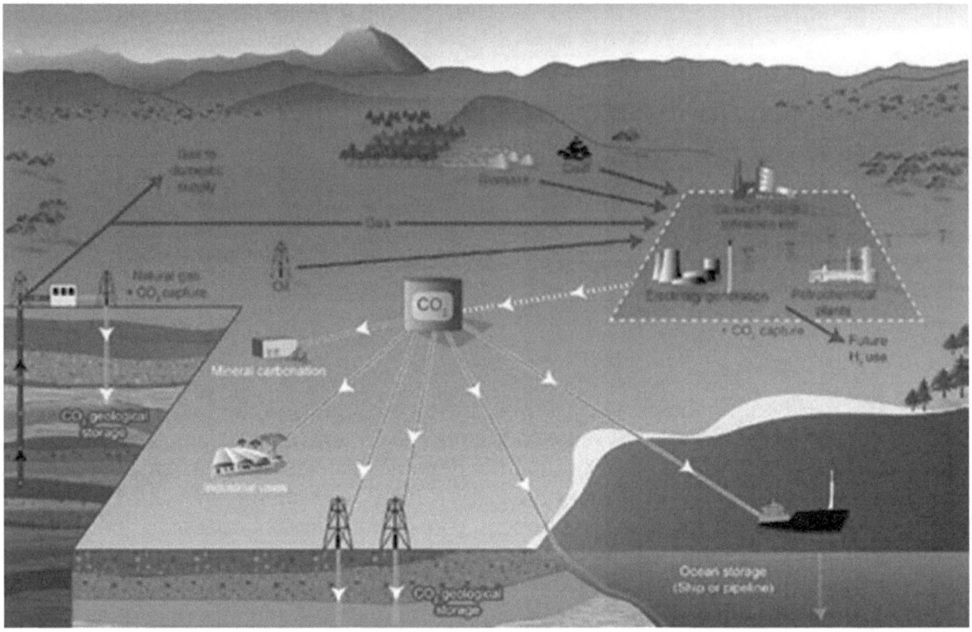

Figure 1. Schematic diagram of possible CCS systems (CO_2CRC).

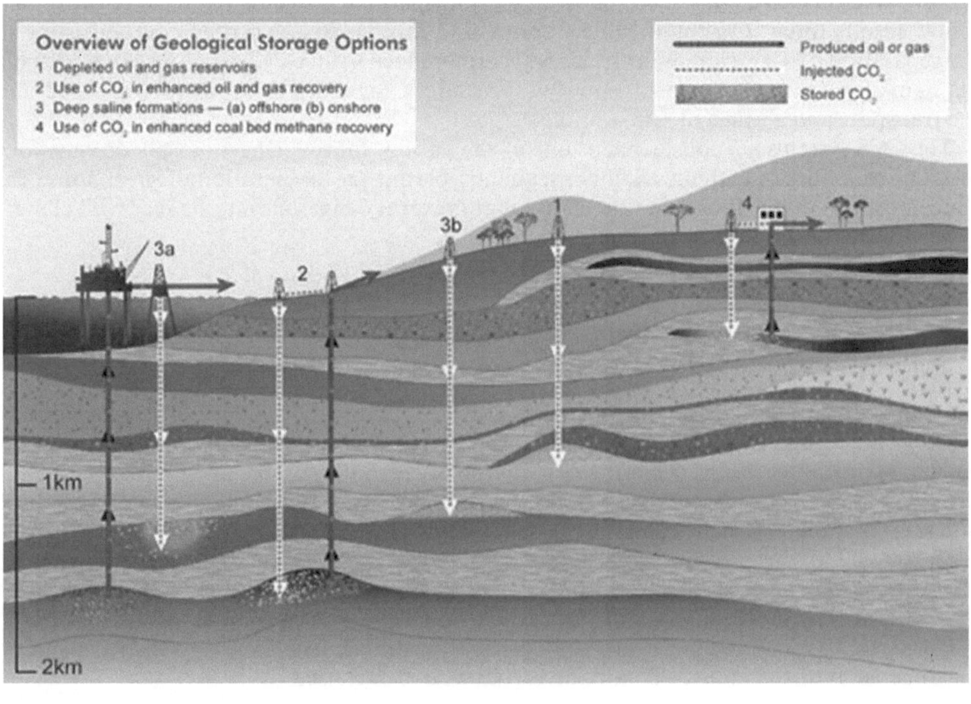

Figure 2. Geological storage options for CO_2 (IPCC, 2005).

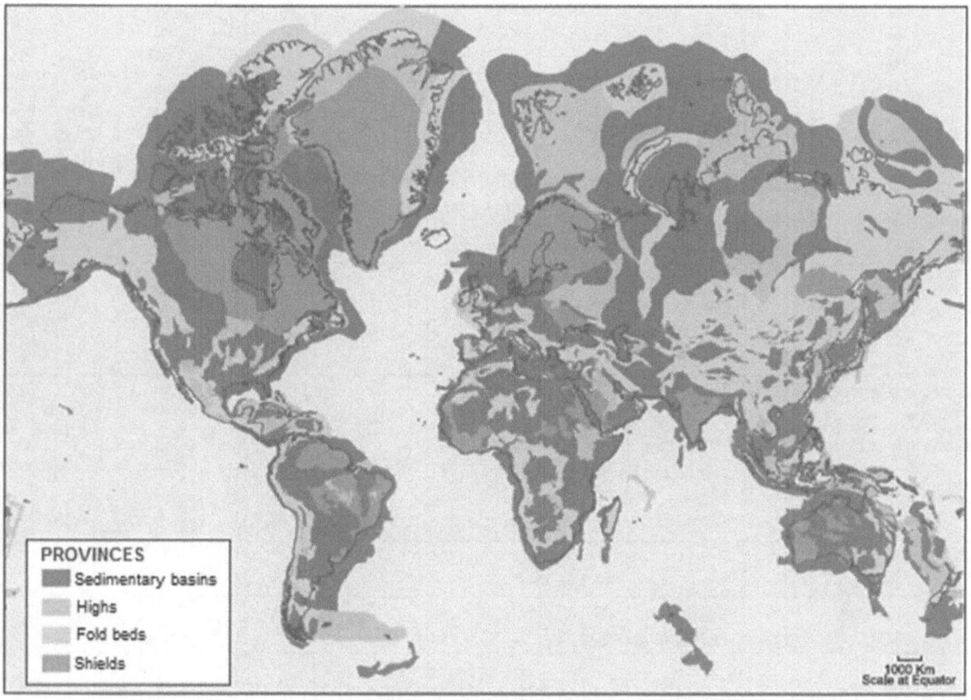

Figure 3. Distribution of sedimentary basins around the world (IPCC, 2005).

Figure 4. Emissions and carbon intensity in China between 1980 and 2006 (Adapted from Zeng et al., 2008).

Large point sources of CO_2 include large fossil fuel or biomass energy facilities, major CO_2-emitting industries, natural gas production, synthetic fuel plants and fossil fuel-based hydrogen production plants.

Potential technical storage methods are: geological storage (in geological formations, such as oil and gas fields, unminable coal beds and deep saline formations), ocean storage (direct release into the ocean water column or onto the deep seafloor) and industrial fixation of CO_2 into inorganic carbonates.

Components of CCS are in various stages of development (see Figure 5). In the Figure, X indicates the highest level of maturity for each component. Complete CCS systems can be

CCS component	CCS technology	Research phase	Demonstration phase	Economically feasible under specific	Mature market
Capture	Post-combustion			X	
	Pre-combustion			X	
	Oxyfuel combustion		X		
	Industrial separation				X
Transportation	Pipeline				X
	Shipping			X	
Geological storage	Enhanced Oil Recovery (EOR)				X
	Gas or oil fields			X	
	Saline formations			X	
	Enhanced Coal Bed Methane recovery (ECBM)		X		
Ocean storage	Direct injection (dissolution type)	X			
	Direct injection (lake type)	X			
Mineral carbonation	Natural silicate minerals	X			
	Waste materials		X		

Figure 5. CCS system components (Adapted from IPCC, 2005).

Reservoir type	Estimate storage capacity ($GtCO_2$)	
	Lower estimate	Upper estimate
Oil and gas field	675	900
Unminable coal seams (ECBM)	3-15	200
Deep saline formations	1000	Uncertain, but possible 10^4

Figure 6. Estimate storage capacity for several geological storage options (IPPC, 2005).

assembled from existing technologies that are mature or economically feasible under specific conditions, although the state of development of the overall system may be less than some of its separate components.

There is relatively little experience in combining CO_2 capture, transport and storage into a fully integrated CCS system. The utilization of CCS for large-scale power plants (the potential application of major interest) still remains to be implemented. The different estimate solutions, as indicated in Figure 6, shown that the estimates of storage in unminable coal seams are more reduced, but the possibility of using abandoned coal mines could be an attractive possibility in countries with a considerable number of mines in this situation, like in China (He, 2011).

Ocean storage can be also considered by two ways (Figure 7): by injecting and dissolving CO_2 into the water column (typically below 1,000 meters) via a fixed pipeline or a moving ship; or by depositing it via a fixed pipeline or an offshore platform onto the sea floor at depths below 3,000 m, where CO_2 is denser than water and is expected to form a "lake" that would delay dissolution of CO_2 into the surrounding environment (IPPC, 2005).

Ocean storage and its ecological impacts are still in the research phase. The dissolved and dispersed CO_2 would become part of the global carbon cycle and eventually equilibrate with

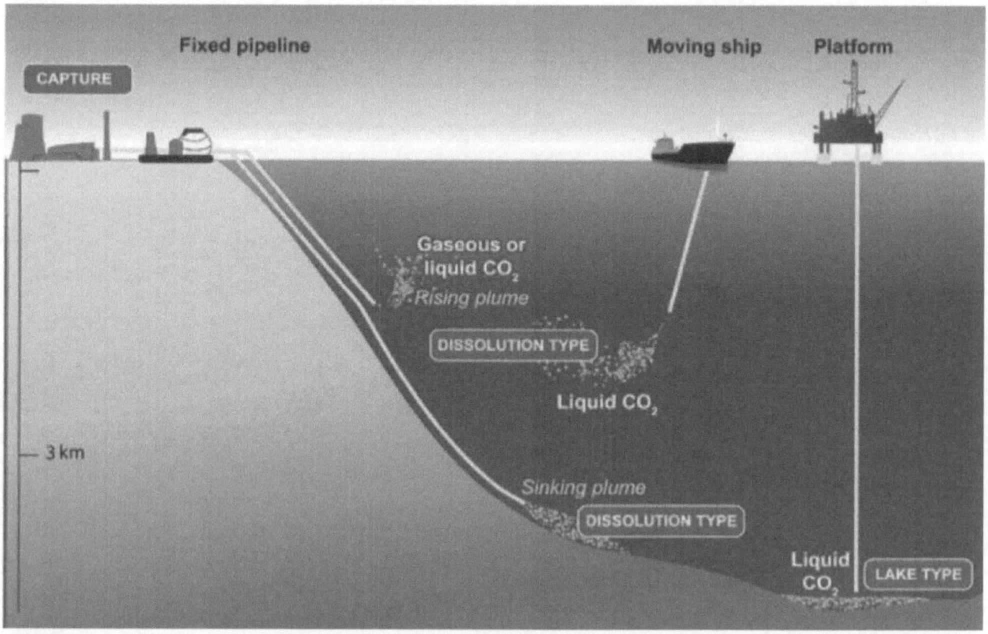

Figure 7. Overview of ocean storage concepts (CO_2CRC).

the CO_2 in the atmosphere. In laboratory experiments, small-scale ocean experiments and model simulations, the technologies and associated physical and chemical phenomena, which include, notably, increases in acidity (lower *pH*) and their effect on marine ecosystems, have been studied for a range of ocean storage options.

2 PROCESSES ASSOCIATED TO THE CAPTURE AND TRANSPORTATION

The purpose of capture of CO_2 is to produce a concentration stream and transport of CO_2 to a storage site. Today, the capture of CO_2 is only technically possible and economically profitable for emission sources of large industries, such as power plants and other industrial plants (Figure 8).

Currently, the CO_2 capture technology is very expensive and energy consuming. Generally speaking, capture systems are divided into three categories: capture the exhaust gases (post-combustion), upstream of the combustion capture (pre-combustion) and combustion with high oxygen content (combustion oxyfuel). In Figure 9, are schematically described the three types of capture, as well as the typology of a traditional fossil fuel power plant. Diagrams of fossil-fuel-based power generation and different capture systems are explicitly represented in Figure 10 (IPCC, 2005).

In the combustion gases from industries and power plants there are in most cases, other gases, such as oxygen, sulfur oxides, nitrogen and water vapor. For economic reasons and risk (hazard levels and different interactions among the various gases), it is impossible to compress and store all the exhaust gases from chimneys. CO_2 is a waste produced in large quantities, which has significant implications for the environment with a capture process more feasible (Gomes, 2010).

Some industrial solutions using different processes of capture are presented at Figures 11 and 12 (IPCC, 2005).

CO_2 is transported as a gas, liquid or solid and commercial transport uses tanks, pipelines and shipping. Gas occupies less if compressed and is normally transported by pipelines

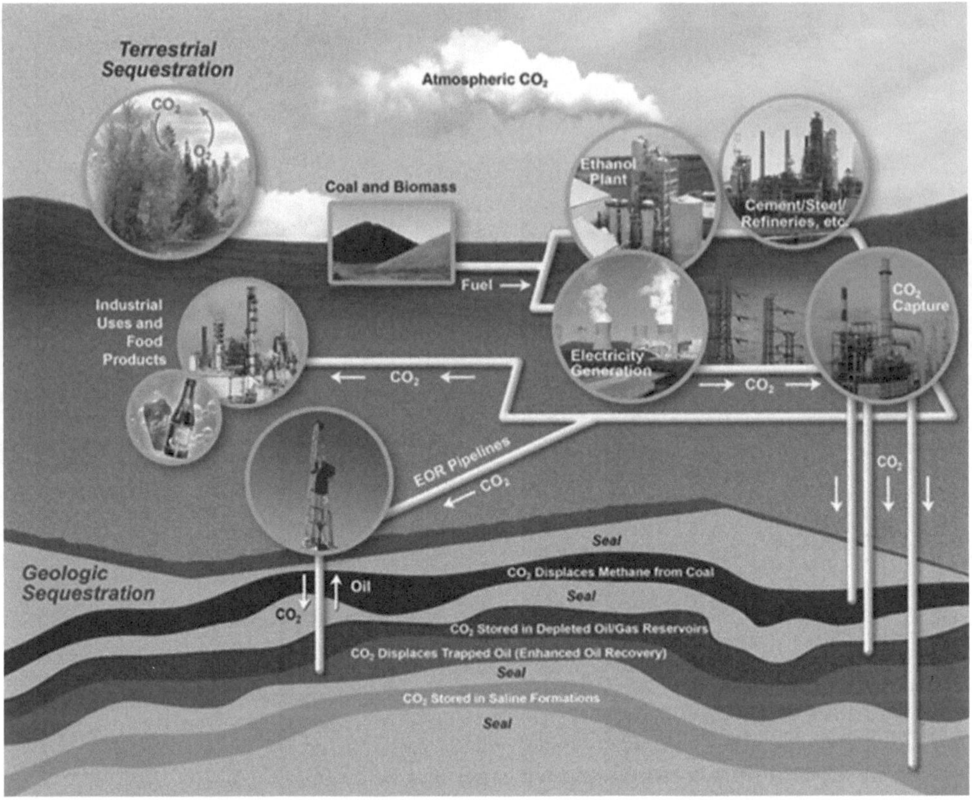

Figure 8. Processes of capturing and storing CO_2 (NETL, 2007).

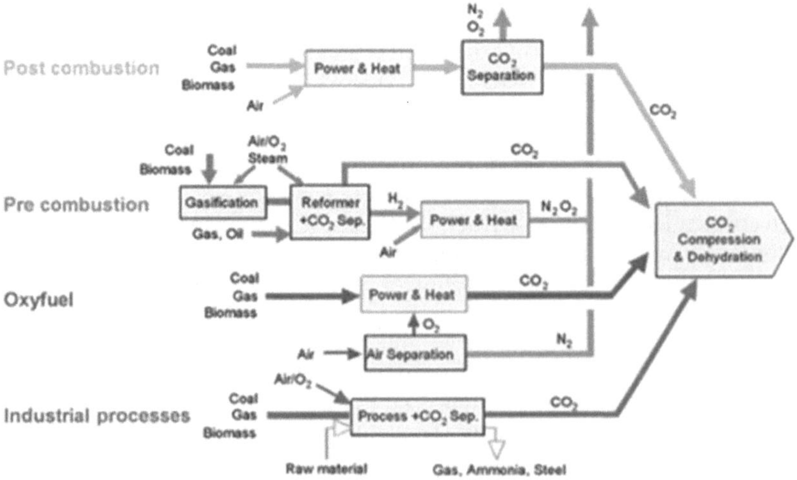

Figure 9. Overview of CO_2 capture processes and systems (IPCC, 2005).

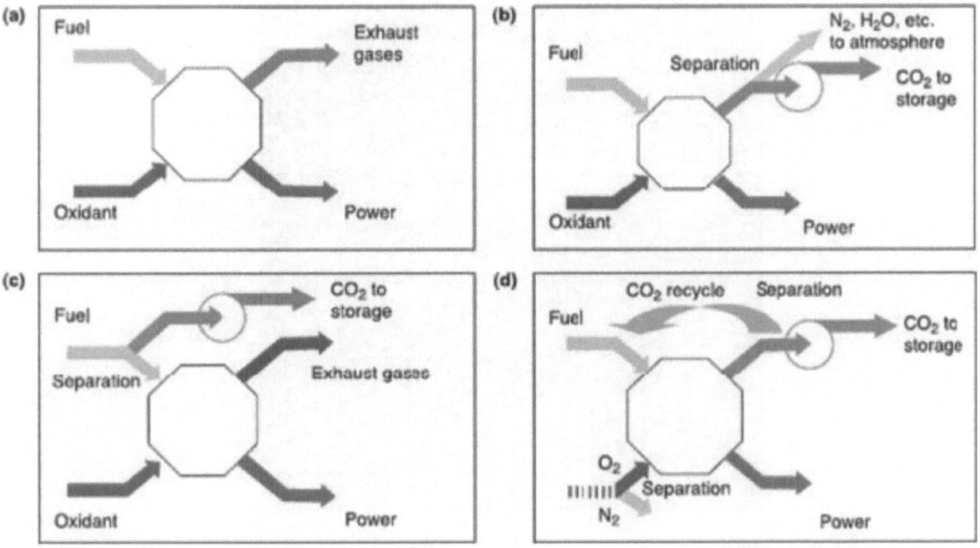

Figure 10. a) Schematic diagram of fossil-fuel-based power generation; b) schematic diagram of post-combustion capture; c) schematic diagram of pre-combustion capture; d) schematic diagram of oxyfuel combustion (IPCC, 2005).

Figure 11. CO_2 post-combustion capture at a plant in Malaysia (IPPC, 2005).

Figure 12. CO_2 precombustion capture at a coal gasification plant in North Dakota (IPCC, 2011).

Figure 13. Transportation by pipelines (Gomes, 2010).

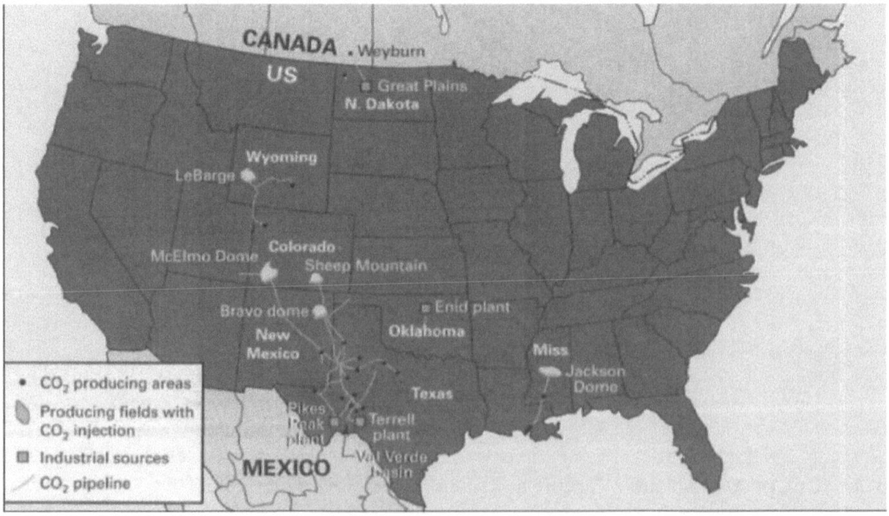

Figure 14. CO_2 pipelines in North America (IPCC, 2005).

Figure 15. Transportation by shipping (Adapted from Gomes, 2010).

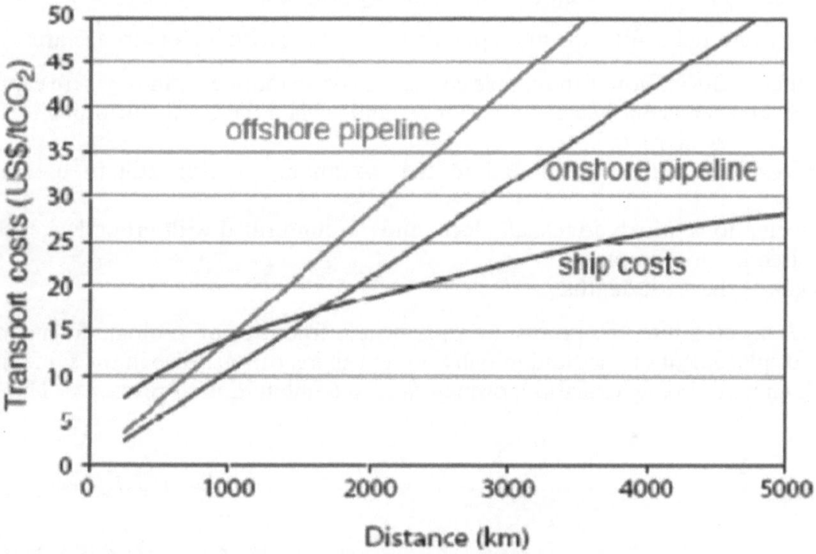

Figure 16. Transport costs associated with pipelines or by shipping (source from Gomes, 2010).

(Figure 13). CO_2 pipelines provide direct route to harmful human exposure or to harmful impacts on animals and plants by producing a local high concentration of CO_2 and generating exposures sufficient to harm or kill people, plants and animals (IPPC, 2005). While an important risk precautions can be taken to minimize the likelihood of a major pipeline rupture.

In the USA existed several long-distance CO_2 pipelines as illustrated in Figure 14, with special emphasis to the Cortez pipeline with 808 km, to the Sheep Mountain pipeline (660 km) and to the Weyburn pipeline with 330 km.

Sometimes shipping can be economically attractive, particularly if the transportation for long distance (Figure 15). In these circumstances the transport of CO_2 is done similarly to other products, such as propane or butane.

Figure 16 compares the costs associated with transport by pipeline and by shipping. It is considered that mass flow of CO_2 is constant at 6 $MtCO_2$/ano, and that the costs of transport by shipping include temporary storage, taxes, fuel costs, and activities of loading and unloading. The curves for the pipelines are actually straight, given the constant slope corresponding to a constant cost, per unit length. It was verified that transport by shipping is an option, and it is more viable for distances over 1,000 km.

3 STORAGE OF CO_2 IN UNDERGROUND

The carbon dioxide captured in order to reduce GHG emissions to the atmosphere, must be stored in proper sites. There are several reservoirs in the nature where the CO_2 is trapped particularly in the pores of sedimentary rocks. Thus geological underground storage can be considered a good option for storage of CO_2.

It is necessary to take into account several aspects in underground storage of CO_2 (Gomes, 2010; Sousa and Sousa, 2011):

- The storage period should be long, preferably hundreds or thousands of years;
- The storage cost, including transportation from the place of capture, injection and monitoring, should be minimized;
- The methodologies used cannot violate any laws or regulations, at national or international levels;
- Environmental impacts should be studied carefully, and should be evaluated ways to make them as minimal as possible;
- The likelihood of accidents should be practically nonexistent.

As already referred before, the main possible types of reservoirs for storage are:

- Carboniferous formations (unminable coal seams or abandoned mines) where exploitation of the coal has no economic benefits (for example, when the depths to which it is located the coal are very high).
- Oil or gas fields already exploited to the maximum, or even still in use in special situations.
- Saline formations, which are usually local underground filled with groundwater that cannot be used for consumption.
- The ocean at the great depths.

The CO_2 injected into the porous and permeable fractures of geologic formations may cause the displacement of the fluid initially present there, can dissolve in the same fluid, can react with minerals in the formation, or may cause a combination of processes. The study of

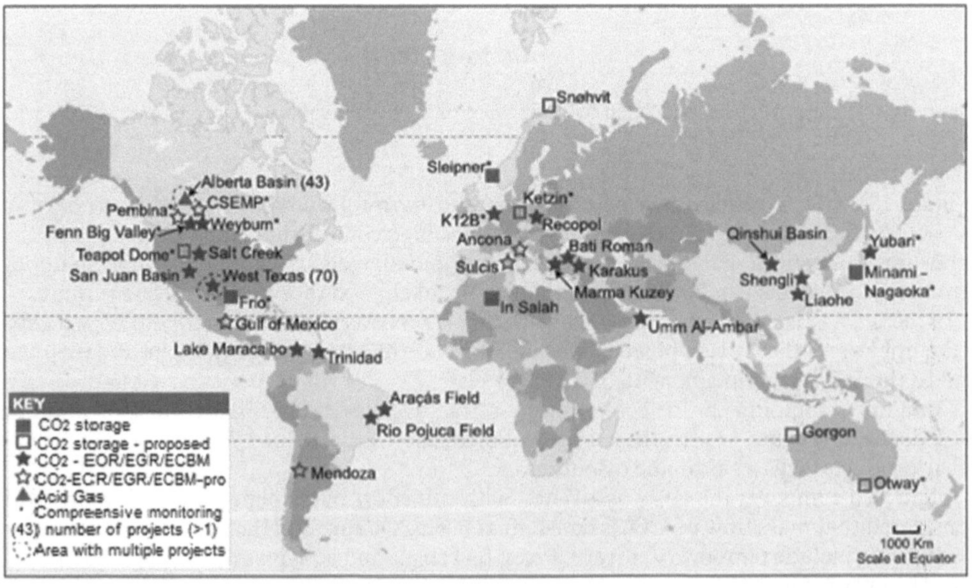

Figure 17. Location of sites where activities relevant to CO_2 storage are planned or under way (IPPC, 2005).

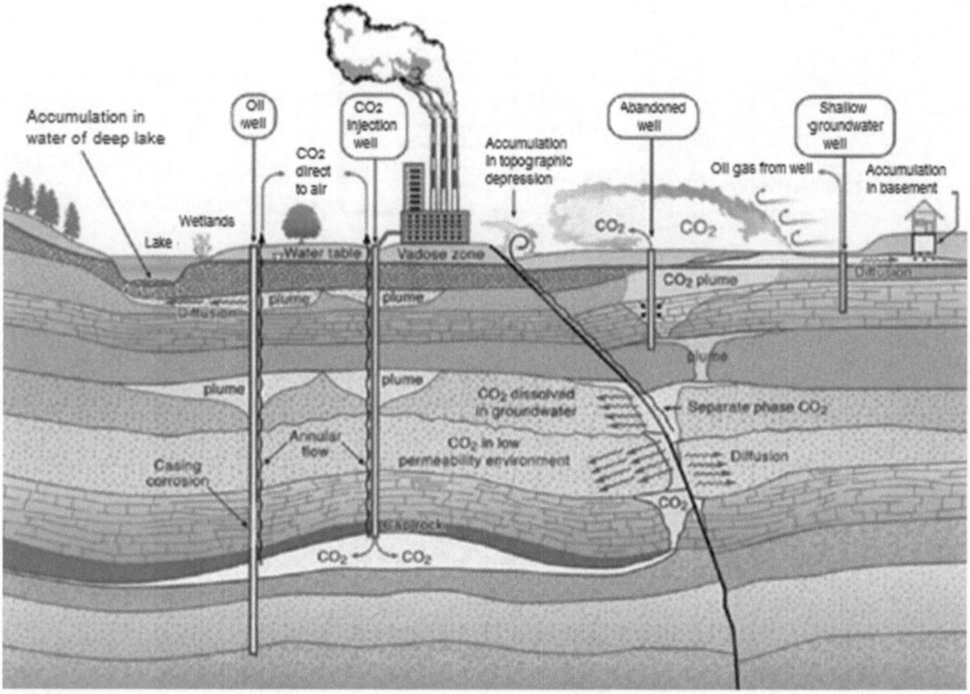

Figure 18. Storage in carboniferous formations with ECBMR (IEA, 2008).

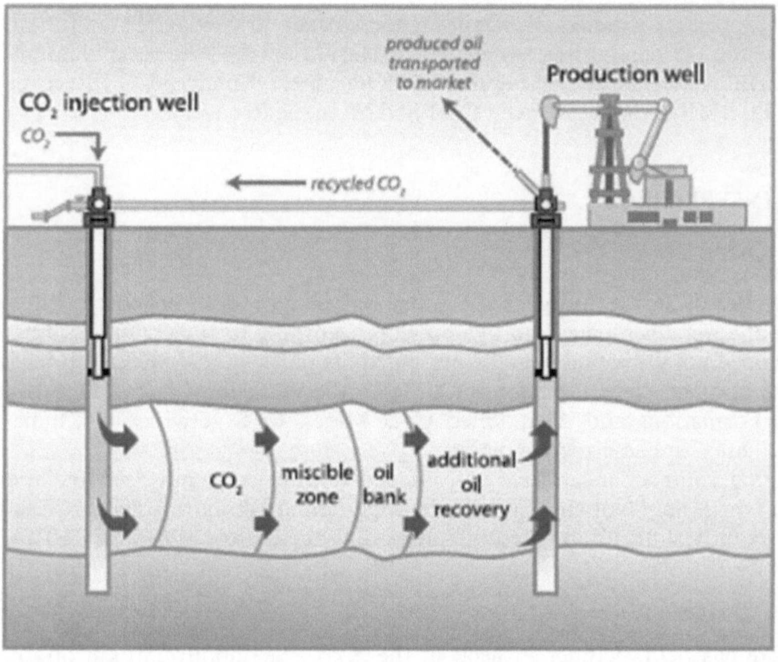

Figure 19. CO_2 storage in oil field reservoir with EOR (CO_2CRC).

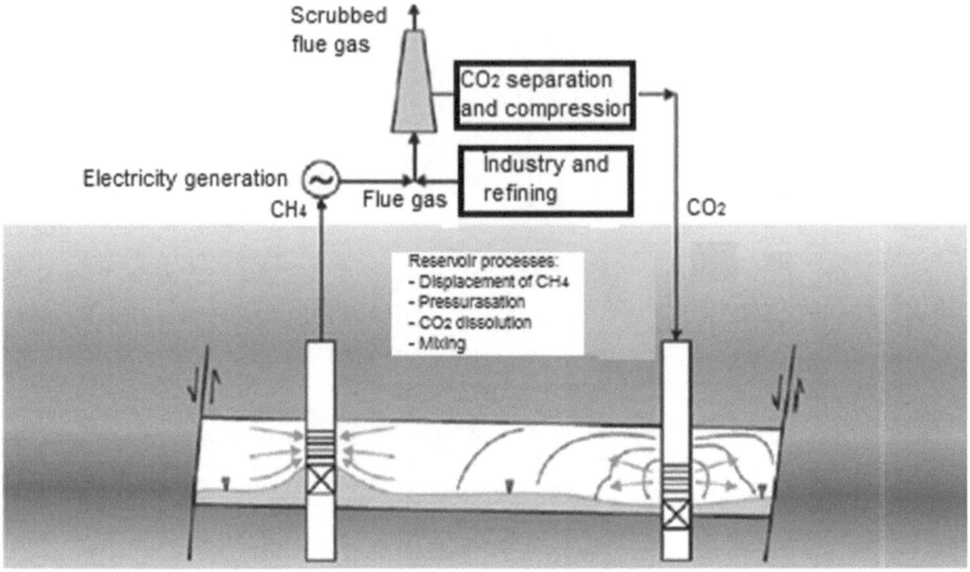

Figure 20. Storage in carboniferous formations with ECBMR (IEA, 2008).

these complex phenomena is presented at the Workshop on CO_2 Storage in Carboniferous Formations and Abandoned Coal Mines at the publications of Elsworth et al. (2011), Qu et al. (2011) and Wang and Liu (2011).

The methodologies for Risk Assessment and resolution of the problems in CCS projects can be seen at the publications of Sousa (2011) and Sousa and Sousa (2011).

Numerical modeling for injection and storage of CO_2 is analyzed in the publication of Vargas et al. (2011).

Figure 17 shows the location of sites where activities relevant to CO_2 storage are planned or under way, according to IPPC (2005).

Figure 18 presents a schematic diagram for storage in carboniferous formations with ECBMR, where the major risks associated is referred. Other schematic diagrams are illustrated in Figures 19 and 20. In Figure 19, EOR means Enhanced Oil Recovery, and, in Figure 20, ECBMR means Enhanced Coal Bed Methane Recovery.

4 EXISTING PROJECTS

4.1 *General*

A great number of storage projects are under way or have been proposed. Figure 21 gives a selection of several geological storage projects according to IPCC (2005). Others are now being developed not yet included in the presented list particularly in China, Brazil and USA. In the publication of Zhang and Sousa (2011) for the Workshop on on CO_2 Storage in Carboniferous Formations and Abandoned Coal Mines, CCS activities in China, including research activities, are described in detail.

Most actual commercial projects are associated with gas production facilities, such as Sleipner in North Sea, Norway, In Salah in Algeria and Weyburn project, Canada. In sections 4.2, 4.3 and 4.4 details are presented regarding these projects (Gomes, 2010).

4.2 *Project Sleipner, Norway*

The offshore gas field Sleipner Project, in the North Sea about 250 km off the coast of Norway, is the first commercial scale project dedicated to geological storage of CO_2 in a

Project	Country	Scale project	Injection start date	Approx. daily rate	Total storage	Storage type	Geol. formation	Age formation	Lithology	Monitoring
Sleipner	Norway	Commercial	1996	3000 t day-1	20 Mt planned	Aquifer	Utsira Formation	Tertiary	Sandstone	4D seismic plus gravity
Weyburn	Canada	Commercial	May 2000	3–5000 t day-1	20 Mt planned	CO_2-EOR	Midale Formation	Mississippian	Carbonate	Comprehensive
Minami-Nagoaka	Japan	Demo	2002	Max 40 t day-1	10,000 t planned	Aquifer (Sth. Nagoaka Gas field)	Haizume Formation	Pleistocene	Sandstone	Crosswell seismic + well monitoring
Yubari	Japan	Demo	2004	10 t day-1	200 t Planned	CO_2-ECBM	Yubari Formation (Ishikari Coal Basin)	Tertiary	Coal	Comprehensive
In Salah	Algeria	Commercial	2004	3–4000 t day-1	17 Mt planned	Depleted Hydrocar. reservoirs	Krechba Formation	Carboniferous	Sandstone	Planned comprehensive
Frio	USA	Pilot	4–13 Oct. 2004	App. 177t day-1 for 9 days	1600 t	Saline formation	Frio Formation	Tertiary	Brine-bearing sandstone shale	Comprehensive
K12B	Netherlands	Demo	2004	100–1000 t day-1 (2006+)	Approx 8 Mt	EGR	Rotleigendes	Permian	Sandstone	Comprehensive
Fenn Big Valley	Canada	Pilot	1998	50 t day-1	200 t	CO_2-ECBM	Mannville Group	Cretaceous	Coal	P, T, flow
Recopol	Poland	Pilot	2003	1 t day-1	10 t	CO_2-ECBM	Silesian Basin	Carboniferous	Coal	–
Qinshui Basin	China	Pilot	2003	30 t day-1	150 t	CO_2-ECBM	Shanxi Formation	Carboniferous Permian	Coal	P, T, flow
Salt Creek	USA	Commercial	2004	5–6000 t day-1	27 Mt	CO_2-EOR	Frontier	Cretaceous	Sandstone	Under development
Snøhvit	Norway	Decided Commercial	2006	2000 t day-1	–	Saline formation	Tubaen Formation	Lower Jurassic	Sandstone	Under development
Gorgon	Australia	Planned Commercial	Planned 2009	Approx. 10,000 t day-1	–	Saline formation	Dupuy Formation	Late Jurassic	Massive sandstone with shale seal	Under development
Ketzin	Germany	Demo	2006	100 t day-1	60 kt	Saline formation	Stuttgart Formation	Triassic	Sandstone	Comprehensive
Orway	Australia	Pilot	Planned late 2005	160 t day-1 for 2 years	0.1 Mt	Saline fm and depleted gas field	Waarre Formation	Cretaceous	Sandstone	Comprehensive
Teapot Dome	USA	Proposed Demo	Proposed 2006	170 t day-1 for 3 months	10 kt	Saline fm and CO_2-EOR	Tensleep and Red Peak Fm	Permian	Sandstone	Comprehensive
CSEMP	Canada	Pilot	2005	50 t day-1	10 kt	CO_2-ECBM	Ardley Fm	Tertiary	Coal	Comprehensive
Pembina	Canada	Pilot	2005	50 t day-1	50 kt	CO_2-EOR	Cardium Fm	Cretaceous	Sandstone	Comprehensive

Figure 21. A selection of geological storage projects according to IPCC (2005).

saline formation (Figure 22). CO_2 content in natural gas varies from 4 to 9.5% and the CO_2 content has to be reduced below 2.5%. CO_2 is injected in salt water containing sand layer, Utsira formation, which lies 1000 m below sea level (IPCC, 2005; Solomon, 2006; Gomes, 2010).

The CO_2 project storage in saline aquifers was established to monitor and research the storage of CO_2. About 1 MtCO_2 is removed annually from produced natural gas and then injected into the formation. The CO_2 injection operation began in October 1996 and in early 2005, more than 7 MtCO_2 had been injected at a rate of about 2700 t/day. A simplified diagram of the Sleipner system is given in Figure 23.

This project is being conducted in three phases. Phase-0 involved the collection and evaluation of data, which was completed in November 1998. Phase-1 consisted in assessing the status of the project after three years of CO_2 injection. In Phase-2, data interpretation and verification of the model began in April 2000.

Characterization of the reservoir and caprock formations was carried out. The reservoir was mapped and characterized by 2D seismic data set and by well data. In more detail the injection set was characterized using 3D seismic dataset and more closely spaced well data. The 2D and 3D seismic data were fundamental for defining the limits of the reservoir and their structure, as shown in Figure 24.

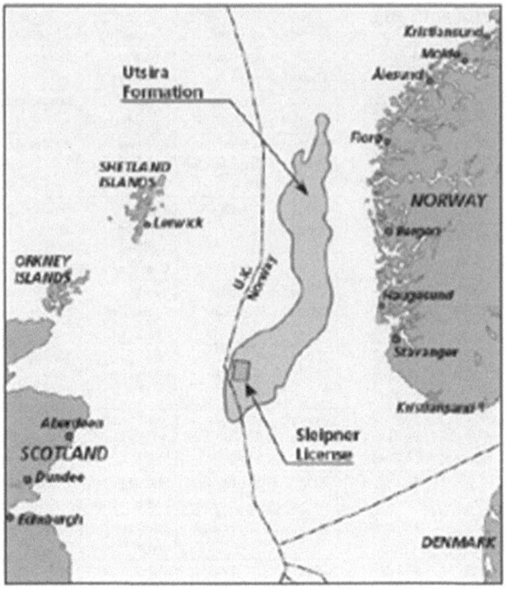

Figure 22. Location map of the offshore gas field Sleipner project (Solomon, 2006).

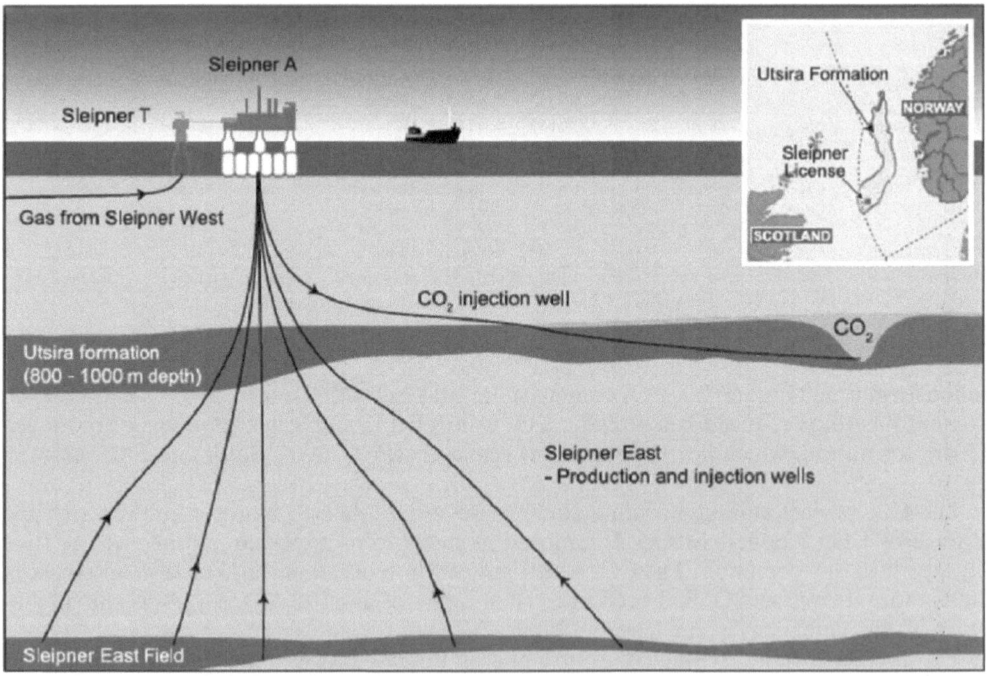

Figure 23. Diagram of the Sleipner CO_2 storage project (IPCC, 2005).

During this project it has been shown that the Utsira formation has good storage quality with respect to porosity and permeability, depth, bedding, pressure and temperature. The formation is an elongated sand reservoir extending to more than 400 km with an area of about 26,100 km^2.

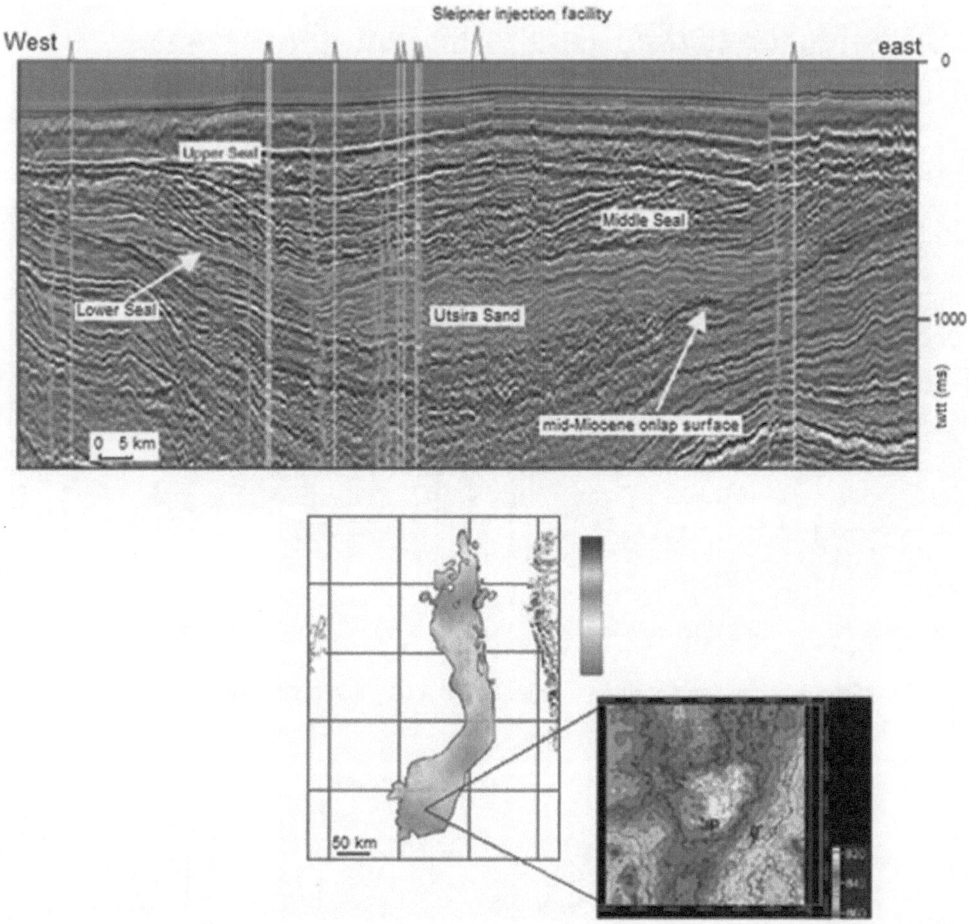

Figure 24. a) Typical 2D seismic reflection profile across the Utsira reservoir; b) Regional depth map to top of Utsira; c) Detailed depth map of Top Utsira sand (from Solomon, 2006).

The dissemination of CO_2 in the storage formation has been monitored successfully by seismic tests (Figure 25). These tests also showed that the caprock is an effective and prevents the migration of CO_2 out of the storage reservoir. Currently, CO_2 at Sleipner spans an area of about 5 km^2. Studies and simulations of reservoirs covering hundreds of thousands of years have shown that the CO_2 will eventually dissolve into pore water, which will become heavier and sink, thus minimizing the potential for long-term leakage (Solomon, 2006).

4.3 *Project In Salah, Algeria*

The In Salah project involves a consortium between Sonatrach, BP and Statoil, and is located in the central Sahara, Algeria (Figure 26). The location of Salah gas project in Algeria is shown in Figure 27.

It is the first project in the world dedicated to large-scale storage of CO_2 in a gas reservoir (Riddiford et al., 2003). The exploration fields Krechba, Teguentour and Reg, in In Salah, produce natural gas with CO_2 concentrations of about 10%, which is higher than the transport specification. So, before delivery to markets in Europe, it is necessary for separation of CO_2 from fuel, which is then re-injected into a sandstone reservoir containing water at a

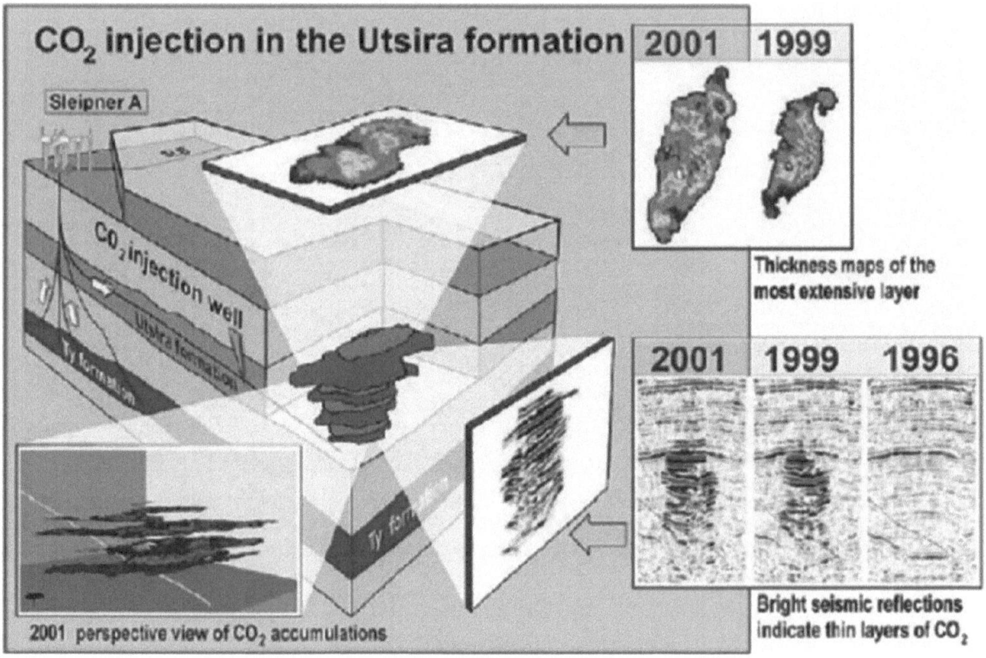

Figure 25. Seismic surveys and position of injected CO_2 (Torp and Gale, 2004).

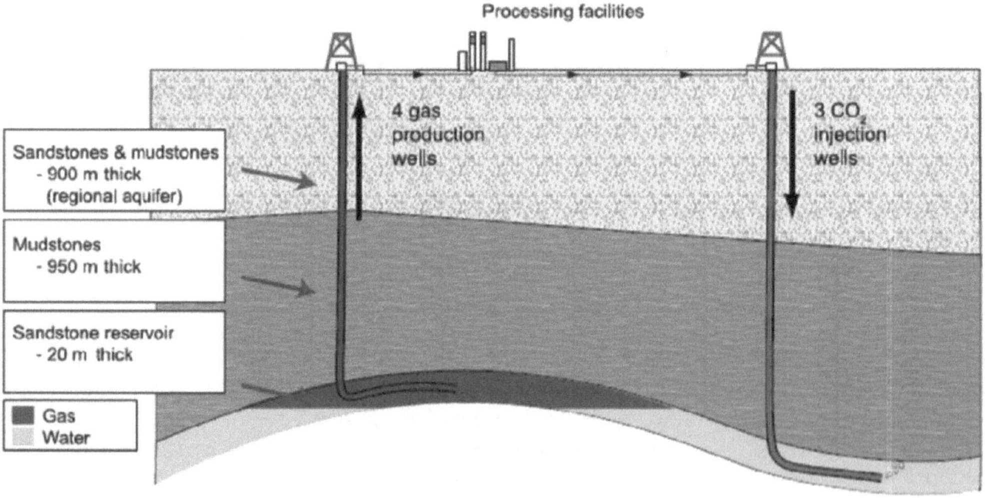

Figure 26. In Salah gas project, Algeria (IPCC, 2005).

depth of 1800 m. It is intended to be injected into storage until $MtCO_2$/year 1.2. The CO_2 injection began in April 2004 and for the duration of the project; it is estimated to store about 17 $MtCO_2$.

The complex consists of four production wells and three injection wells (Figure 26). Horizontal wells at long range (up to 1.5 km) are used to inject CO_2 in the areas of the reservoir with high permeability.

The Krechba field is a relatively simple anticline. The injected CO_2 is expected to migrate into the area of the current gas field after depletion of the gas zone. The area was mapped

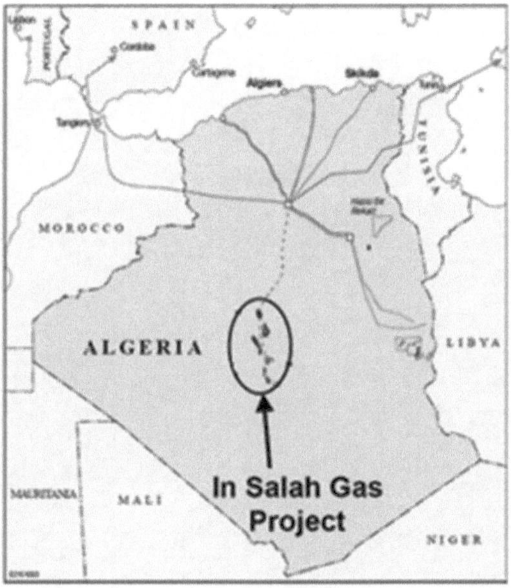

Figure 27. Location of Salah gas project in Algeria.

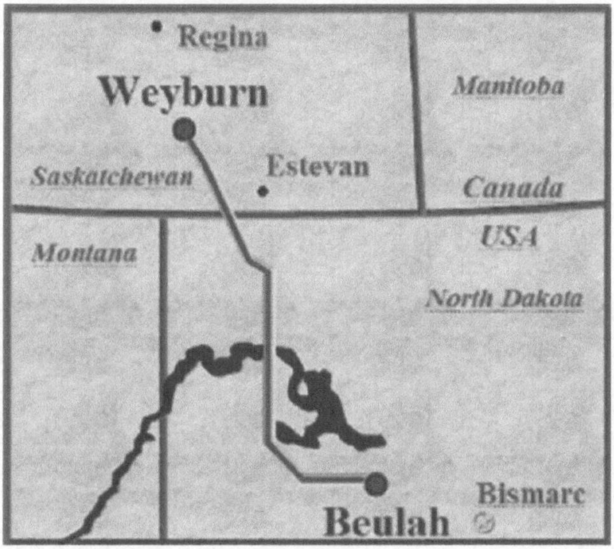

Figure 28. Location of the Weyburn CO_2-EOR and storage reservoir at Canada.

with three-dimensional seismic analysis and surveys. Were located deeply flawed, but the most superficial levels in the rock matrix is presented intact (IPCC, 2005).

4.4 *Project Weyburn, Canada*

The Weyburn Project, located at Canada (Figure 28), is an enhanced oil recovery (EOR) project located at Williston Basin, a geological structure that extends from south-central

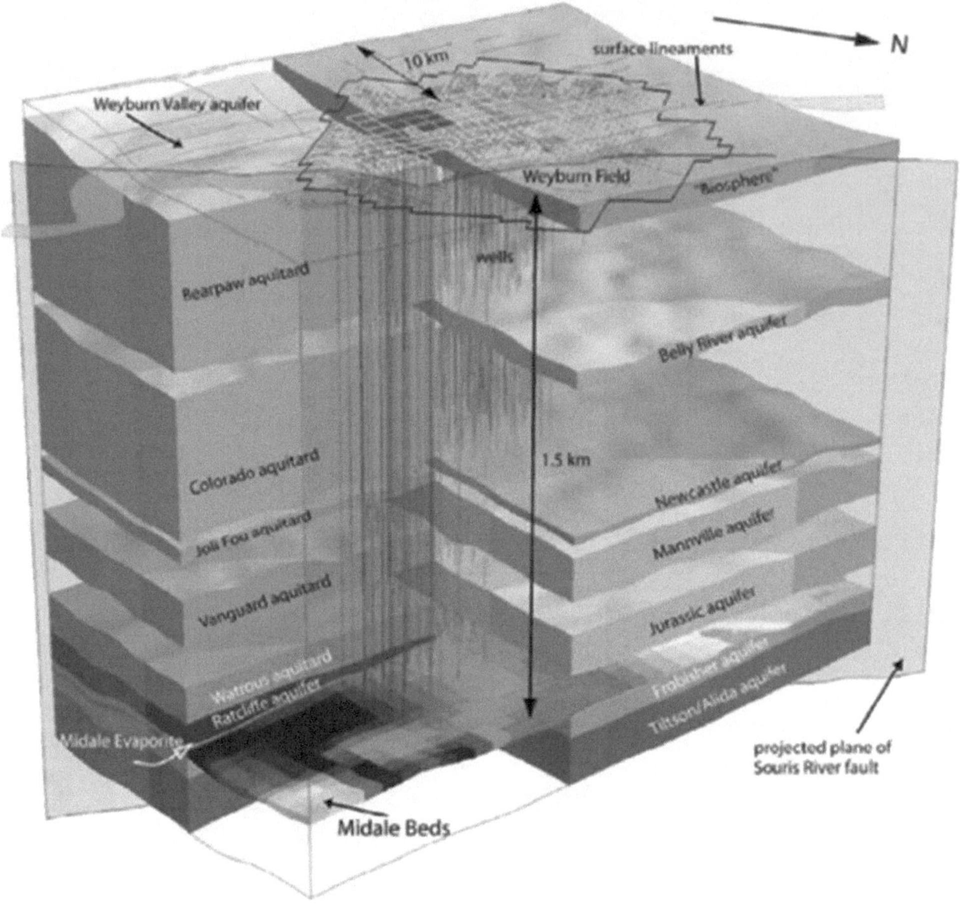

Figure 29. Block diagram of the geological model for the Weyburn project (Chalaturnyk and Whittaker, 2006).

Canada to the north-central United States. The project aims to permanently store almost all of the injected CO_2, eliminating the CO_2 that would normally be released during the end of the field life.

The source of CO_2 for this project is the installation of Dakota Gasification Company facility, located approximately 325 km south of Weyburn, in North Dakota, USA. At the plant, coal is gasified to make syngas (methane), with a relatively pure stream of CO_2 as a byproduct. This flow of CO_2 is dried, compressed and piped to Weyburn.

The Weyburn field covers an area of 180 km^2, with an initial volume of oil of about 222 million m^3 (1,396 million barrels). It is expected that over the life of the project (20 to 25 years) about 20 MtCO_2 being stored in the field, according to current economic conditions and technology of oil recovery.

The oil reservoir is a fractured carbonate, with 20–27 m thick. A plan and expressed it above the reservoir bed is a good barrier to leakage of CO_2.

Since the CO_2 injection began in late 2000, the EOR project has been implemented as planned. Currently, about 1,600 m^3/day (10,063 bpd) of oil is being produced from the field. All the CO_2 produced is captured and recompressed for reinjection in the production area. About 1,000 tCO_2/day are now reinjected, which grows as the project matures. Monitoring is extensive, with high resolution seismic and surface monitoring to determine potential leaks. The monitoring includes surface sampling and analysis of potable groundwater, and the collection and analysis of soil gases (Moberg et al., 2003).

The geological characterization was one of the largest and most diverse aspects of the project. The Weyburn reservoir is at an average depth of 1.5 km and includes a upper dolostone unit, with an average thickness of about 6 m, and a lower limestone unit that averages around 15 m in thickness. More details about the geological aspects can be described in paper to International Symposium on Site Characterization for CO_2 Geological Storage presented by Chalaturnyk and Whittaker (2006). A block diagram of the geological model for the Weyburn project is presented in Figure 29.

5 STORAGE IN CARBONIFEROUS FORMATIONS

The CO_2 injection and sequestration in carboniferous reservoirs can be performed either in deep unminable coal seams or abandoned coal mines. Coal formations contain cleats that impart some permeability to the system. Between cleats coal has a large number of micropores into which gas molecules can diffuse and be tightly absorbed. Gaseous CO_2 injected through wells will flow through the cleat system, diffuse the coal matrix and be absorbed onto the coal micropore surfaces (Figure 30). The cleat spacing is very uniform and ranges from the order of millimeters to centimeters (Shi and Durucan, 2005). If CO_2 is injected into coal seams it can display gas methane enhancing coal bed methane recovery.

The coals can be classified according to their content of fixed carbon, whose proportion increases as the ore is formed. In ascending percentage of carbon, the main coal types are peat, lignite, bituminous coal and anthracite. In Figure 31, there are photographs of various types of coal followed by a sketch of its stratigraphic position.

The gas storage mechanism in coal seams is distinctively different from that in oil and gas reservoirs and aquifers, where injected CO_2 occupies the pore space as a separate phase or is dissolved in water or oil. Over the last two decades coalbed methane (CBM) has become an important source of (unconventional) natural gas supply in the United States. Carbon dioxide enhanced coalbed methane recovery (CO_2-ECBM) is an emerging technology, which has the potential to store large volumes of CO_2 in deep unminable coal formations (coalbeds).

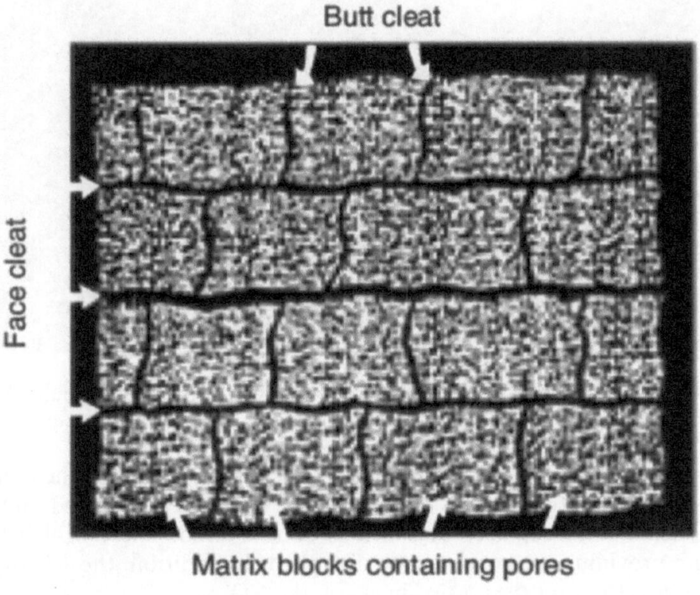

Figure 30. A schematic of coal structure (Shi and Durucan, 2005).

Figure 31. Several types of coal (Coal, 2010).

Figure 32. Coal bed methane test well at Sealand, Chester, UK (Evergreen Resources).

When listing coal seams also contain gases such as methane. The gas is held in pores on the surface of the coal and in fractures in the seams. If CO_2 could be injected into a coal seam, then the methane could be replaced and in turn be recovered. The CO_2 will remain stored within the seam, providing the coal is never disturbed. In addition, the sale of the methane produced could help to offset the cost of injecting the CO_2.

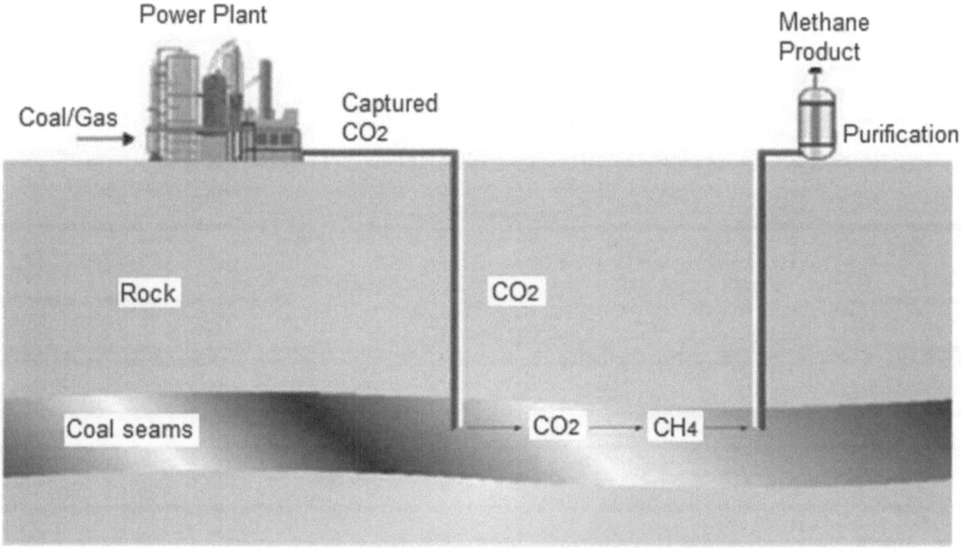

Figure 33. Schematic diagram of the CO_2-ECBM process (Adapted Gomes, 2010).

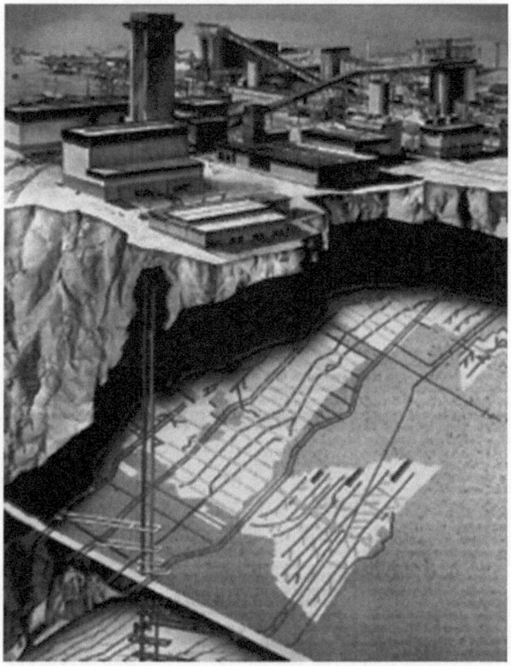

Figure 34. Coal mine in China.

In the most favorable coal basins it is estimated that 15 Gtonnes of CO_2 could be stored in unminable coal seams. In the best sites, the operating income from increased methane production would compensate for the additional costs associated with CO_2 injection. The more CO_2 (perhaps 20 to 50 times as much) could be stored in less favorable coal basins, the higher cost would be. A key factor determining the attractiveness of a particular site is the permeability of the coal.

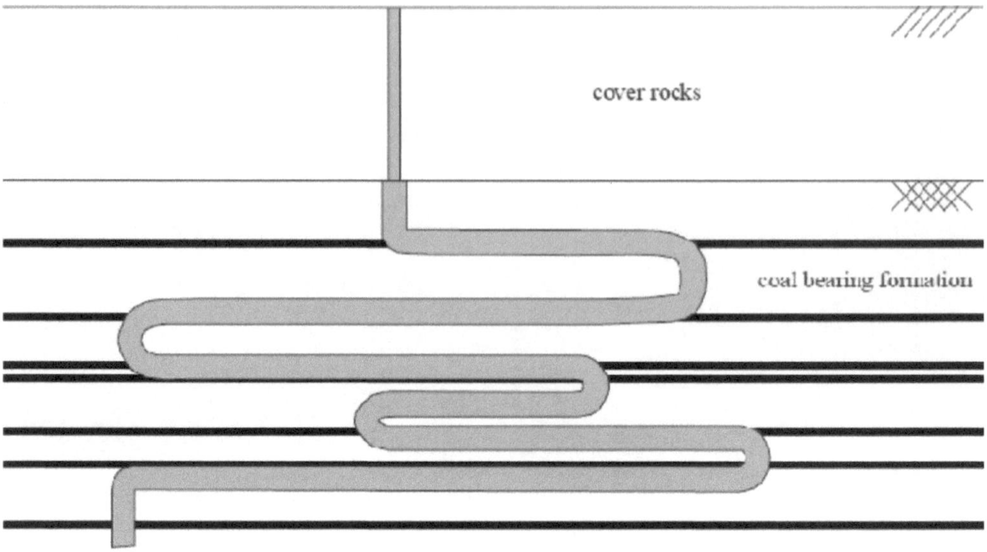

Figure 35. Schematic representation of pressure evolution in an abandoned mine (Piessens and Dusar, 2003).

One pilot project has been underway for the past three years. Burlington Resources, together with BP, is operating a 13-well CO_2-ECBM pilot unit in the San Juan Basin in southwestern USA. Initial results show that increased methane production can be achieved by CO_2 injection. Also, no CO_2 has been found in the produced gas indicating that the CO_2 is being stored in the coal seam as predicted. Figure 32 presents a coal bed methane test at Sealand, UK.

A schematic diagram of the CO_2-ECBM process is presented at Figure 33.

CO_2 can be stored in abandoned coal mines. Figure 34 gives an idea of the complexity of these reservoirs. The mine can be represented as a long gallery as shown in Figure 35. An explanation of the problems and constrains associated to these type of reservoirs can be obtained in the publication of Piessens (2011).

6 CONCLUSION

Underground geological storage of carbon dioxide has a considerable potential for mitigating global climate changes. CO_2 can be safely injected and stored at well characterized and properly managed sites, where the deeply geological formations could be stored for a long-term. Depleted oil and gas reservoirs, saline aquifers and carboniferous formations can be used for storage of CO_2. Particularly, the abandoned coal mines can be a good option for China due to the existence of a large number of existing mines in this situation.

This chapter is an introductory work and intends to present the present day conditions of CCS projects in the world describing with more detail the processes associated to capture and transportation of CO_2, the different options for the storage of CO_2 in underground geological formations, the existing commercial projects and the possibilities and problems associated to the storage in carboniferous formations, including abandoned coal mines.

REFERENCES

Chalaturnyk, R.; Whittaker, S. 2006. Geological characterization of the Weyburn field for geological storage of CO_2. Proc. of Int. Symposium on Site Characterization for CO_2 Geological Storage, Berkely, pp. 101–105.

Coal, Underground 2010. Available in WWW: <http://www.undergroundcoal.com.au/>
Elsworth, D.; Wang, S.; Izadi, G.; Kumar, H.; Liu, J.; Lee, D.-S.; Mathews, J.; Pone, D. 2011. Complex process couplings in systems pushed far-far-from equilibrium: Applications to deep geologic sequestration and energy recovery. *Int. Workshop on CO_2 Storage in Carboniferous Formations and Abandoned Coal Mines*, Ed. He, Sousa, Elsworth and Vargas. Beijing, pp. 55–68.
Gomes, A. 2010. CO_2 injection processes in carboniferous formations (in Portuguese). MSc Thesis, University of Porto, Porto, 116p.
He, M.; Sun, X. 2011. Considerations on CO_2 storage in abandoned coal mines in China. Int. Workshop on CO_2 Storage in Carboniferous Formations and Abandoned Coal Mines, Beijing, pp. 25–36.
IEA—International Energy Agency. 2008. CO_2 Capture and Storage—A Key Carbon Abatement Option. Head of Communication and Information Office, Paris.
IPCC 2005. Carbone dioxide and storage. IPCC Special Report, Cambridge University Press, 431p.
Moberg, R.; Stewart, D.; Stachniak, D. 2003. The IEA Weyburn CO_2 monitoring and storage project. Proc. of the 6th Int. Conf. on Greenhouse Gas Control Technologies (GHGT-6), Gale and Kaya (eds.), Kyoto, pp. 219–224.
NETL—National Energy Technology Laboratory 2007. Carbon Sequestration Technology Roadmap and Program Plan 2007. United States Department of Energy (http://www.netl.doe.gov/technologies/carbon_seq/refshelf/project%20portfolio/2007/2007Roa dmap.pdf).
Piessens, K. 2011. The conceptual model for an abandoned coal mine reservoir. *Int. Workshop on CO_2 Storage in Carboniferous Formations and Abandoned Coal Mines*, Ed. He, Sousa, Elsworth and Vargas. Beijing, pp. 179–200.
Piessens, K.; Dusar, M. 2003. CO_2-Sequestration in Abandoned Coal Mines. Royal Belgian Institute for Natural Sciences, Geological Survey of Belgium.
Price, P.; McKone, T.; Sohn, M. 2008. Carbon sequestration risks and risk management. Lawrence Berkeley National Laboratory, Environment Energy Techniques Division, Berkeley, 19p.
Qu Hongyn, Liu Jishan, Zhong Wei, Pan Zhejun & Connell Luke. 2011. Multiphysics of coal-gas interactions. Something old, something new and something very new. *Int. Workshop on CO_2 Storage in Carboniferous Formations and Abandoned Coal Mines*, Ed. He, Sousa, Elsworth and Vargas. Beijing, pp. 69–92.
Riddiford, F.; Tourqui, A.; Bishop, C.; Taylor, B.; Smith, M. 2003. A cleaner development: The In Salah Gas project, Algeria. Proc. of the 6th Int. Conf. on Greenhouse Gas Control Technologies (GHGT-6), Gale and Kaya (eds.), Kyoto, pp. 601–606.
Shi, J.; Durucan, S. 2005. CO_2 Storage in deep unminable coal seams. Oil & Gas Science and Technology—Rev. IFP, Vol. 60 (2005), No. 3, pp. 547–558.
Solomon, S. 2006. Carbon dioxide storage: Geological security and environmental issues—case study on the Sleipner gas field in Norway. Bellone Foundation, Oslo, 20p.
Sousa, L.R.; Sousa, R.L. 2011. Risk associated to storage of CO_2 in carboniferous formations. Application of Bayesian Networks. *Int. Workshop on CO_2 Storage in Carboniferous Formations and Abandoned Coal Mines*, Ed. He, Sousa, Elsworth and Vargas. Beijing, pp. 153–178.
Sousa, R.L. 2011b. Methodologies for Risk Analysis and Decision Making. *Int. Workshop on CO_2 Storage in Carboniferous Formations and Abandoned Coal Mines*, Ed. He, Sousa, Elsworth and Vargas. Beijing, pp. 125–152.
Torp, T.; Gale, J. 2004. Demonstrating storage of CO_2 in geological reservoirs: The Sleipner and SACS projects. Energy, 29, pp. 1361–1369.
Vargas, E.A. Jr.; Velloso, R.Q.; Ribeiro, W.N.; Muller, A.L.; Vaz, L.E. 2011. Considerations on the numerical modeling of injection processes of CO_2 in geological formations with emphasis on carboniferous formations and abandoned coal mines. *Int. Workshop on CO_2 Storage in Carboniferous Formations and Abandoned Coal Mines*, Ed. He, Sousa, Elsworth and Vargas. Beijing, pp.107–123.
Wang, J.G. & Liu, J.S. 2011. Micro-scale modeling of gas-coal interaction in coalbed seam – heterogeneity effect. *Int. Workshop on CO_2 Storage in Carboniferous Formations and Abandoned Coal Mines*, Ed. He, Sousa, Elsworth and Vargas. Beijing, pp. 93–104.
Zeng, N. et al. 2008. Climate change—The Chinese challenge. Science, Vol. 319, 8/Feb/2008, pp. 730–731. The American Association for the Advancement of Science, Washington DC.
Zhang N.; Sousa, L.R. 2011. Carbon Capture and Storage (CCS) activities in China. Int. Workshop on CO_2 Storage in Carboniferous Formations and Abandoned Coal Mines, Ed. He, Sousa, Elsworth and Vargas. Beijing, pp. 37–53.

CO_2 Storage in Carboniferous Formations and Abandoned Coal Mines –
He, Ribeiro E Sousa, Elsworth & Vargas Jr. (eds)
© 2012 Taylor & Francis Group, London, ISBN 978-0-415-62079-6

Considerations on CO_2 storage in abandoned coal mines in China

M.C. He
State Key Laboratory for GeoMechanics and Deep Underground Engineering of China University of Mining and Technology, Beijing, China

ABSTRACT: Due to economic development, China has surpassed the U.S. as the world's largest emitter of greenhouse gases. Underground geological storage of carbon dioxide has great potential for the mitigation of climate change. Carboniferous formations and underground abandoned coal mines can be used to store carbon dioxide. CO_2 storage in abandoned underground coal mines (CSAC), especially, provides a new way for the realization of China's CO_2 emission reduction targets. Three types of underground space are defined for underground coal mines in China. As an example, abandoned coal mines from Xuzhou Mining Group were analyzed with respect to the available underground space for CO_2 storage. Taking Qingshanquan coal mine as a potential CO_2 storage site, ideas about in-situ tests are proposed according to its specific geological conditions. Key issues related to CO_2 injection and storage process are discussed. Although there is still some scientific issues to be studied in depth, the geological conditions and the technology are available on a feasibility basis.

1 INTRODUCTION

With the increasing economic development, China has surpassed the U.S. as the world's largest emitter of greenhouse gases. In the context of global warming, China faces increasing challenges in the international climate negotiations.

The Chinese government has been devoting itself to promoting energy conservation and emissions reduction to address global climate change. The Chinese President announced that China wants to cut 1.5 billion tons of CO_2 emissions in the future 5 years. Also it was announced a significant reduce of carbon emission intensity per unit of GDP on the basis of 2005 by 2020, at the Summit on Climate Change held in New York, September 23, 2009.

During the past decades, China has carried out several activities to promote carbon capture and storage (CCS) development. A detailed description of these activities is presented in the Chapter of Zhang and Sousa (2011). In this Chapter emphasis is given to the possibilities of CO_2 Storage in Abandoned underground Coal mines (CSAC) in China.

Abandoned coal mines from Xuzhou Mining Group, located at Jiangsu province, are analyzed with respect to the available underground space for CO_2 storage. The Qingshanquan coal mine was selected as a potential CO_2 storage site, and some ideas are proposed for in-situ pilot tests according to the specific geological conditions of the mine.

2 MAJOR SOURCES AND REDUCTION PATHWAYS

Long-term development of China's economy mainly depends on coal-based fossil fuels, among which coal accounts for about 70% of China's total primary energy consumption (Figure 1).

Compared with U.S., CO_2 emission in China has been increasing very quickly since 2004 (Figure 2). By 2009, it reached to 1.86 billion ton, accounting for a quarter of global fossil

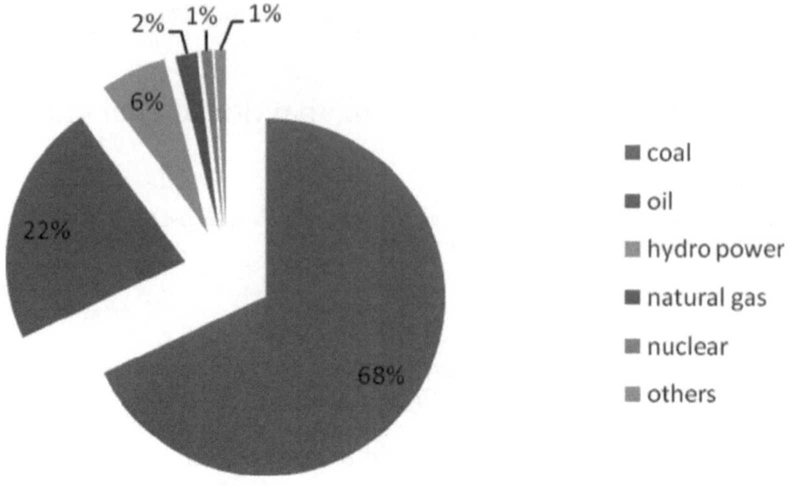

(a) 2004

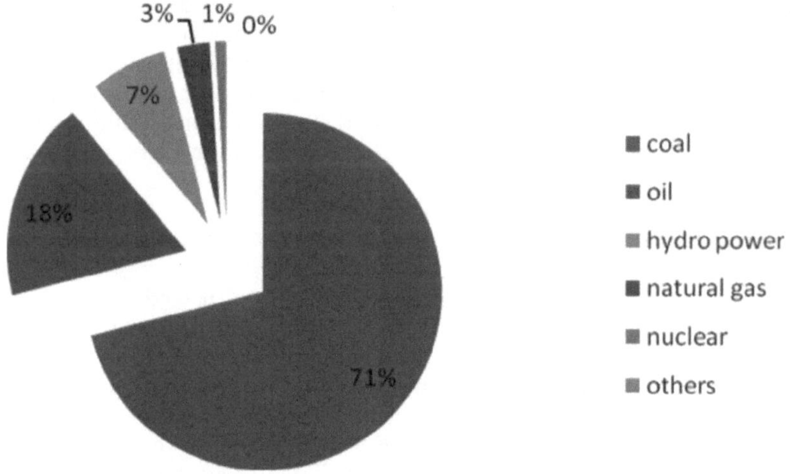

(b) 2008

Figure 1. China's primary energy composition in 2004 (a) and 2008 (b) (National Development and Reform Commission, 2004 & 2008).

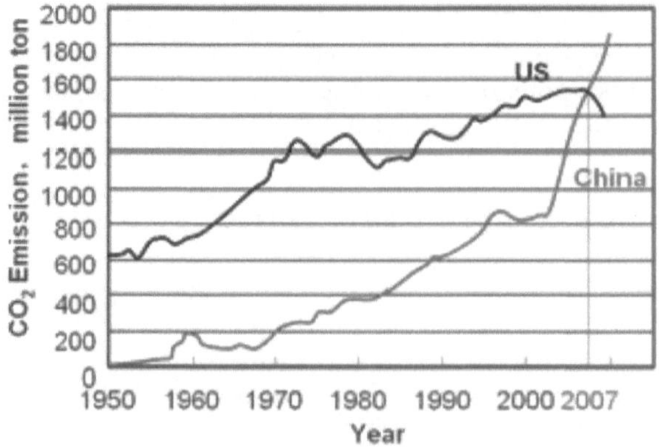

Figure 2. Carbon dioxide emissions from coal-fired power (CDIAC and BP).

fuel combustion emissions, and this part of the carbon dioxide emissions are mainly from coal-fired generation.

CCS has become the primary approaches to reduce carbon dioxide emissions. A lot of research work has been carried out around the world and some pilot explorations were performed (Sousa, 2011; Shenhua, 2011; Peng, 2010; Climit, 2010; NZEC, 2007; Petrochina, 2007).

Geological sites for carbon dioxide storage include the depleted oil and gas reservoirs, deep saline aquifers, deep unmineable or abandoned coal seams and ocean (Sousa, 2011). However, this CCS technology is difficult to apply in China for two main reasons: (1) high energy consumption—the CO_2 capture process will consume a large amount of energy; (2) high cost—CCS project commercial operation costs is about $ 70/t.

Underground storage of CO_2 in abandoned mines is a new technology proposed in recent years (Paria et al., 2011; Shi and Durucan, 2005; Piessens and Dusar, 2004; Piessens, 2011) and has a promising application prospect in China. Analysis of China's carbon dioxide emissions from major industrial point source (Figures 3 and 4) showed that China has more than 1620 large stationary sources for carbon dioxide emissions, which include coal-fired power plants, cement production, iron and steel industry, oil refineries and other industrial facilities.

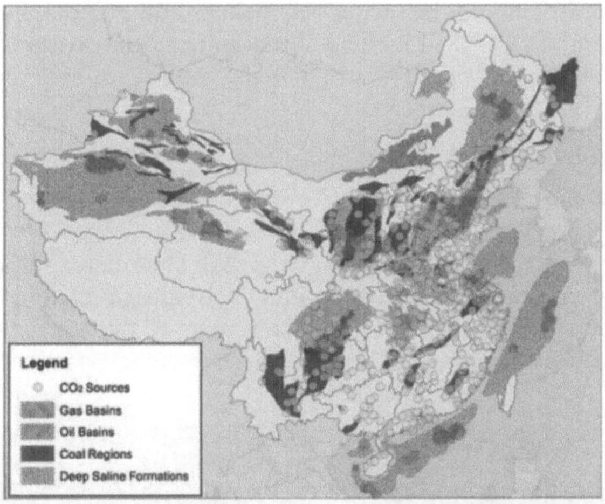

Figure 3. Distributions of CO_2 point sources and coal resources in China (Li, 2009).

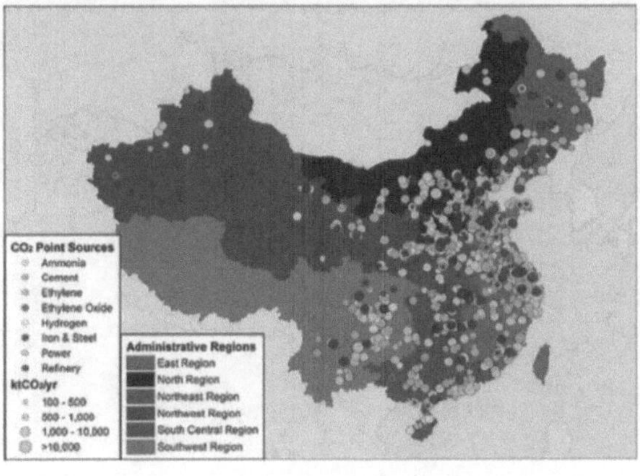

Figure 4. Distribution of CO_2 industrial point sources in China (Li, 2009).

Of these, about 50 large power plants located near coal mines and some abandoned mine, which provide adequate underground space for storage of CO_2 emitted from coal-fired power generation. China has a total of 15,000 coal mines and the number of abandoned mines continue to increase. Meanwhile, a large amount of underground space is being produced everyday due to coal mining. Therefore, CSAC is an important technical means of carbon storage, which is more economical and feasible to China.

3 UNDERGROUND SPACE IN COAL MINES

3.1 *Introduction*

Underground space in coal mines is produced by four general developing approaches, i.e. vertical shaft development, inclined shaft development, adit development and mining excavation development.

The access to the mines can be provided by vertical shafts (Figure 5) that are widely used in China. This approach can be divided into single level and multi-level developments. Inclined shaft development is another option to the access, as shown in Figure 6.

Adit development as shown in Figure 7 is another type of coalfield development for the underground space. In general, a main adit is built for transportation of coal, materials and gangue, drainage, ventilation and the laying of pipelines and other tasks.

Base on above approaches of minefield development, the integrated development of underground space is shown in Figure 8, which is the combination of them.

3.2 *Types of underground space in coal mines*

According to the different approaches, underground space can be divided into three types: SCP space, GR space and WP space.

SCP space includes the space of shafts, underground chambers and the permanent roadways. Those spaces are not affected by coal mining activities and should remain stable after mining. GR space includes the space of Gate Roadways of the working face in coal mines,

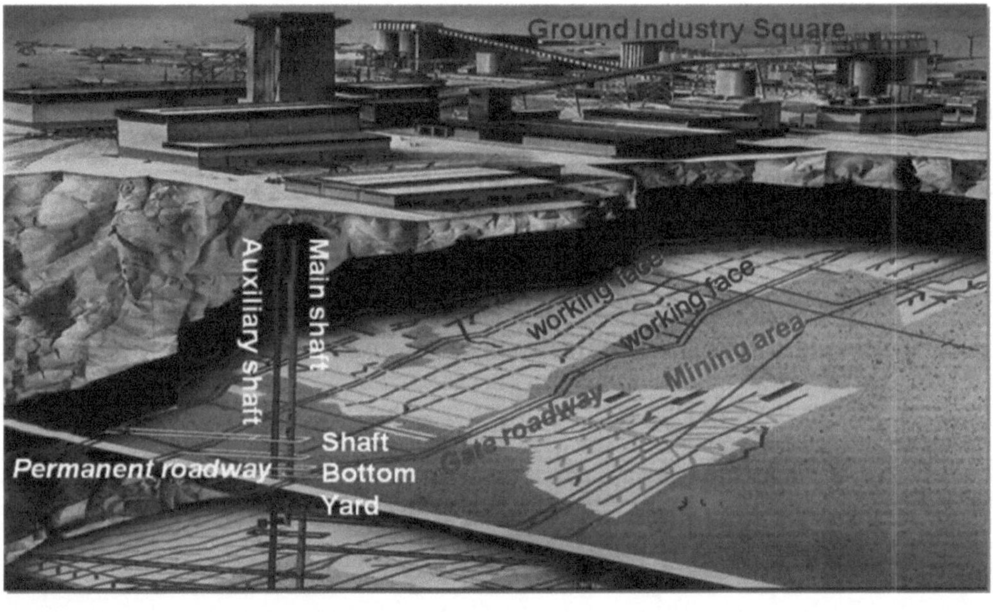

Figure 5. Vertical shaft development.

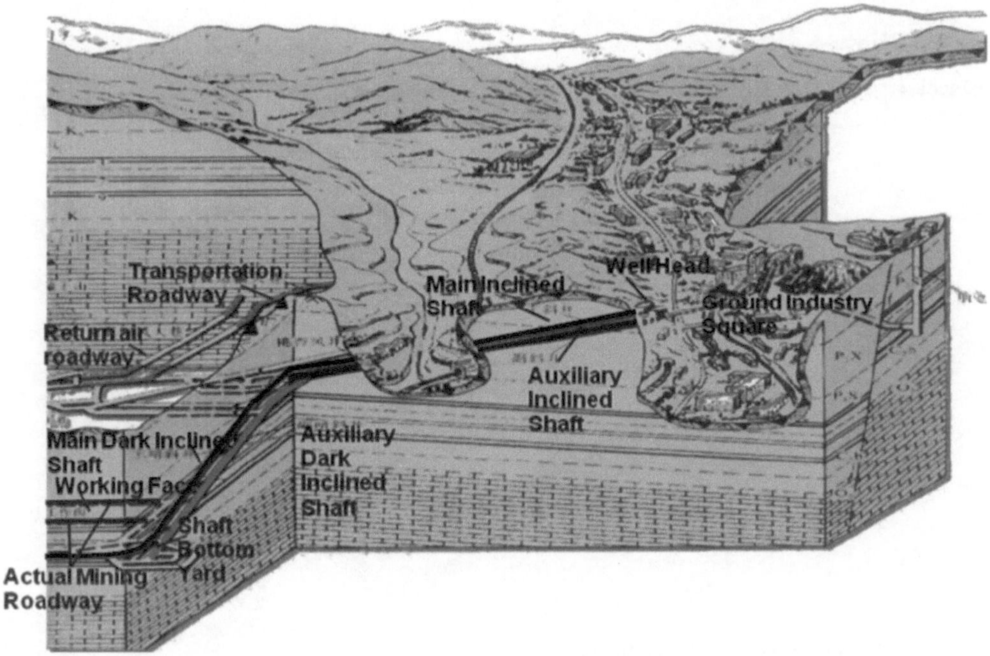

Figure 6. Inclined shaft development.

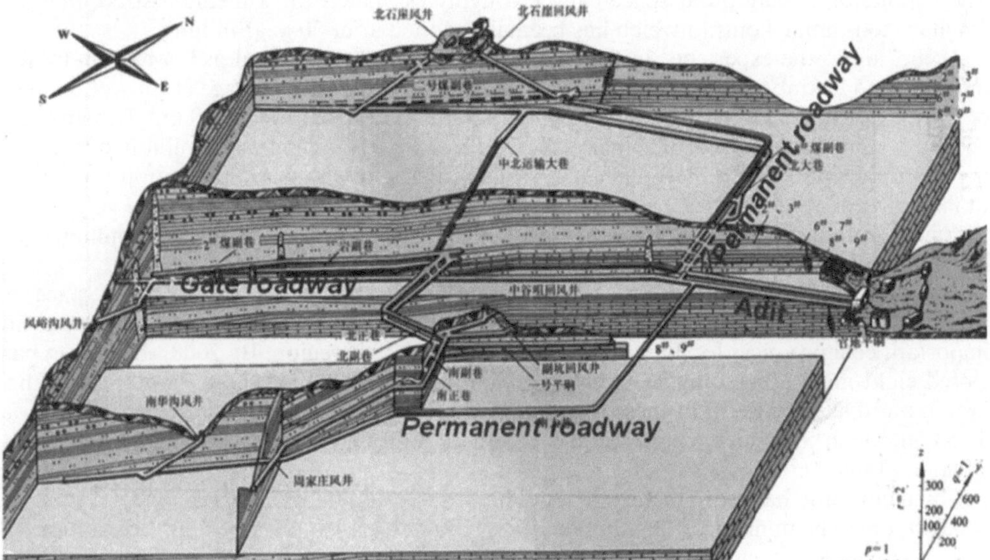

Figure 7. Adit development.

partly affected by coal mining, and remains stable after mining in the working face, but some relevant deformations can happen. WP space includes the space of Working Face of Panel, strongly affected by coal mining and collapse of overlying strata could happen after mining, depending on the excavation mining techniques.

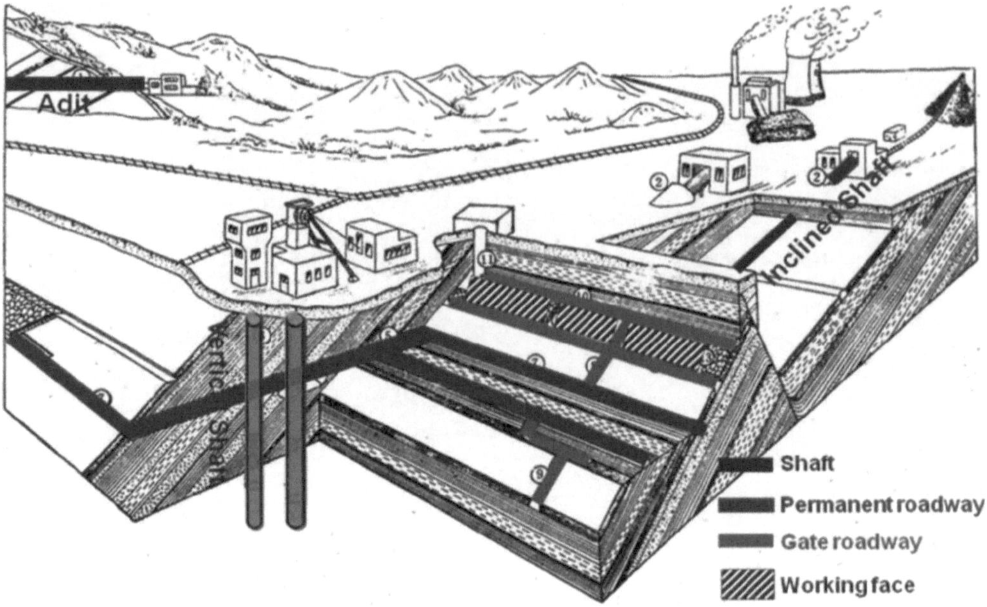

Figure 8. Mining excavation development.

3.3 *Underground space quantity*

The volume of underground space can be roughly estimated for a medium-sized mine of 2 million tons annual output which has been abandoned after 30-year mining.

According to the experience in China, SCP space can be estimated as 0.4 million m^3 for underground chambers and a permanent roadway of 20 km with average cross-section area of 20 m^2. GR space can be estimated as 0.6 million m^3 for a roadway of 40 km with an average cross-section area of 15 m^2. Finally, WP space can be estimated as 15 million m^3 according to a total amount of mined coal of 60 million tons, with a rock sedimentation coefficient of 0.7, and an average coal density of 1200 kg/m^3. Therefore, the total estimated volume for the underground space of an abandoned coal mine with an annual output of 2 million tons is roughly for 16 million m^3.

The Xuzhou Mining Group was selected in order to quantify the underground space of abandoned coal mines (Figure 9). The Group has 120 years of coal mining history and is an important coal producer for Jiangsu Province in East China region. By 2008, the Group has closed eight mines, including Dongzhuang, Hanqiao, Mapo, Yian, Qingshanquan, Xinhe, Woniu and Dahuangshan. In these abandoned mines, the shafts were intact, and a total of 46.53 million m^3 of underground space can be used for CO_2 storage according to above calculation (Table 1).

China currently has more than 2000 medium-large scale coal mines with annual output of 400,000 tons. The number of abandoned coal mines is increasing fast. In our opinion, there is adequate underground space which can be used for CSAC in China.

4 IDEAS FOR AN IN-SITU TESTS FOR CSAC

4.1 *Introduction*

The feasibility of CO_2 injection into underground space of coal mines, the CO_2 sealing performance and long-term stability of CO_2 geological sequestration are key issues related to CSAC.

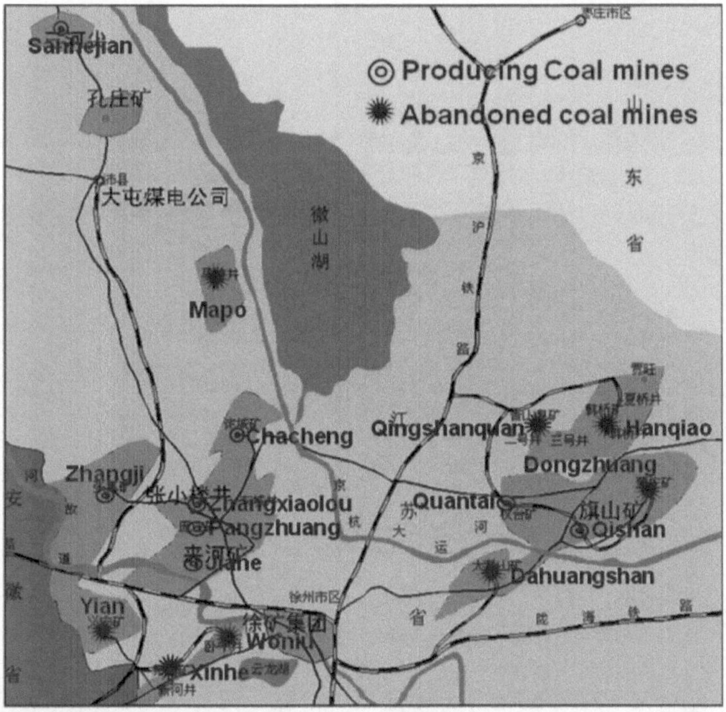

Figure 9. Distribution of underground coal mines of Xuzhou Mining Group.

Table 1. Estimated underground space of abandoned coal mines of Xuzhou Mining Group.

Coal mine	Depth (m)	Output (kt/a)	Production period	Underground space (10^3 m^3)
Dongzhuang	100~650	600	1958–2000	7300
Hanqiao	100~450	500	1950–2007	8100
Mapo	200~400	200	1972–1999	2350
Yian	100~700	450	1963–2007	5950
Qingshanquan	200~1121	300	1950–2000	4750
Xinhe	300~1100	450	1960–2007	6280
Woniu	100~300	300	1960–2007	4500
Dahuangshan	100~600	600	1958–2000	7300
Total				46530

The above key issues can be overcome by the special features of abandoned coal mines. The Qingshanquan coal mine was selected as a potential CO_2 storage site and the design ideas for in-situ tests for CSAC is proposed according to the specific geological conditions of this mine.

Key issues with respect to CO_2 injection and storage process are discussed below.

4.2 *CO_2 injection*

Coal-bearing strata of Qingshanquan mine has two main coal seams, at depths of 800 and 1120 m. Seam roofs are respectively mudstone and sandstone, and floors are sandstones (Figure 10).

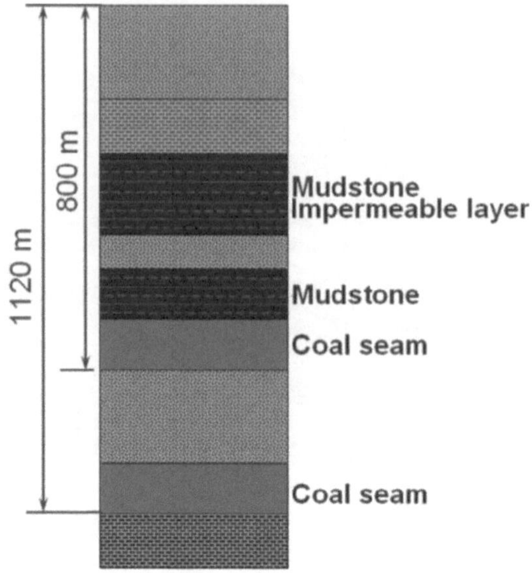

Figure 10. Rock profile.

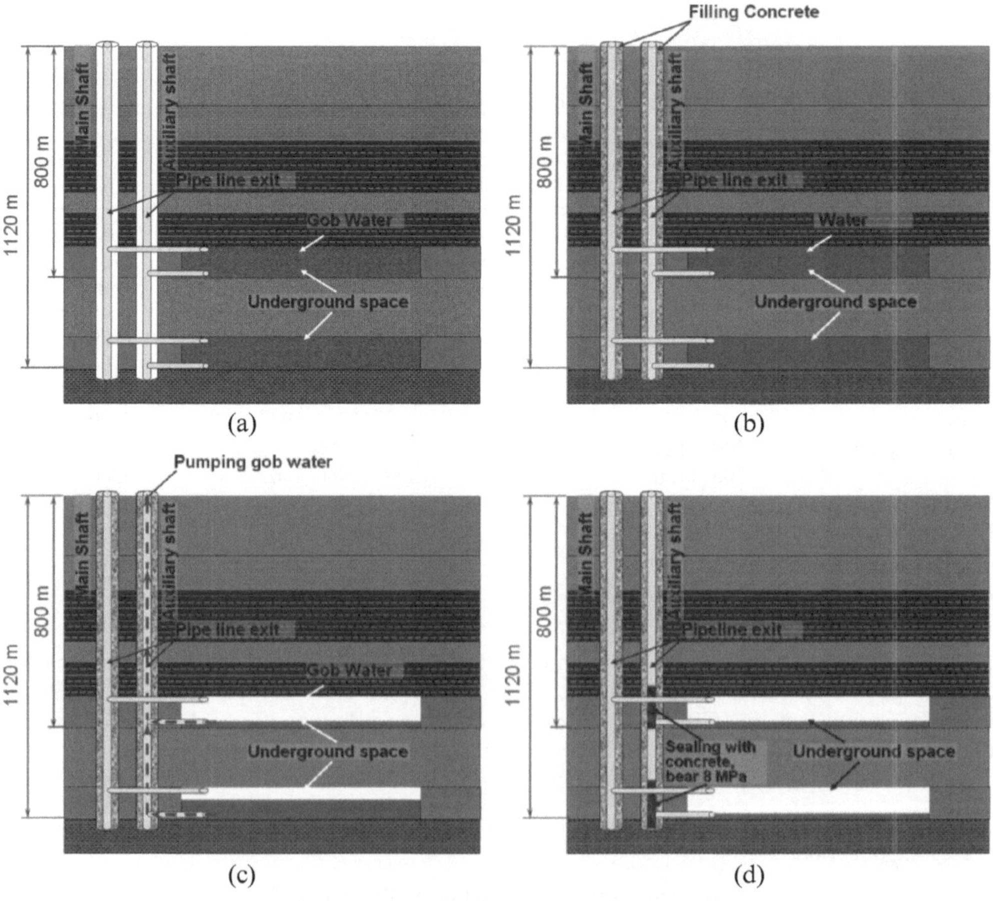

Figure 11. Underground water pumping processes: a) After mining; b) Filling concrete; c) Pumping underground water; and d) Water pipeline sealing.

Pipelines constructed in the mining process to ensure underground production were kept after abandonment of the mine. Therefore, these pipelines after treatment can be used to pump water and new pipelines should be designed for the injection of CO_2.

1. Underground water extraction
After coal mining activities, mined-out area is filled with water (Figure 11a), so underground water needs to be pumped in order to inject CO_2. Firstly, the main and auxiliary shafts should be closed by grouting as illustrated in Figure 11b. Secondly, the auxiliary shaft will be used for pumping underground water in order to create the WP space (Figure 11c). Finally, water pumping pipes should be closed by grouting (Figure 11d).

2. CO_2 injection
CO_2 can be injected to the deep underground space using another set of pipelines, and then when the injection process ends, the injection pipelines should be closed by grouting (Figure 12).

4.3 *Prevention of shaft leakage*

The closure of the shafts and the pipelines used for water pumping and CO_2 injection by means of grouting can effectively prevent leakage problems during CO_2 injection and follow-up processes of CSAC. According to the depth of underground spaces of the mine,

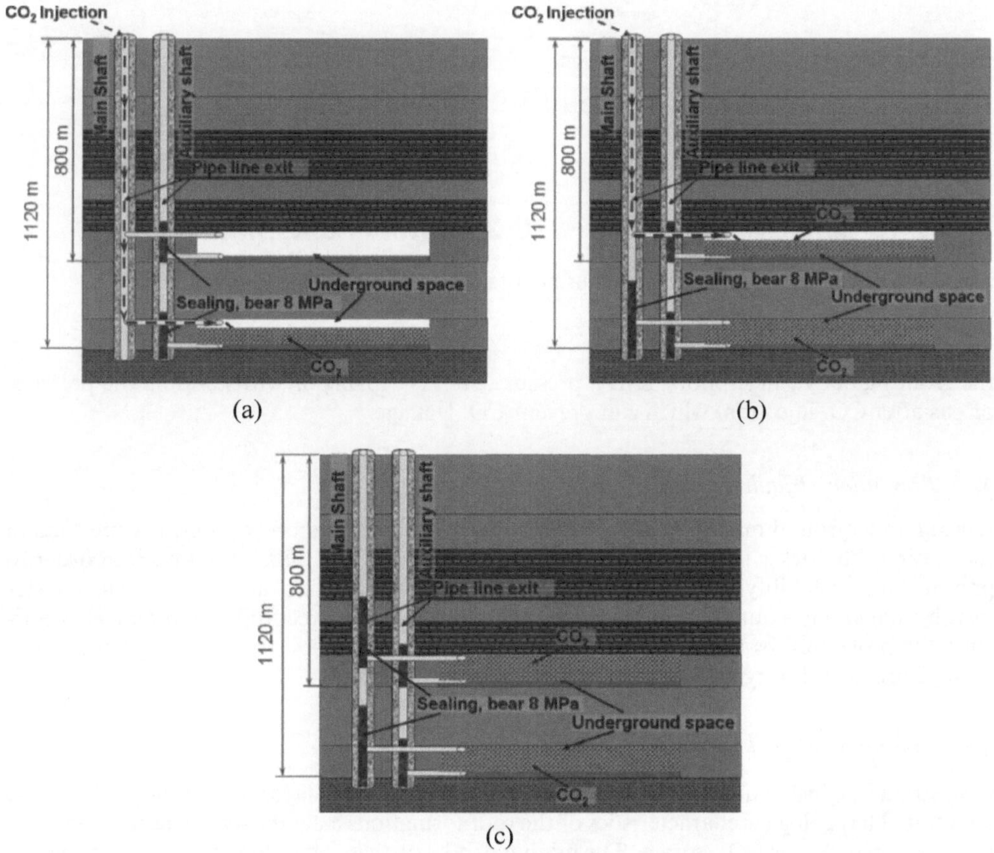

Figure 12. CO_2 injection processes: (a) Lower underground space injection; (b) Gas pipeline sealing and upper underground space injection; and (c) Gas pipeline sealing.

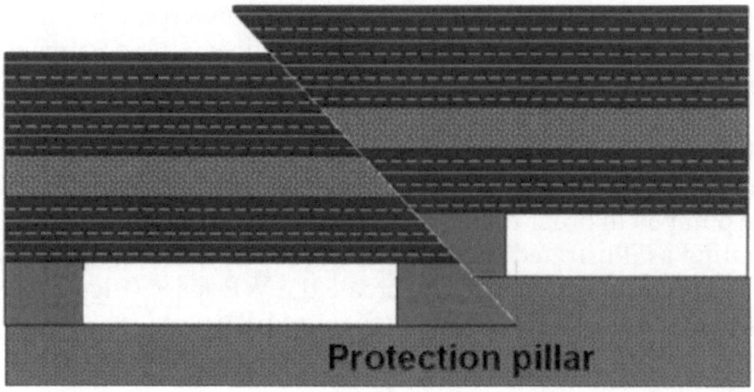

Figure 13. Protection pillar.

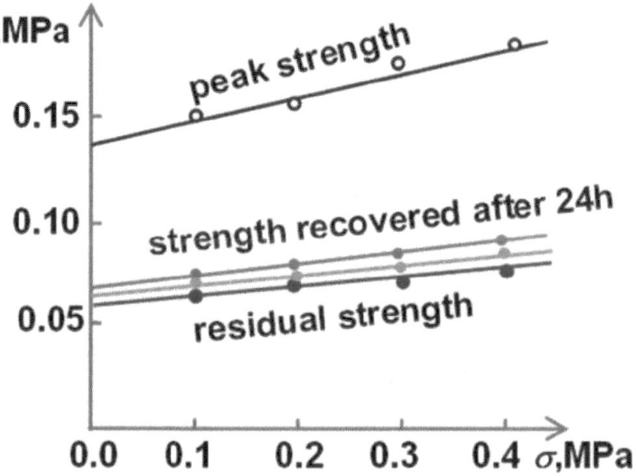

Figure 14. The shear strength properties of mudstone.

the grouting body can support 8 MPa pressure, so it is capable of withstanding the pressure of gas after CO_2 injection, which can prevent CO_2 leakage.

4.4 *Prevention of fault leakage*

During underground mining, when a fault is detected, the common countermeasure used in the mining activities is to keep the fault in the protection of the pillar (Figure 13) in order to prevent fault instability which will affect the advance of the work face and can cause water inrush, coal and gas outburst and other engineering disasters. According to practical experience, the protection by the pillar can sustain the rock strata stability, and also prevent fault leakage problem during CO_2 storage in underground space.

4.5 *Prevention of roof leakage*

Coal strata in Qinshanquan mine have mudstone and sandstone layers at the roofs with dense structure. The geological characteristics of the roof formations meet the sealing requirements of regional cap rocks for CO_2 storage. The mining depths of these abandoned mines are relatively deep, around 800 and 1120 m, which determines a sufficient thickness for regional sealing.

After coal mining, the residual tectonic stresses in the rock geological structures will release during the process of overlying rock breaking and settlement. With the stabilization of overlying rock, the entire rock structure tends to stabilize and can maintain stabilization for a long period of time. In addition, the overlying sedimentary strata in the coal seams, contains layers of mudstone and sandstone rocks with impervious characteristics. Another important aspect to be considered is the fact that the mudstone recovers strength properties after damaging as shown by laboratory tests (Figure 14). Therefore, after stable settlement of the overlying rocks, cracks can be filled, which can prevent the leakage of CO_2 and meet the long-term stability requirements for CO_2 geological sequestration.

5 ADVANTAGES OF CSAC

CO_2 storage needs to take full account of the geographical matching between CO_2 emission sources and geological storage sites, in order to reduce carbon transport costs and environmental risks.

According to CEC's 2009 Annual Report of China Electric Power Industry statistics data, in 2009 the national power generation capacity was 3.6812 trillion kwh, of which thermal power generation accounts for more than 81.8%, and 60,000 kilowatts and above the national coal (including coal gangue) unit power generation accounts for 95.2% of thermal power generation capacity. Electricity supply dependence on coal in China has been as high as 70% (Figure 1). On this basis, CO_2 emissions of thermal power generation reached to 3 billion ton (calculated by 0.997 kg CO_2 emissions per kWh).

Distribution of large-scale thermal power plants is concentrated at regions including the Pearl River Delta, Yangtze River Delta region, eastern North China Plain, the Yellow River region, the Northeast region, the Yunnan-Guizhou Plateau and Sichuan-Chongqing region. Among them, the eastern North China Plain, the middle reaches of Yellow River region and northeast China's are major coal production bases and there are a large number of abandoned coal mines, which can provide favorable conditions for the storage of CO_2 emissions from thermal power generation.

In addition, conventional thermal coal transport has many disadvantages, such as huge scale, long distance, and overburden. These unfavorable elements caused a series of problems including environmental pollution, rising prices and other social issues. Thus, construction of coal mine pit power plants has become an economical and secure measure to improve energy transmission. At present, China has built about 50 large-scale pit power plants with generating capacity of 300 million kWh/a. Total generating could be 150 million kwh per year. CO_2 emissions can be reduced by 150 million t/year if CO_2 emissions from these pit power plants are stored in underground space of coal mine directly after they were emitted. The benefits of CSAC can be significant.

6 CONCLUSIONS

Underground geological storage of carbon dioxide for the mitigation of climate change has great potential. Carboniferous formations and abandoned mines can be used to store carbon dioxide in underground space.

CO_2 storage in abandoned underground coal mines, especially, provides a new way for the realization of China's CO_2 emission reduction targets. Although this technology has some key scientific issues to be studied, the geological conditions and the technology are available on a feasibility basis.

In particular, with national plans of increasing pit coal-fired power generation up to 40%, conditions can be created for storing CO_2 in underground space of abandoned coal mines. Therefore, CSAC technology research and application will have broad prospects for China to achieve emission reduction targets.

REFERENCES

Climit 2010. CCS projects in China. Available at http://www.climit.no/frontend/files/CONTENT/CCSWorld/Japan/China_summary-issue5.pdf.

Li Guiju 2009. China-US cooperative research report shows that China has great potential carbon dioxide storage. *Science Research Dynamic Monitoring Express* 20, 6–8p.

NZEC 2007. CCS activities in China. Available at http://www.nzec.info/en/assets/Reports/CCS-Activities-in-China.pdf.

Paria J., Serkan S. & Cinar Y. 2011. CO_2 Storage in Abandoned Coal Mines. Underground Coal Operators' Conference. Paper 378, 354–360p.

Peng B. 2010. CO_2 Storage and Enhanced Oil Recovery in Jilin Oil Field. Enhanced Oil Recovery Research Center, China. University of Petroleum, Beijing, China. Available at http://www.ga.gov.au/image_cache/GA16242.pdf.

PetroChina 2007. CCS Activities and Developments in China. Available at http://www.un.org/esa/sustdev/sdissues/energy/op/ccs_egm/presentations_papers/li_ccs_china.pdf.

Piessens K. 2011. The conceptual model for an abandoned coal mine reservoir. *Int. Workshop on CO_2 Storage in Carboniferous Formations and Abandoned Coal Mines*, Ed. He, Sousa, Elsworth and Vargas. Beijing, pp. 179–200.

Piessens K. & Dusar M. 2003. CO_2-sequestration in abandoned coal mines. Proc. *Int. Coalbeb Methane Symposium, Tusaloosa*, Alabama, 346, 10p.

Piessens K. & Dusar M, 2004. Feasibility of CO_2 sequestration in abandoned coal mines in Belgium. *Geologica Belgica*, 7(3–4), 165–180p.

Shenhua 2011. Shenhua CCS Demonstration Project. Available at http://stock.sohu.com/20110104/n278670341.shtml.

Shi J.Q., Durucan S.. CO_2 storage in caverns and mines. Oil & Gas Science and Technology—Rev. IFP, 2005, 60(3): 569–571.

Sousa L.R. 2011. Present day conditions in the world of carbon capture and storage (CCS) projects. *Int. Workshop on CO_2 Storage in Carboniferous Formations and Abandoned Coal Mines*, Ed. He, Sousa, Elsworth and Vargas. Beijing, pp. 1–23.

Su, W. 2009. GreenGen Project in China. Available at http://www.iea.org/work/2009/bergen/Wenbin.pdf.

CO_2 Storage in Carboniferous Formations and Abandoned Coal Mines –
He, Ribeiro E Sousa, Elsworth & Vargas Jr. (eds)
© 2012 Taylor & Francis Group, London, ISBN 978-0-415-62079-6

Carbon Capture and Storage (CCS) activities in China

N. Zhang
State Key Laboratory for GeoMechanics and Deep Underground Engineering of China University of Mining and Technology, Beijing, China

L. Ribeiro e Sousa
State Key Laboratory for GeoMechanics and Deep Underground Engineering of China University of Mining and Technology, Beijing, China
University of Porto, Portugal

ABSTRACT: Carbon Capture and Storage (CCS) is an effective way to mitigate the potentially harmful effects caused by global warming. During the past decades, China has carried out a series of activities to promote CCS development. This paper provides a comprehensive description of CCS activities in China based on the documents collected from a variety of credible resources. A summary of CCS projects undertaken by Chinese organizations of science and technology is presented, as well as actions developed by Chinese enterprises. An outline of international science and technology cooperation on CCS is included. Particular emphasis is made to the EU projects and to the pilot tests conducted to Qinshui mine under the China–Canada projects. Finally, reference is made to laboratory experiments and numerical modeling studies performed or ongoing in China.

1 INTRODUCTION

Carbon capture and storage (CCS) is a very important and potentially effective approach to mitigate climate change caused by the emission of greenhouse gas CO_2 (Glessner and Young, 2008; IPCC, 2005; Su and Fletcher, 2010). CCS has started to develop in China mainly in the last decade with the increasing concern on the threat of climate change under the global circumstances. China has become the largest emitter of CO_2 in 2007 (He, 2011; He et al., 2011). Therefore, "CCS, as an option in the portfolio of mitigation actions to combat climate change, is expected to have far-reaching implications for China" (Climit, 2010).

In order to better identify the status quo of CCS activities in China, existing publications were collected and stored in a database. This permits one to improve the knowledge on existing CCS activities in China. Around 90 pieces of information including papers, publications and website information were collected. The topics cover CO_2 Enhanced Coal Bed Methane (ECBM), numerical modeling, capture technology, transportation, risk assessment, capacity estimation, and status quo of CCS in China. The distribution of documents by topics is illustrated in Figure 1 and distribution by sources in Figure 2.

In this chapter, we are intended to summarize the overall CCS activities in China with reference to several aspects as following: CCS projects supported by Chinese government (or institutions) and enterprises; international collaboration projects; typical demonstration project—ECBM Project at Qinshui; activities under development at State Key Laboratory for GeoMechanics and Deep Underground Engineering in Beijing; and numerical modeling and other theoretical studies.

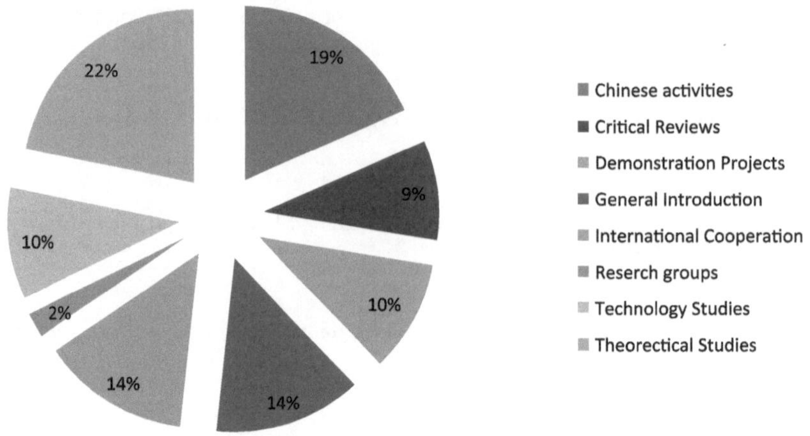

Figure 1. Distribution of collected documents by topics.

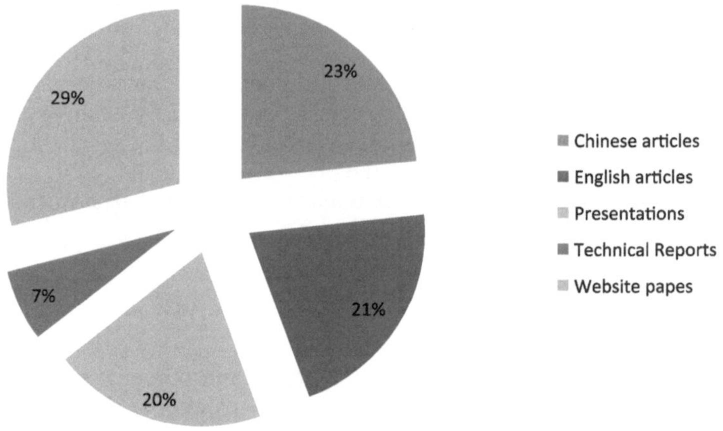

Figure 2. Distribution of collected documents by sources.

2 CCS ACTIVITIES IN CHINA

2.1 *Activities by Chinese institutions*

Chinese government has made great efforts to combat climate change. Strong support on technology research and development programs on climate change, in particular on mitigation technologies was initiated by governmental organizations and ministries headed by the Ministry of Science and Technology of the People's Republic of China (MOST).

2.1.1 *National Key Technology R&D program (KTRD)*

KTRD is one of the four Programs funded by MOST. During the 10th and 11th five-year periods, the KTRD program has supported strategic studies on CCS by Chinese research institutions. In particular, studies have addressed the applicability of CCS in China and its potential impacts on the energy system and Greenhouse Gas (GHG) emissions reduction.

2.1.2 *National Basic Research Program (NBRP, 973 program)*

NBRP program is the second key program supported by MOST. There are at least four CCS related projects supported by this program. One of the most important projects started in 2006 is Utilizing Greenhouse Gas as Resources in Enhanced Oil Recovery (EOR) and Geological Storage, which was led by Professor Shen Pinpin (Chief Scientist). This project addresses four key scientific issues: Geological issues associated with CO_2 storage; Physical and chemical issues during CO_2 injection and storage; Nonlinear flow mechanics during the CO_2 injection process; Issues of CO_2 capture and corrosion prevention. PetroChina, Institute of Geology and Geophysics Chinese Academy of Sciences, Peking University, China University of Petroleum (Beijing), Huazhong University of Science and Technology, Tsinghua University and the Jilin oilfield are jointly responsible for subtopics attached to this project. The overall 8 topics are as follows (NZEC, 2007):

- Topic 1—Standard setting and evaluation of the potential for CO_2 storage to fit China's geological characteristics (PetroChina).
- Topic 2—Geological theory of CO_2 subsurface storage, Institute of Geology and Geophysics (Chinese Academy of Sciences).
- Topic 3—Theory and technology of monitoring and predicting CO_2 storage processes (Peking University).
- Topic 4— Research on multiphase and multicomponent phase theory during CO_2 injection with EOR (PetroChina).
- Topic 5—Research on non-linear flow mechanisms and principles of multiphase and multicomponent flow during the CO_2 injection process (China University of Petroleum, Beijing).
- Topic 6—Principles of O_2/CO_2 circulating combustion for coal and the mechanism of synergetic removal of pollutants (Huazhong University of Science and Technology).
- Topic 7—Theory and technology of CO_2 separation and concentration from coal combustion (Tsinghua University).
- Topic 8—Theory and method of preventing CO_2 corrosion (Jilin Oilfield).

The second project is *Fundamental Research on Syngas Production through Coal Gasification and Pyrolysis*. Research into syngas production based on coal gasification and pyrolysis has been proposed in connection to the problem of gas emission and pollution in China's coal industry. The following key technology problems are addressed: Fundamentals of the large-scale gasification and syngas production based on coal gasification and pyrolysis; Fundamentals of catalyst synthesis for the fuel slurry reactor, of engineering of the reactor and of combustion of relief gas and crude gas; Special conditions of dual gas multi-generation systems and design and optimization theory for complex systems. Several participants were included in the project (NZEC, 2007).

The third project *Fundamental Research on the High Efficiency Transfer of Natural Gas and Syngas* aims to develop dynamic research on catalyst theory and concepts, and research into optimum catalyst reaction conditions, starting with the development of a new catalytic reaction and new catalytic materials followed by in-depth study of the optimum micro-structure and macro-structure for natural gas and syngas; and to extend the results to relevant coal conversion processes by catalyst-centered science and technology innovation. The project comprises eight research subtopics as follows (NZEC, 2007):

- Topic 1—Key problems of syngas production from natural gas, large scale hydrogen production and CO_2 handling.
- Topic 2—Obtaining high quality liquid fuels from syngas.
- Topic 3—Oxygen containing compound production from syngas.
- Topic 4—High temperature fuel cells based on syngas and natural gas.
- Topic 5—Direct conversion of methane into methanol.
- Topic 6—Novel processes for direct conversion of methane into methanol.
- Topic 7—The relationship between structure and activity and dynamic properties of catalysts and catalyst systems.
- Topic 8—Micro-mechanisms and the identification of intermediates during the catalytic process.

The fourth project *A Study of High Efficiency Heat Transfer in Gas Turbines* researches the energy transmission, conversion and utilization and CO_2 removal process of the system. Eight subjects were incorporated in the project (NZEC, 2007):

- Topic 1—Research on the macromolecular structure of biomass and response mechanisms under thermo-chemical circumstances.
- Topic 2—Fundamental research on the biomass gasification.
- Topic 3—Fundamental research on biomass selective pyrolysis.
- Topic 4—Research on the selective control law of the complex bio-oil catalytic process.
- Topic 5—Research on the characteristic molecular properties and separation of bio-oils.
- Topic 6—Study of reaction mechanisms and fundamental laws of hydrogen production by reformation of bio-oil.
- Topic 7—Research on the mechanisms involved in the catalytic synthesis of liquid fuel by rich-CO_2 biomass.
- Topic 8—Optimization theory and modeling of biomass liquefaction processes.

These subjects are jointly undertaken by ten Chinese organizations (NZEC, 2007).

2.1.3 *National High-Tech Program (NHTP, 863 program)*

The third key program is NHTP. The project *Development of CCS* has been supported by this program. Three major topics under this project include:

- Topic 1—CO_2 capture technologies based on absorption.
- Topic 2—CO_2 capture technologies based on adsorption.
- Topic 3—CO_2 geological storage.

2.1.4 *Other programs*

Special reference is made to the following projects:

a. *Risk Assessment of CO_2 Injection Processes and Storage in Carboniferous Formations*
The project consists of risk assessment of CO_2 injection and sequestration in carboniferous reservoirs, either in abandoned mines or deep unminable coal seams. The importance of the project is based on the fact that China is the major producer of coal in the world. Therefore, there are several possibilities for selecting appropriate sites for the reservoirs, and in abandoned coal mines.

The project is funded by the State Administration of Foreign Experts Affairs and is being conducted by the State Key Laboratory for GeoMechanics and Deep Underground Engineering of China University of Mining and Technology, Beijing (GDUE). Five technical tasks are planned in the project, namely:

- Regional characterization and determination of relevant formation rocks properties.
- Creation of a 3D geological-transport-geomechanical model of proposed CO_2 reservoirs.
- Numerical simulation of the injection processes.
- Risk Assessment of different escape scenarios.
- Pilot tests.

An International Workshop on *CO_2 Storage in Carboniferous Formations and Abandoned Coal Mines* was held in January 8–9, 2011, Beijing. Several publications were included in the Proceedings of the Workshop, as referred in Table 1.

b. *Experiments on CO_2 and CH_4 Sorption*
Besides the *Risk Assessment of CO_2 Storage in Carboniferous Formations*, GDUE is also carrying out another research program about gas sorption of soft rock. The purposes of the studies are as following:

- Conduct CO_2 and CH_4 sorption experiments of different types of soft rock under different temperature and humidity conditions
- Summarize CO_2 and CH_4 sorption characteristics of various soft rocks
- Explore CO_2 and CH_4 sorption mechanisms and affecting factors of various soft rocks.

Table 1. List of publications of the workshop.

Title	Authors	Reference
Present day conditions in the world of CCS projects	L. Sousa	Sousa (2011a)
Considerations on CO_2 storage in abandoned coal mines in China	M.C. He	He (2011)
Carbon capture and storage (CCS) activities in China	N. Zhang & L. Sousa	Present paper
Complex process couplings in systems pushed far-from-equilibrium: Applications to CO_2 sequestration in carboniferous formations	D. Ellsworth, S. Wang, G. Izadi, H. Kumar, J. liu, D. Lee, J. Mathews & D. Pone	Ellsworth et al. (2011)
A fully coupled gas flow, coal deformation and thermal transport model for the injection of carbon dioxide into coal seams	Hongyn Qu, Jishan Liu, Zhongwei, Zhejun Pan & Luke Connell	Qu et al. (2011)
Micro-scale modeling of gas-coal interaction in coalbed seam—Heterogeneity	Wang G. & Jishan L.	Wang and Liu (2011)
Considerations on the numerical modeling of injection processes of CO_2 in geological formations with emphasis on carboniferous formations and abandoned coal mines	E.A. Vargas Jr., R.Q. Velloso, W.N. Ribeiro, A.L. Muller & L.E. Vaz	Vargas et al. (2011)
Methodologies for risk analysis and decision making	R. Sousa	Sousa (2011b)
Risk associated to storage of CO_2 in carboniferous formations. Application of Bayesian Networks	L. Sousa & R. Sousa	Sousa and Sousa (2011)
The concetula model for an abandoned coal mine reservoir	K. Piessens	Piessen (2011)

c. *Integrated Research on CO_2 Emission Reduction, and its Resource Recycling, Low NOx Combustion, SOx Control and Multi-pollutants Removal*

The project funded by MOST is an application study on integrated control of multiple pollutants in coal-fired power plants. The project comprises three parts: research on coal combustion, control and analysis of CO_2 emissions reduction and multiple pollutants. An integrated system for eliminating flue gas, CO_2 and multiple pollutants had been designed, manufactured and installed. The system has reached the design requirements after successful adjustment and tests (NZEC, 2007).

2.2 *Activities by Chinese enterprises*

2.2.1 *Programs by PetroChina*

PetroChina, China's largest oil and gas producer, has performed several basic research projects funded by MOST, which include the following programs:

- Standard stipulation and potential evaluation for CO_2 storage that meets China geologic characteristic.
- Research on phase theory of multiphase and multicomponent during CO_2 flooding process.
- Theory and method of engineering for CO_2 corrosion prevention and antiscale, Jilin Oilfield.

In addition to the theoretical researches mentioned above, PetroChina conducted a CO_2-EOR pilot test in Jilin Oilfield (Peng, 2010). Jilin Oilfield is located in the southern Songliao basin of Songyuan district, Jilin province (Figure 3) and its exploration began in 1955. In 2007, a project *Exploitation of CO_2-containing Natural Gas, CO_2 storage and* Utilization in Jilin Oil

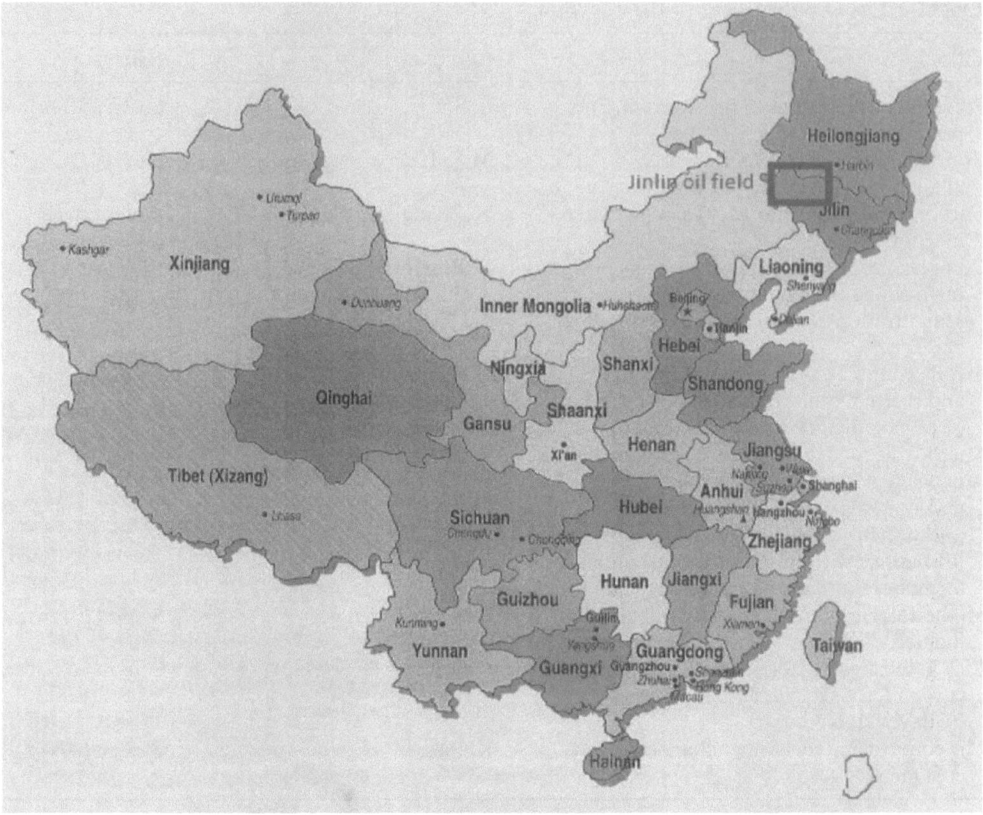

Figure 3. Location of Jilin Oilfield (Peng, 2010).

Field was established by PetroChina. The duration of the project is from April, 2007 to 2009. The primary goal of the project was to develop the bottleneck technology of the exploitation of CO_2 containing natural gas and to provide technical support for CO_2 storage (PetroChina, 2007). Natural source of CO_2 is from natural gas (about 22.5%) and reserve of CO_2 for Jilin Oilfield is 12.5 Gm^3. CO_2-EOR pilot tests of block Hei79, Hei59, Hong87-2, Qian'an and other 6 blocks had been accomplished by the end of 2009. Figure 4 shows the distribution of test regions in Jilin Oilfield. Test results indicated that recovery rate can be enhanced by 10~20% for miscible flooding and by 5~10% for immiscible flooding (Peng, 2010).

2.2.2 *Programs by Huaneng Group*

a. *Grengen project*

China Huaneng Group (CHNG) is China's largest power generation company. GreenGen project was initiated by CHNG in 2004 based on the support of National Development and Reform Commission (NDRC). The Greengen Corporation is a joint venture with 8 shareholders which are the largest power and coal companies in China (NZEC, 2007; Su, 2009).

Greengen project aims at the development, demonstration and popularization of hydrogen production by coal gasification, hydrogen gas turbine combined power generation, fuel cell power generation and coal based energy system with CCS in order to elevate coal generation efficiency and realize near zero emissions of pollutant and CO_2, and to prepare technology for addressing climate change to realize the sustainable development of coal-based power generation (NZEC, 2007).

The project was designated for three stages (Figure 5).The first stage was to construct an Integrated Gasification Combined Cycle (IGCC) power plant (Figure 6). In this stage, a

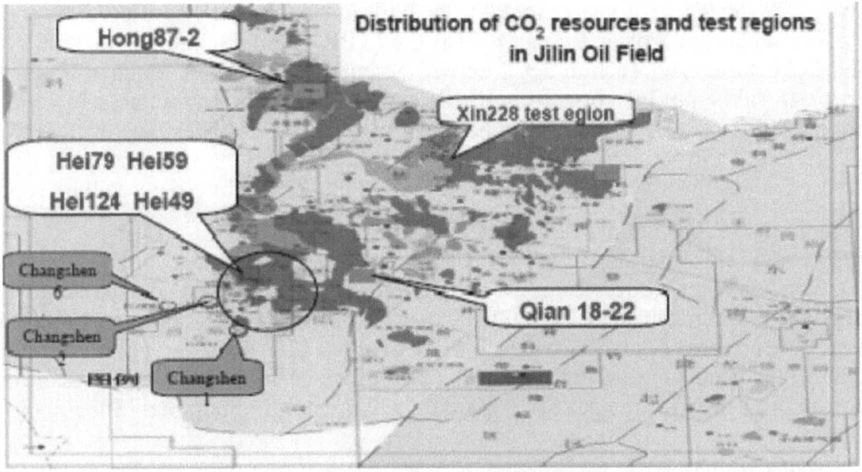

Figure 4. Distribution of test regions in Jilin Oilfield (Peng, 2010).

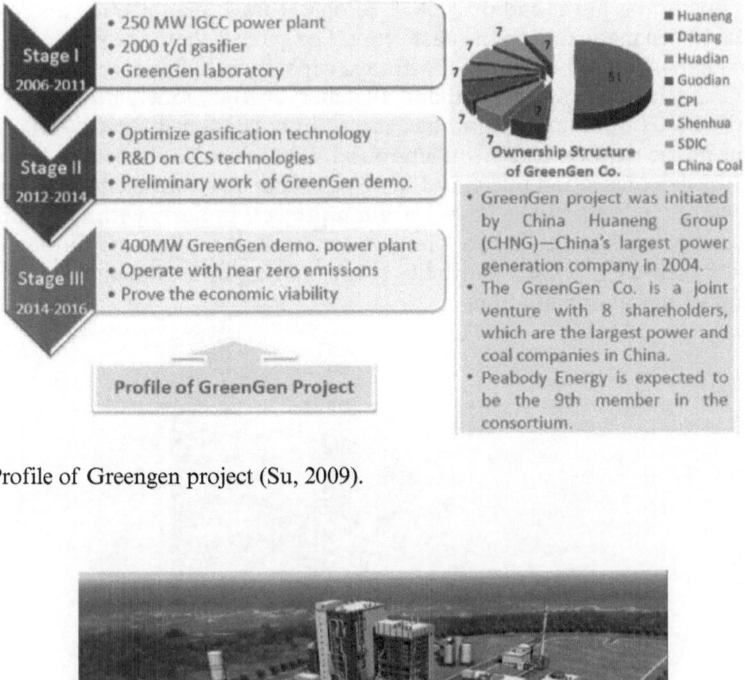

Figure 5. Profile of Greengen project (Su, 2009).

Figure 6. Design sketch of Greengen IGCC plant to be finished in 2011 in Tianjing (Su, 2009).

250 MW IGCC power plant was planned to be built in Tianjin with gasifier (2000 tpd) supplied by Xi'an Thermal Power Research Institute (TPRI) and E-class gas turbine between 2006 and 2011. Greengen Technology Lab will be built at the IGCC plant in parallel. The IGCC project will be put into operation by the end of 2011. The second stage includes optimization of a TPRI gasifier and development of the main technologies. The third stage is the implementation of the 400MW Greengen near-zero emissions demonstration power plant in which hydrogen gas turbine power generation and other CCS technologies will be verified.

b. *Post-combustion CO_2 Capture (PCC) demonstration project*

Huaneng Group and Australia's Commonwealth Scientific and Industrial Research Organization (CSIRO) started cooperation on a PCC pilot plant deployed in Huaneng's Beijing thermal power plant in 2008 (NZEC, 2007; SciDev, 2008). The post-combustion CO_2 capture demonstration project (Figure 7) with capacity of 3000 t/a located in Beijing Gaobeidian Power Plant has been successfully put into operation since June, 2008. A post-combustion CO_2 capture project with capacity of 100,000 t/a will be built in Shanghai (Su, 2009).

2.2.3 *Shenhua Group*

Shenhua Group as China's largest coal producer launched a CCS project at its 1-million-ton direct coal-to-liquids plant in Ordos, Inner Mongolia autonomous region in 2004. This project intended to build China's first commercial CCS facility. Estimated investment of the project is about 210 million RMB. Shenhua completed a first commissioning with 303 hours of successful operations at the end of 2008. Geologic storage site selection and characterization efforts have been made during the past years. The geology in the region looks promising, with an estimated 4.4 billion tons of CO_2 storage capacity in saline formations in the Ordos basin (NZEC, 2007; Reuters, 2009; Su and Fletcher, 2010). Liquefaction and Purification Equipment of the CCS demonstration has successfully produced a purity of 99.2% liquid CO_2 which meets the requirement of underground storage on December 30, 2010 (Shenhua, 2011). The CCS demonstration injecting 100,000 tons per year started commissioning in the early of 2011. On the basis of the demonstration project, projects with CO_2 collection and storage capacity of 1 million tons and 3 million tons will be built in the future in 2 steps, of which 1 million tons CCS project feasibility study is being done, but there is no clear timetable for its full-scale implementation.

Figure 7. Demonstration project of post-combustion CO_2 capture in Beijing Gaobeidian power plant (Su, 2009).

2.2.4 *ENN Group*

ENN, China's big global clean energy company, is working on capturing carbon emissions based on algae, a tiny plant organism, and a process called bio-absorption (ENN, 2010). The algae can turn CO_2 into oil product that is a clean burning fuel by utilizing solar energy. This biological technology prevents carbon from escaping into the atmosphere. ENN's work in this area is being conducted at its experimental center in Langfang, Hebei Province near Beijing and supported by China's national high-tech development program known as Program 863, which is designed to stimulate development of advanced technologies in China.

3 INTERNATIONAL COLLABORATION PROJECTS

In addition to the projects committed by the Chinese government, institutions and enterprises, China has taken active part in a number of international CCS projects and programs during past decade. Table 2 lists the projects jointly undertaken by China and other countries. The topics cover a wide range in CCS research field (Table 3).

3.1 *China–EU cooperation projects*

3.1.1 *Cooperation action within CCS China-EU (COACH)*

The China-EU COACH project was launched in November 2006 as a project under the theme Sustainable Energy Systems and ended in December 2009. Major activities of the project include application of carbon capture technology to coal-fired power plants, evaluation of China's CO_2 storage potential and the study of relevant regulations and financing

Table 2. International projects with China's involvement.

		Planning		Major topics[5]			
Project	Origin	Beg.	End	T1	T2	T3	T4
MOVECBM	EU	2006	2008	1	2	3	4
COACH	EU	2006	2009	5	6	7	–
STRACO$_2$	EU	2008	2009	7	8	–	–
CACHET (I (II)	EU[4]	?	?	11	8	5	–
CAPRICE	EU	2007	2008	5	11	8	–
GeoCapacity	EU[2]	2006	2008	8	11	–	–
China–UK Near Zero Emissions Coal (NZEC)	UK[3]	2010	2015				
China–Australia CAGS Project	Australia	2009	?				
China–Canada ECBM Project01	Canada	2002	2006	9	10	1	–
China–Canada ECBM Project02	Canada	2008	2013	9	10	1	–
Tianjin Dagang CCS Project	USA?	2008	?	11	3	8	–
Sino-Japan CCS Cooperation	Japan	2008	2011	11	9	5	8
Chinese Advanced Power Plant Carbon Capture Options (CAPPCCO)	UK	?	?	11	5	8	7
Developing CCS Guidelines for China	USA	?	?	7	8	11	–
US-China Study on Regional Opportunities for CCS	USA	?	?	11	8	6	–
APP (Asia-Pacific Partnership on Clean Development and Climate	Several[1]	?	?	8	11	3	–

[1] Australia, Canada, China, India, Japan, Korea and USA.
[2] With China, India and Russia (Assessing European Capacity for Geological Storage of Carbon Dioxide).
[3] Also with EU.
[4] Cachet II started in 2010.
[5] See Table 3.

Table 3. Designation of the research topics of international CCS projects.

Topic	Designation
1	Understanding CO_2 injection into coal seams
2	Migration of methane
3	Long-term reliable and safe storage
4	Modeling and laboratory work
5	Application of carbon capture technology in coal-fired power plants
6	Evaluation of China CO_2 storage potential
7	Study of relevant regulations and financing mechanisms
8	Facilitate and increase S&T cooperation
9	Coal bed methane technology
10	In situ pilot tests
11	CCS technologies

mechanisms. Activities were also be carried out to improve public awareness and to build CCS capacity. 8 Chinese organizations including Tsing Hua University, Zhejiang University, Institute of Geochemistry, Chinese Academy of Sciences (IGC/CAS), The Administrative Centre for China's Agenda 21 (ACCA21), Research Institute of Petroleum Exploration and Development, PetroChina (RIPED), TPRI, GREENGEN were involved in the project. Details about COACH, please see http://www.co2-coach.com/.

3.1.2 *China–UK Near Zero Emissions Coal (NZEC)*

The Joint China–UK initiative addresses the challenge of rising CO_2 emissions through joint research, development and demonstration of near zero emissions coal technology. UK envisages an ambitious three-phase approach to achieve NZEC demonstration objective. Phase 1 has explored options for demonstration and builds capacity for CCS in China. Phase 2 will carry out further development work on storage and capture options leading to Phase 3, which will construct a demonstration plant by 2015. NZEC is funded by the UK's Department of Energy and Climate Change (DECC), and is being taken forward in partnership with MOST. 19 Chinese partners were included in the project. Progress of the project can be seen at http://www.nzec.info/en/.

3.1.3 *Monitoring and Verification of CO_2 Storage and ECBM in Poland (MOVECBM)*

The MOVECBM project started in November 2006 and has a duration of 2 years. The objective of MOVECBM project was to improve understanding on CO_2 injection into coal seams and the migration of methane, ensuring long-term reliable and safe storage. In the project, modeling and laboratory work will be based on the parameters of the test site in Kaniów, Poland (previously investigated by EU RECOPOL project). Chinese Organizations including State Key Laboratory of Coal Conversion (SKLCC), RIPED and China United Coalbed Methane Company (CUCBM) were involved in this project. Details about MOVECBM can be found at its official website http://www.movecbm.eu/.

3.1.4 *Assessing European Capacity for Geological Storage of Carbon Dioxide (GeoCapacity)*

The GeoCapacity project is a 3 year project and started on January 2006 and ended on December 2008. The main objective of the project was to assess the European Capacity for Geological Storage of Carbon Dioxide (EU GeoCapacity). The project included full assessments of a number hitherto not covered countries, and updates of previously covered territory. Also a priority was the further development of innovative methods for capacity assessment, economic modelling and site selection criteria. Finally, an important mission was to initiate scientific collaboration with China, a member of the CSLF and Tsinghua University is the only one partner from China of the total 26 partners from 21 countries. A website has been established that contains available public information (www.geocapacity.eu).

3.1.5 *Support to Regulatory Activities for Carbon Capture and Storage ($STRACO_2$)*

The $STRACO_2$ program started in January 2008 and ran for 18 months. The major objective of the project is to support the ongoing development of the comprehensive regulatory of China framework of EU for CCS technologies for zero emissions applications. Its specific objectives were: Identify science and technology development priorities in multiple aspects of technical applications and regulations for specific themes; Facilitate and increase S&T cooperation at the EU level; Identify through research, pivotal issues for regulatory development; Establish the EU regulatory framework as the basis for dialogue and priority setting with regulatory authorities in China with a view to further joint activities; Facilitate and increase S&T cooperation at the international level with China. The $STRACO_2$ Project work plan is divided into eight Work Packages (WP) to ensure that all partners work in their area of expertise. 3 Chinese organizations including ACCA21, Institute of Engineering Thermophysics, CAS (IET/CAS), and the Institute of Policy and Management of the Chinese Academy of Sciences (IPM) were involved in the project. Progress of $STRACO_2$ can be obtained at http://www.euchina-ccs.org/index.php.

3.1.6 *Other EU projects*

Some Chinese organizations also participated in some funded EU projects such as CACHET and CAPRICE. These projects mainly focus on research areas of interest in Europe, and only 1 or 2 Chinese organizations have joined each project.

3.2 *China Australia Geological Storage (CAGS) project*

The China–Australia Geological Storage of CO_2 (CAGS) project is a collaborative, bilateral project between Australia and China. The CAGS project started in 2009 after meetings including China, Australia and EU/UK and will be ended in 2011. The GACS project aims to deliver the technology transfer to facilitate China's own assessment of sites for geological storage of CO_2 across different regions. The work program is divided into the following areas:

- Capacity building events, including a series of technical workshops, a summer school, and a technical symposium towards the end of the project;
- Building awareness of CCS in China through a study tour for policymakers and business leaders;
- Research into geological storage in saline formations, oil and gas fields, and CO_2-enhanced oil recovery, as well as research into environmental impacts and risk assessment;
- Researcher and student exchanges between China and Australia, and support for Chinese researchers and professionals working in CCS to attend international CCS conferences.

More details about this project are located at http://www.cagsinfo.net/.

3.3 *China–Canada ECBM projects*

A project agreement on coal bed methane technology development and CO_2 storage was signed between China and Canada on 2002 until 2006. CUCBM and Canada Alberta Research Council finished a micro-pilot field test at Qinshui Basin, Shanxi province. The details about this micro-pilot test at Qinshui are given in the following chapter. In 2008, a project focuses on the development of EOR technology through a multi-well pilot CO_2 injection in deep unmineable coal seams and geological storage has been launched as a continuation of the micro-pilot test in Qinshui Basin. The north block of Shizhuang at Qinshui Basin was selected as the pilot site (NZEC, 2007).

3.4 *Tianjin Dagang CCS project*

In 2008 EESTech and Tianjin Dagang Huashi Power Generation signed a memorandum of understanding. Dagang plans to use one of its two 330 MW power units to demonstrate the capture of CO_2 from its flue stream and then transport the CO_2 for geo-sequestration and EOR.

3.5 *Sino–Japan CCS Cooperation*

In May 2008, Japan and China announced their intention to jointly develop a CCS and EOR project that aims to recover 3–4 million tonnes of CO_2 per annum from two coal-fired power plants in China. For China, the NDRC is the lead government department with input from: CNPC-PetroChina, Daqing Oil Field Ltd. (local oil field partner), Harbin district government, Harbin Utilities Company and China Huadian Corporation.

3.6 *Chinese Advanced Power Plant Carbon Capture Options (CAPPCCO)*

The CAPPCCO project aims to produce a carbon capture characteristics database for existing and planned plants; to develop and assess capture options for planned new pulverized coal (PC) plants and existing PC plants. CAPPCCO is sponsored by the UK Department of Energy & Climate Change (DECC) and China's Ministry of Science and Technology (MOST).

3.7 *Developing CCS guidelines for China*

This is a collaborative project by Tsinghua University and the World Resources Institute (WRI), funded by the US Department of State and the Asia Pacific Partnership on Clean Development and Climate. The main outcomes will be: 1) a publication of a set of guidelines for safe and effective CCS (including capture, transport, and sequestration) and 2) a framework for the post-closure stewardship.

3.8 *US–China study on regional opportunities for CCS*

The US Department of Energy and MOST are funding a joint study examining Regional Opportunities for Carbon Dioxide Capture and Storage in China. Partners include the U.S./China Energy and Environmental Technology Center, Tsinghua University, Institute of Rock and Soil Mechanics: Chinese Academy of Sciences, Leonardo Technologies, Inc., Battelle Pacific Northwest National Laboratory and Montana State University.

4 MICRO-PILOT TEST AT QINSHUI BASIN

The ECBM Micro-pilot Field Test at Qinshui Basin was one of the largest cooperation programs between China and Canada (Figure 8). The objective of the micro-pilot test deployed at Qinshui was to measure and evaluate data to obtain estimates of reservoir properties and sorption behavior and to calibrate a simulation model for estimation of the enhancement of CBM recovery in a larger-scale pilot or full field development (Gunter et al., 2005).

Location of demonstration site at Qinshui Basin is shown in Figure 9. The attractive characteristics of the Qinshui Basin includes: 1) large areal extent—24,000 km^2; 2) abundant CBM resource—5.5 trillion m^3; 3) high rank semi-anthracite and anthracite coal; 4) laterally thick continuous coal seams; 5) relatively shallow depths of coal seams; 6) reasonable access to local and distant markets (west to east pipeline is 30–90 km from south Qinhsui at Jincheng); 7) relatively more explored than other basins (Gunter et al., 2005).

Major tasks of the micro-test at Qinshui included: 1) potential pilot site selection; 2) geological-engineering-environmental characterization and ranking of selected 3 pilot sites; 3) design of micro-pilot field test to evaluate CBM reservoir properties; 4) carry out a single well micro-pilot field test at the best suitable site: selecting of existing wells or drilling new wells and up to 3 micro-pilot tests will be performed if first 2 tests do not show commercial potential; 5) micro-pilot test evaluation and numerical model calibration and fine tuning; 6) large-scale pilot design leading to commercial production; 7) training and technology transfer to be conducted in Canada and China.

Figure 10 shows the operation scene of liquid CO_2 injection at the project site. Goal of CO_2 injection was to inject 200 tons into reservoir over a 12 day period. Injection rate was

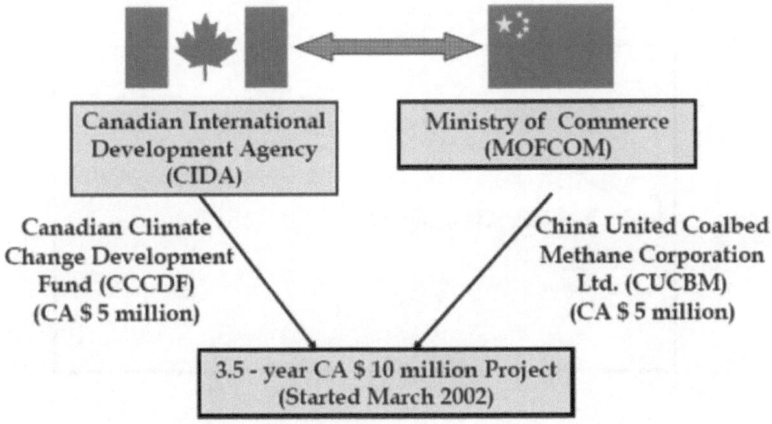

Figure 8. China–Canada ECBM project (NZEC, 2007).

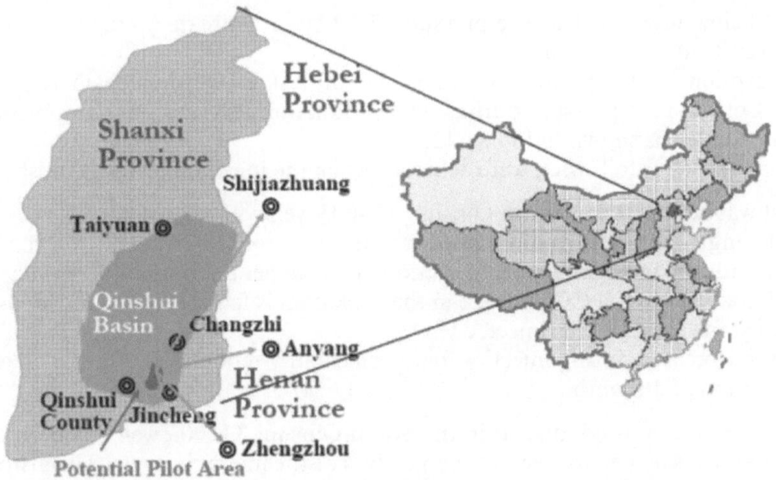

Figure 9. Location of demonstration site at Qinshui Basin (Gunter et al., 2005).

Figure 10. Injection of liquid CO_2 at the project site of micro-pilot test at Qinshui (Gunter et al., 2005).

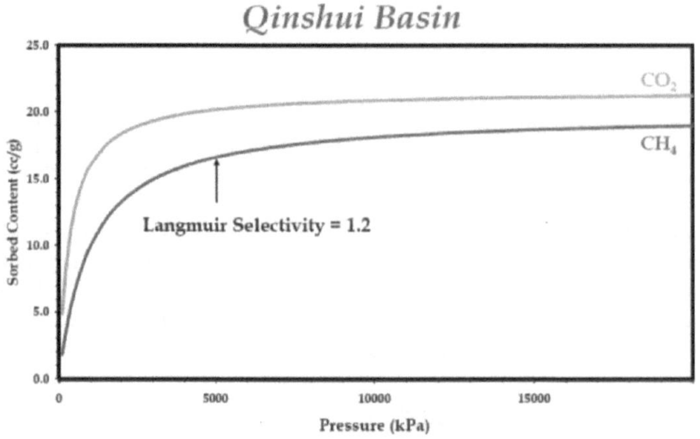

Figure 11. Adsorption isotherms for CO_2 and CH_4 of test at Qinshui basin (Gunter et al., 2005).

maintained below reservoir fracture pressure (7.5 MPa). Estimate average injection rate of 30 L/min over 8–10 hour period.

The adsorption isotherms for CO_2 and CH_4 of test at Qinshui basin is presented in Figure 11. Bottom-hole pressure during final production test and gas composition during final production test is shown in Figure 12.

Final conclusions as following were achieved when the injection test was finished:

- Pilot test was conducted in South Qinshui TL-003 well with success.
- The matching of dataset from the pilot test and the simulation prediction for the multi-well pilot indicated a significant production enhancement compared to primary production and that substantial CO_2 storage in the coal seam is feasible in a multi-well-project.
- 192 metric tones of CO_2 was injected.
- Infectivity decreased during injection but permeability rebounded after an extended production period of 1 month.

The micro-pilot test as conducted in the South Qinshui TL-003 well has been completed successfully and has met all the technical objectives of the micropilot test. The history matching of the dataset from the micro-pilot and the simulation prediction for the multi-well pilot indicated a significant production enhancement compared to primary production, and that substantial CO_2 storage in the coal seam is feasible in a multi-well project.

5 LABORATORY EXPERIMENTS AND NUMERICAL MODELING

In addition to the demonstration tests that have been jointly carried out by Chinese organizations and foreign institutions, a number of fundamental researches have also been addressed by Chinese scientists and engineers with or without the cooperation of foreign experts through numerical simulation methods and laboratory experiments. Only the following publications are referred: Only the following publications are referred: Hu et al. (2010); Wu et al. (2010); Yu et al. (2008); and Zhu et al. (2009).

Based on the configuration of one-injection well/one-production well under ideal conditions, Yu and his colleagues (2008) constructed an innovative experimental apparatus (Figure 13) for CH_4 displacement with CO_2 injection to investigate the basic procedure of coalbed methane (CBM), ECBM, and CO_2 sequestration at driven-conditions. The results of their laboratory experiments were successfully implemented in a small scale experiment of CH_4 displacement with CO_2 injection. Furthermore, on the basis of this primary apparatus, a new experimental setup developed to simulate ECBM and CO_2 sequestration on field scale based on the ideal apparatus was under consideration of Yu's research team.

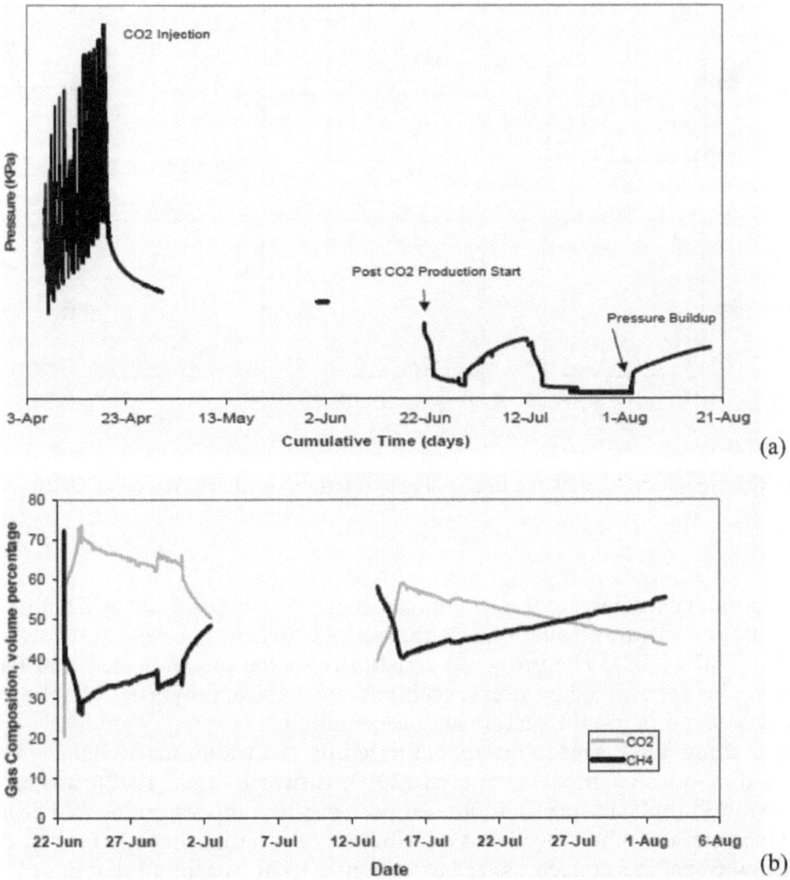

Figure 12. Bottom-hole pressure during final production test (top); gas composition during final production test (bottom) (Gunter et al., 2005).

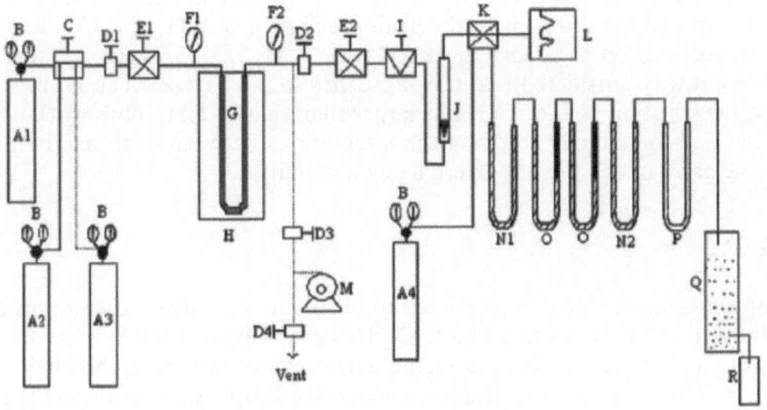

Figure 13. Schematic diagram of experimental apparatus used in Wu's study. A1, A2, A3 or A4: CH_4, CO_2, He or H_2 cylinder; B: manometer and regulator on gas cylinder; C: switch valve; D1, D2: shut-off valve of front-column and back-column; D3, D4: shut-off valve used to detect leak; E1, E2: regulator of front-column and back-column; F1, F2: manometer of front-column and back-column; G: coal sample column; H: thermostatic chamber; I: needle valve; J: flowmeter; K: six-port valve; L: GC; M: vacuum pump; N1, N2: moisture absorbent; O1, O2: CO_2 absorbent; P: protective tube; Q: lower-opening bottle; R: graduated cylinder (Yu et al., 2008).

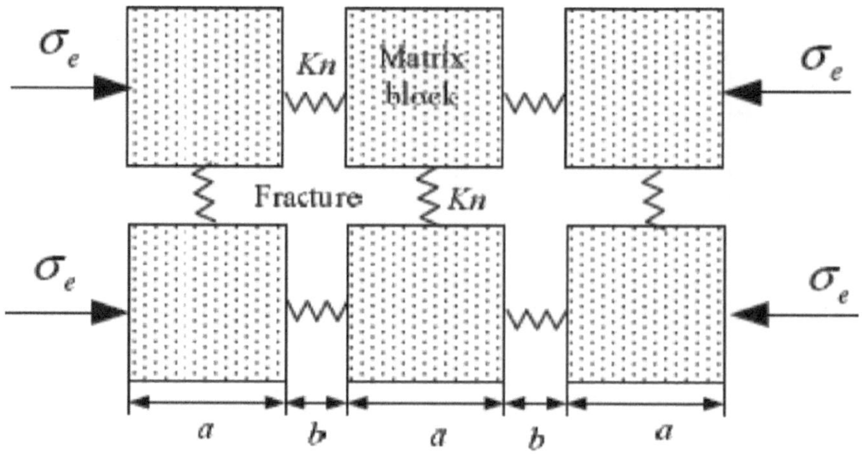

Figure 14. Representation of coal as a dual-porosity fractured medium (Wu et al., 2010).

Besides laboratory studies, numerical simulations also have been done by Chinese researchers. A new set of constitutive equations for the saturate mixtures of gas-filled coal was developed by Zhu et al. (2008). The proposed constitutive equations indicated that mechanical response of coal is determined by stress condition, mechanical properties of solid phase, gas pressure, gas adsorption, pore structure and its evolution.

In a recent study, the permeability model based on the traditional general porosity and permeability dual-porosity model (Bai et al., 1993; Elsworth et al., 1992), was extended to define the evolution of gas sorption-induced permeability anisotropy by Wu et al. (2010). The model incorporates a heterogeneous distribution of fractures in coal. In their derivation of the model, the coal was conceptualized as in Figure 14. It was found that under the condition of constant reservoir volume the interaction between fracture and matrix is controlled by the sorption-induced strain only, while under stress controlled conditions, the differential stress may elevate the gas sorption-induced coal permeability anisotropy.

In another recent work (Hu et al., 2010), based on molecular simulation, Hu et al. investigated and compared the diffusion and sorption behavior of CH_4 and CO_2 in coal. It was indicated that coal tended to adsorb more CO_2 than CH_4 at a given temperature and pressure. Moreover, it was considered that CH_4 adsorbed in the coal seam could be replaced by CO_2 because the sorption heat of CO_2 was larger than that of CH_4. Hu's work proposed an alternative method for directly studying the interactions between coal macromolecule and small-molecule gases under various external environments.

6 SUMMARY

Climate change is a major threat to global environment and sustainable development. China is carrying out a series of activities to promote CCS (Carbon Capture and Storage) development. A summary of CCS projects funded through the national science and technology programs is presented, as well as the actions by Chinese enterprises. Emphasis is made to the risk assessment and CO_2 & CH_4 adsorption programs conducted at GDUE. An outline of international science and technology cooperation on CCS is included. Particular emphasis is made to the EU projects and to the pilot tests conducted to Qinshui mine under the China-Canada ECBM projects. Finally, reference to laboratory experiments and numerical modeling studies newly performed in China. In summary, China has been making great efforts in developing CCS technologies and promoting CCS technology implantations in industry home and abroad.

REFERENCES

Bai, M., Elsworth, D. & Roegiers, J.C. 1993. Multiporosity/Multipermeability approach to the simulation of naturally fractured reservoirs. *Water Resources Research* 29, 1621–1633.

Climit 2010. CCS projects in China. Available at http://www.climit.no/frontend/files/CONTENT/CCS-World/Japan/ China_summary-issue5.pdf.

Elsworth, D. & Bai, M. 1992. Flow-deformation response of dual-porosity media. *Journal of Geotechnical Engineering* 118, 107–124.

Elsworth D., Wang, S., Izadi, G., Kumar, H., Liu, J., Lee, D.-S., Mathews J. & Pone, D. 2011. Complex process couplings in systems pushed far-far-from equilibrium: Applications to deep geologic sequestration and energy recovery. *Int. Workshop on CO_2 Storage in Carboniferous Formations and Abandoned Coal Mines*, Ed. He, Sousa, Elsworth and Vargas. Beijing, pp. 55–68.

ENN 2010. Available at http://www.enn.cn/en/about/profile.html.

Glessner, M.W. & Young, J.E. 2008. Carbon Capture and Storage. *Chemical Engineering* (NY) 115, 28–30.

Gunter, B., Wong, S., Law, D., ARC, Sanli, F., Jianping, Y., Zhiqiang, F. & CUCBM 2005. Enhanced Coalbed Methane (ECBM) Field Test at South Qinshui Basin, Shanxi Province, China. *GCEP workshop* August 22–23, 2005 Beijing. Available at http://gcep.stanford.edu/pdfs/wR5MezrJ2SJ6NfFl5sb5Jg/15_china_gunter.pdf.

He, M.C. 2011. Considerations on CO_2 storage in abandoned coal mines in China. *Int. Workshop on CO_2 Storage in Carboniferous Formations and Abandoned Coal Mines*, Ed. He, Sousa, Elsworth and Vargas. Beijing, pp. 25–36.

He, M., Sousa, L., Sousa, R., Gomes, A., Vargas, Jr., E. & Zhang, N. 2011. Risk assessment of CO_2 injection processes and storage in carboniferous formations. *Journal of Rock Mechanics and Geotechnical Engineering*, 3 (1): 39–56.

Hu, H., Li, X., Fang, Z., Wei, N. & Li, Q. 2010. Small-molecule gas sorption and diffusion in coal: Molecular simulation. *Energy* 35, 2939–2944.

IPCC 2005. IPCC Special Report on Carbon Dioxide Capture and Storage. Cambridge University Press.

NZEC 2007. CCS activities in China. Available at http://www.nzec.info/en/assets/Reports/CCS-Activities-in-China.pdf.

OilChina Jilin Oilfield Introduction. Available at http://www.oilchina.com/eng/Service-Center/oilfileds/Jilin.htm.

Peng, B. 2010. CO_2 Storage and Enhanced Oil Recovery in Jilin Oil Field. Enhanced Oil Recovery Research Center, China. University of Petroleum, Beijing, China. Available at http://www.ga.gov.au/image_cache/GA16242.pdf.

PetroChina 2007. CCS Activities and Developments in China. Available at http://www.un.org/esa/sustdev/sdissues/energy/op/ccs_egm/presentations_papers/li_ccs_china.pdf.

Piessens K. 2011. The conceptual model for an abandoned coal mine reservoir. *Int. Workshop on CO_2 Storage in Carboniferous Formations and Abandoned Coal Mines*, Ed. He, Sousa, Elsworth and Vargas. Beijing, pp. 179–200.

Qu Hongyn, Liu Jishan, Zhong Wei, Pan Zhejun & Connell Luke. 2011. Multiphysics of coal-gas interactions. Something old, something new and something very new. *Int. Workshop on CO_2 Storage in Carboniferous Formations and Abandoned Coal Mines*, Ed. He, Sousa, Elsworth and Vargas. Beijing, pp. 69–93.

Reuters 2009. Shenhua to launch China's first carbon capture project. Available at http://uk.reuters.com/article/2009/04/08/us-shenhua-carboncapture-idUKTRE5370EY20090408.

SciDev 2008. China ventures into carbon capture. Available at http://www.scidev.net/en/news/china-ventures-into-carbon-capture.html.

Shenhua 2011. Shenhua CCS Demonstration Project. Available at http://stock.sohu.com/20110104/n278670341.shtml.

Sousa, L.R. 2011a. Present day conditions in the world of carbon capture and storage (CCS) projects. *Int. Workshop on CO_2 Storage in Carboniferous Formations and Abandoned Coal Mines*, Ed. He, Sousa, Elsworth and Vargas. Beijing, pp. 1–23.

Sousa, L.R. & Sousa, R.L. 2011. Risk associated to storage of CO_2 in carboniferous formations. Application of Bayesian Networks. *Int. Workshop on CO_2 Storage in Carboniferous Formations and Abandoned Coal Mines*, Ed. He, Sousa, Elsworth and Vargas. Beijing, pp. 153–178.

Sousa, R.L. 2011b. Methodologies for Risk Analysis and Decision Making. *Int. Workshop on CO_2 Storage in Carboniferous Formations and Abandoned Coal Mines*, Ed. He, Sousa, Elsworth and Vargas. Beijing, pp. 125–152.

Su, H. & Fletcher, J.J. 2010. Carbon Capture and Storage in China: Options for the Shenhua Direct Coal Liquefaction Plant. *International Association for Energy Economics*, 29–31.

Su, W. 2009. GreenGen Project in China. Available at http://www.iea.org/work/2009/bergen/Wenbin.pdf.
Vargas, E.A. Jr., Velloso, R.Q., Ribeiro, W.N., Muller, A.L. & Vaz, L.E. 2011. Considerations on the numerical modeling of injection processes of CO_2 in geological formations with emphasis on carboniferous formations and abandoned coal mines. *Int. Workshop on CO_2 Storage in Carboniferous Formations and Abandoned Coal Mines*, Ed. He, Sousa, Elsworth and Vargas. Beijing, pp. 105–123.
Wang, J.G. & Liu, J.S. 2011. Micro-scale modeling of gas-coal interaction in coalbed seam—heterogeneity effect. *Int. Workshop on CO_2 Storage in Carboniferous Formations and Abandoned Coal Mines*, Ed. He, Sousa, Elsworth and Vargas. Beijing, pp. 95–106.
Wu, Y., Liu, J. & Elsworth, D. 2010. Development of Permeability Anisotropy during Coalbed Methane Production. 2010 *ARMA*, American Rock Mechanics Association.
Yu, H., Yuan, J., Guo, W., Cheng, J. & Hu, Q. 2008. A preliminary laboratory experiment on coalbed methane displacement with carbon dioxide injection. *International Journal of Coal Geology* 73, 156–166.
Zhu, J., Jiang, Y., Zhao, Y., Lu, Y. & Zhao, Y. 2009. Constitutive equations for coal containing gas considering gas adsorption. *Procedia Earth and Planetary Science* 1, 425–431.

CO_2 Storage in Carboniferous Formations and Abandoned Coal Mines –
He, Ribeiro E Sousa, Elsworth & Vargas Jr. (eds)
© 2012 Taylor & Francis Group, London, ISBN 978-0-415-62079-6

Complex process couplings in systems pushed far-from-equilibrium: Applications to CO_2 sequestration in carboniferous formations

D. Elsworth, S. Wang, G. Izadi, H. Kumar & J. Mathews
G^3 Center, EMS Energy Institute, and Energy and Mineral Engineering, Pennsylvania State University, University Park, PA, USA

J.S. Liu
Petroleum Engineering Program, Mechanical Engineering, University of Western Australia, Nedlands, Western Australia

D.-S. Lee
Petroleum and Marine Resources, Korea Institute of Geoscience and Mineral Resources, Daejeon, South Korea

D. Pone
ConocoPhillips, Bartlesville, Oklahoma, USA

ABSTRACT: We explore important couplings between the infiltration of CO_2 into fractured coals and into shale caprocks to examine changes in the storage and transport properties that may be helpful in ensuring long term sequestration. These effects include the potential sorption of gas into organic matter in both the coals and shales with the effect of long term fixation and also the swelling-induced reduction in permeability in shales that may create a self-sealing capability in caprocks. We show how changes in permeability in both coals and shales are indexed to sorption-induced swelling, even for conditions of zero mechanical restraint. A model of a discretely-fractured medium is used to identify modes and magnitudes of permeability change, as a function of gas pressure and effective stress and related gas content. This model is used to explain observations of non-monotonic changes in permeability of both coal and shales infiltrated with CO_2.

1 INTRODUCTION

Gas transport in coal seams is significantly different from that of other rock types because of the phenomena of gas sorption and coal swelling. The relative roles of stress level, gas pressure, gas composition, fracture geometry of coal, and water content are intimately connected to the processes of gas sorption, diffusion, transport, and coal swelling. Scientific research on coal-gas interactions has been conducted for more than a century, but the physicochemical and hydro-thermodynamic phenomena are still not fully understood. Understanding and quantifying these interactions is essential in developing the best possible process-based models of behavior that incorporate key parameters observed in response. This characterization is important in preventing gas outbursts in coal mines, in predicting CO_2 injectability into coal seams and in estimating the long-term stability of CO_2 sequestered in coal beds.

Significant experimental effort has been applied to investigate gas permeability and its evolution in coal. Laboratory measured permeabilities of coal to sorbing gases such as CH_4 and CO_2 are known to be lower than permeabilities to nonsorbing or lightly sorbing gases such as argon and nitrogen (N_2) (Somerton et al., 1975; Siriwardane et al., 2009). Permeabilities may decrease by as much as five orders of magnitude for confining pressures increasing from 0.1 to 70 MPa (Somerton et al., 1975; Durucan and Edwards, 1986). Under constant total stress, sorbing gas permeability decreases with increasing pore pressure due to coal swelling (Robertson and Christiansen, 2005; Mazumder and Wolf, 2008; Pan et al., 2010; Wang et al., 2010, 2011), and increases with decreasing pore pressure due to matrix shrinkage (Harpalani and Schraufnagel, 1990; Seidle and Huitt, 1995; Harpalani and Chen, 1997; Cui and Bustin, 2005). Rebound pressure, which corresponds to the minimum permeability, has been observed for CO_2 injection at 1.7 MPa (Pini et al., 2009), and at 7 MPa (Palmer and Mansoori, 1996; Shi and Durucan, 2004). Permeability of sorbing gas in coal is found to be a function of gas exposure time (Siriwardane et al., 2009). Pemeability is also influenced by both the presence of water and the magnitude of water saturation (Han et al., 2010). Based on field and laboratory experimental results, several permeability models have been developed for coal seams (Seidle and Huitt, 1995; Palmer and Mansoori, 1998; Pekot and Reeves, 2002; Shi and Durucan, 2005; Cui and Bustin, 2005; Zhang et al., 2008; Wang et al., 2009; Liu and Rutqvist, 2010; Izadi et al., 2010).

The sorption capacities of coal to N_2, CH_4 and CO_2 have been explored using a variety of measurement methods. Experiments have shown that CO_2 is adsorbed preferentially relative to CH_4 in most instances, and the ratios of the sorption capacities (in molar units) are between 1.15 and 3.16 (Clarkson and Bustin, 1999; Busch et al., 2003; Mastalerz et al., 2004; Harpalani et al., 2006; Bae and Bhatia, 2006; Ottiger et al., 2008; Battistutta et al., 2010; Li et al., 2010; Wang et al., 2011). This ratio decreases with increasing temperature (Bae and Bhatia, 2006; Li et al., 2010). The sorption capacity ratios of CO_2 to N_2 are found to be between 2:1 and 8.5:1 (Shimada et al., 2005; Saghafi et al., 2007; Battistutta et al., 2010). However, some coals under certain conditions show a larger sorption capacity to CH_4 than to CO_2 (Busch et al, 2003; Busch et al., 2006; Majewska et al., 2009). The presence of water reduces the sorption capacity to gases by around 30% (Siemons and Busch, 2007; Gruszkiewicz et al., 2009; Kelemen and Kwiatek, 2009; Wang et al., 2011), due to the competition between water molecules and the sorbing gas for sorption sites on the coal surface (Busch et al., 2007; van Bergen et al., 2009). Applied stress can reduce the sorption capacity of coal by 5%–50% (Hol et al., 2010). Sorption and swelling processes have been shown to be heterogeneous in coal (Karacan and Okandan, 2001; Karacan, 2003; Karacan, 2007; Day et al., 2008) as apparent from quantitative X-ray CT imaging and from optical methods.

Previous studies have shown that coal swells when exposed to N_2, CH_4, and CO_2, with volumetric strain ranging from 0.1% to 15%, under pressures up to 20 MPa and temperatures up to 55°C (Harpalani and Schraufnagel, 1990; Seidle and Huitt, 1995; Levine, 1996; Robertson and Christiansen, 2005; Cui et al., 2007; Karacan, 2007; Day et al., 2008; Mazumder and Wolf, 2008; Kiyama et al., 2010; Wang et al., 2010, 2011). Coal swelling strain increases with increasing pore pressure and strain induced by CO_2 is commonly larger than that induced by CH_4. The swelling strain is either reversible (Levine, 1996; Day et al., 2008; Battistutta et al., 2010) or irreversible (Czerw, 2010; Majewska et al., 2010; Wang et al., 2011). The relation between swelling strain and the amount or volume of gas sorbed is found to be either linear (Levine, 1996; Chikatamarla et al., 2004; Robertson and Christiansen, 2005; Cui et al., 2007; Czerw, 2010) or non-linear (Day et al., 2008; Kelemen and Kwiatek, 2009; Majewska et al., 2010; Wang et al., 2011). Coal swelling strain of wet coals is less than that of dry coals (Mazumder and Wolf, 2008; van Bergen et al., 2009; Kiyama et al., 2010; Wang et al., 2011), illustrating the effect of water on the swelling strain.

The effect of sorption on the mechanical strength and structure of coal has also been previously investigated. Weakening due to the introduction of CO_2 to a coal is found in uniaxial compression tests (Viete and Ranjith, 2006; Ranjith et al., 2010) and in triaxial compression tests (Wang et al., 2011). In some cases, no evidence of coal weakening is found to result following gas sorption (Ates and Barron, 1988; Day et al., 2008; Pan et al., 2010) even though the sorption process may alter the pore structure of the coal (Larsen, 2004; Goodman et al., 2005; Liu et al., 2010).

Gas transport in coal seams is commonly represented as a dual porosity system accommodating two serial transport mechanisms: diffusion through the coal matrix then laminar flow through the cleat system (Elsworth and Bai, 1992; Bai and Elsworth, 2000). The permeability is primarily determined by the cleat aperture (Zhang et al., 2008; Wu et al., 2010a, b). The change in cleat aperture is a function of effective stress through poroelasticity (Izadi et al., 2010; Wu et al., 2010a, b). Meanwhile, coal swelling and shrinkage under a confining stress may also change the cleat aperture (Izadi et al., 2010; Wu et al., 2010a, b). Thus, the net change in coal permeability is a function of both the poroelastic response and the coal swelling or shrinkage. To explore the evolution of permeability due to coal swelling we evaluate rates and magnitudes of expected swelling and how these effects translate into mechanisms of permeability evolution. We then compare these predictions with observed magnitudes of permeability change in both coal and in shales, with application to CO_2 storage in Carboniferous formations.

2 ANALYSIS

Experimental observations have shown that swelling response to the infiltration of CO_2 exhibits two key features. As CO_2 infiltrates the rock, potentially at the expense of CH_4, the matrix material will swell and permeability will reduce if the gas pressure is below the Langmuir pressure. This occurs regardless of the mechanical constraint applied to the cracked sample. Once the Langmuir pressure is exceeded, the permeability will recover as effective stress effects dominate in the absence of swelling-induced closure. A homogeneous continuum with non-interacting flaws (Figure 1) can be shown incapable of replicating this observed response (Izadi et al., 2011). We have shown that permeability decreasing with an increase in swelling may be accommodated (Izadi et al., 2011) either for a cracked continuum containing a zone of damage around the crack (Figure 1 (right)) or for multiple interacting cracks (Figure 2). We choose the geometry of an interacting elliptical fractures within a swelling medium as the intrinsic mechanistic component to describe permeability response, as shown in Figure 1.

2.1 *Response of a cracked continuum with interacting flaws*

The selected geometry is for a regular array of interacting cracks. We examine the influence of effective stress and swelling response for an elliptical crack where the deformation modulus and swelling coefficient are constant with radius. We consider two models to represent this behavior. The first is a single component part removed from the array where the appropriate boundary conditions are for uniform displacement along the boundaries. This

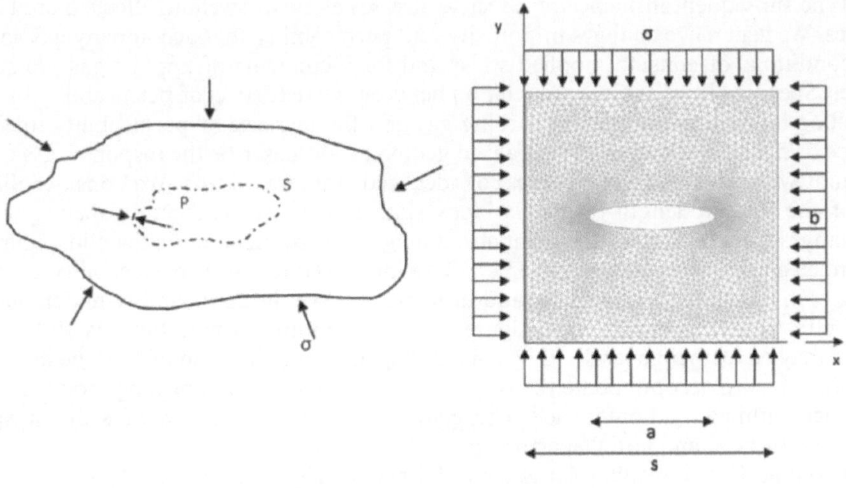

Figure 1. A homogeneous aggregate with pore. The line S represents the surface of the pore that is subject to a pore pressure that is equal to the confining pressure.

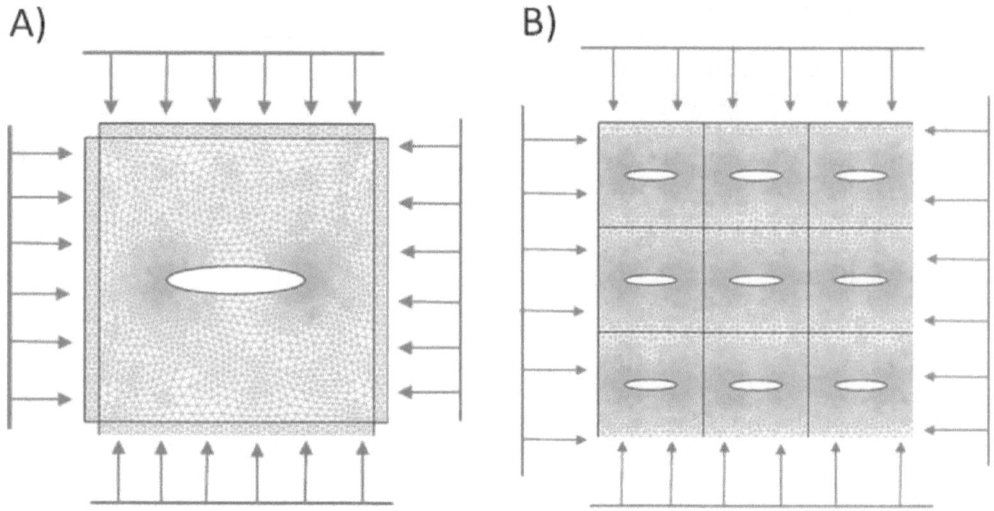

Figure 2. Simulation model.

represents the symmetry of the displacement boundary condition mid-way between flaws as shown in Figure 2(A). An alternative method to represent this geometry is an equivalent model that incorporates multiple flaws and automatically accommodates the appropriate displacement boundary condition as shown in Figure 2(B). We examine the change in aperture due to the combined influence of swelling and effective stress. For each of the two representations of an interacting network of flaws the evolution of aperture is identical as apparent in Figure 3(A) and Figure 3(B) and conforms to the anticipated behavior where swelling effects are staunched at higher gas pressures. Similarly, where these changes in apertures are converted to permeabilities, identical permeability evolution trends result for the two models. These responses are represented in Figure 3(C) and Figure 3(D) for initial permeabilities of 10^{-13} to 10^{-15} m^2.

2.2 *Generalized response*

We generalize our understanding of the reduction of permeability due to unconstrained swelling and the subsequent influence of effective stresses alone as swelling effects halt at higher pressures. We generalize changes in porosity and permeability that accompany gas sorption under conditions of constant applied stress and for increments of applied gas pressure for fractures. Specifically we explore the relation between the reduction of permeability by applying swelling-induced sorption of a sorbing gas and the increase of permeability due to the influence of effective stress for a generalized geometry. We describe the response of a cracked continuum with various fracture sizes for idealized fracture widths. We consider different ratios of the fracture length to matrix block size, defined as the fracture spacing. As this ratio changes from 0.025 to 1 the normalized magnitude of closure is shown in Figure 4(A) for a dimensionless pressure of $(P/E)\varepsilon_L$. The initial reduction in permeability caused by swelling of the unconstrained block and ultimate increase in permeability under the influence of effective stress is shown in Figure 4(B), where initial permeability is 10^{-13} m^2. The response may be further generalized by considering an approximation of the true mechanical behavior. This is to accommodate the swelling of a soft medium (vanishing modulus) that is constrained within a rigid outer shell. This geometry contains a fracture of width, a, spacing between fractures, s, and initial aperture, b.

Total volume V for a rectangular crack is defined as (Figure 1)

$$V = s^3. \tag{1}$$

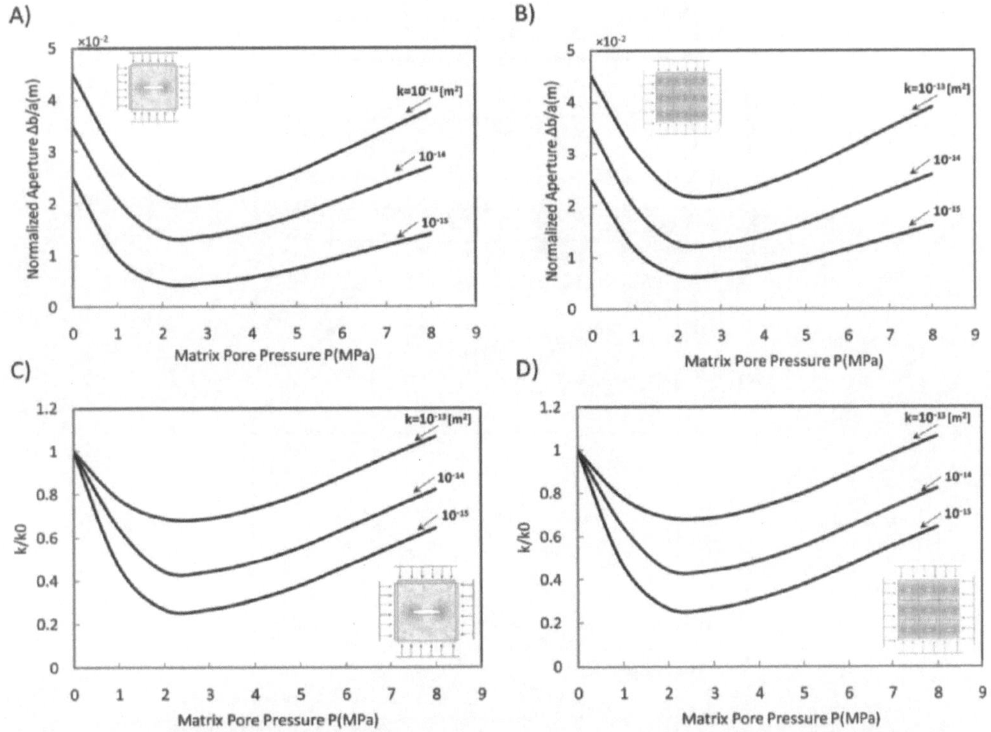

Figure 3. A) The relation between ratio of aperture change to initial aperture and pressure for one fracture. B) The relation between ratio of aperture change to initial aperture and pressure for multiple fractures. C) The relation between matrix permeability ratio and matrix pore pressure for one fracture. D) The relation between matrix permeability ratio and matrix pore pressure.

The corresponding change in volume, ΔV is defined in terms of the swelling strain ε_s as

$$\Delta V = s^3 \varepsilon_s. \tag{2}$$

The change in volume of the fracture, ΔV_f, depends on the length, a, and width, b, of the fracture and also the size of the matrix when the external displacement of the body is null. Thus the volumetric change in the fracture is defined as

$$\Delta V_f = a \cdot s \cdot \Delta b. \tag{3}$$

Substituting equation (1) into (2) and equating the result with equation (3) yields

$$V \varepsilon_s = a \cdot s \cdot \Delta b. \tag{4}$$

where the external boundary has zero displacement, the swelling strain is defined as

$$\varepsilon_s = \frac{\Delta b \cdot a}{s \cdot s} \tag{5}$$

And also the resulting change in aperture may be recovered by substituting the strain of equation (5) into the volume constraint relationship of equation (4) as

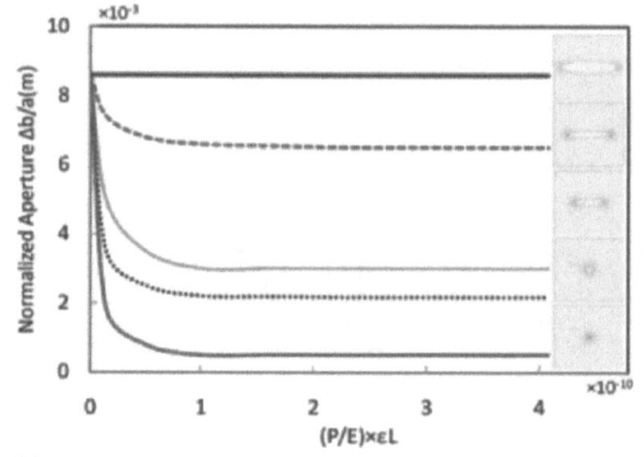

A)

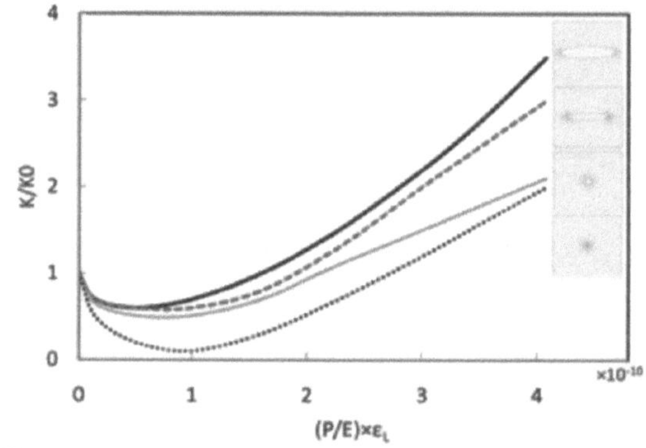

B)

Figure 4. A) The relationship between the ratio of aperture change and dimensionless pressure. B) The relationship between matrix permeability ratio and dimensionless pressure.

$$\Delta b = \frac{\varepsilon_s s^2}{a}. \tag{6}$$

Finally, from equation (6) we can determine

$$\frac{\Delta b}{b_0} = \frac{\varepsilon_s s^2}{a b_0}. \tag{7}$$

Where initial aperture size is evaluated from $b_0 = \sqrt[3]{12 k_0 s}$ we can determine the change in permeability as

$$\frac{k}{k_0} = \left(1 + \frac{\Delta b}{b_0}\right)^3 = \left(1 + \frac{\varepsilon_s s^3}{a b_0}\right)^3 = \left(1 + \left(\frac{\varepsilon_L s^2}{a b_0}\right)\frac{p}{p_m + p_L}\right)^3. \tag{8}$$

Thus, permeability scales according to the physical parameters controlling response as

$$\frac{k}{k_0} = f\left(\frac{\varepsilon_L s^2}{a b_0}; p; p_L\right). \tag{9}$$

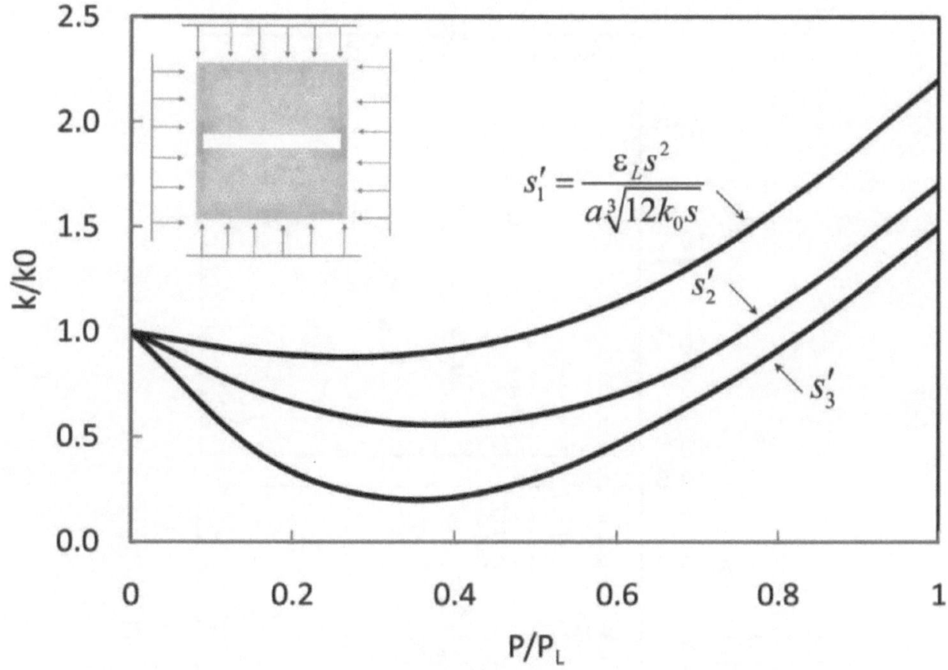

Figure 5. The relationship between fracture permeability ratio and dimensionless pressure.

This enables the evolution of relative permeabilities to be determined as a function of the non-dimensional parameter s_n' incorporating initial permeability, swelling strain and geometric constraints, alone as

$$s_n' = \frac{\varepsilon_L s^2}{a\sqrt[3]{12k_0 s}}. \tag{10}$$

This relation is used to represent the families of permeability evolution for constant magnitudes of s_n'. The resulting permeability evolution for different initial permeabilities of 10^{-13} to 10^{-15} m^2 is shown as a function of dimensionless pressure (p/p_L) in Figure 5.

3 PERMEABILITY EVOLUTION IN COAL

To evaluate the utility of such a model, we compare laboratory measurements of permeability at different gas saturations on a sample of fractured anthracite. Details of the experimental method are documented elsewhere (Wang et al., 2011). The standard cylindrical sample is confined under invariant total stress and the gas pressure to different permeants is augmented for different water contents. Results for the evolution of permeability of a ubiquitously fractured sample under dry and water-saturated conditions at a total stress 6 MPa is shown in Figure 6(A). Under water-saturated conditions, He has the largest permeability, and is followed progressively by CH_4 and CO_2. This progression of permeability magnitudes is the same as in tests under dry conditions. However, the initial permeabilities of wet samples are uniformly reduced by two orders of magnitude over the dry samples—indicating the role of water in occluding microfractures (Han et al., 2010).

Measured permeability with He increases with a higher rate with pore pressure compared with the results under dry condition. However, different from the results from dry samples, no permeability reduction is seen for CH_4 and CO_2 with increasing pore pressure. The pres-

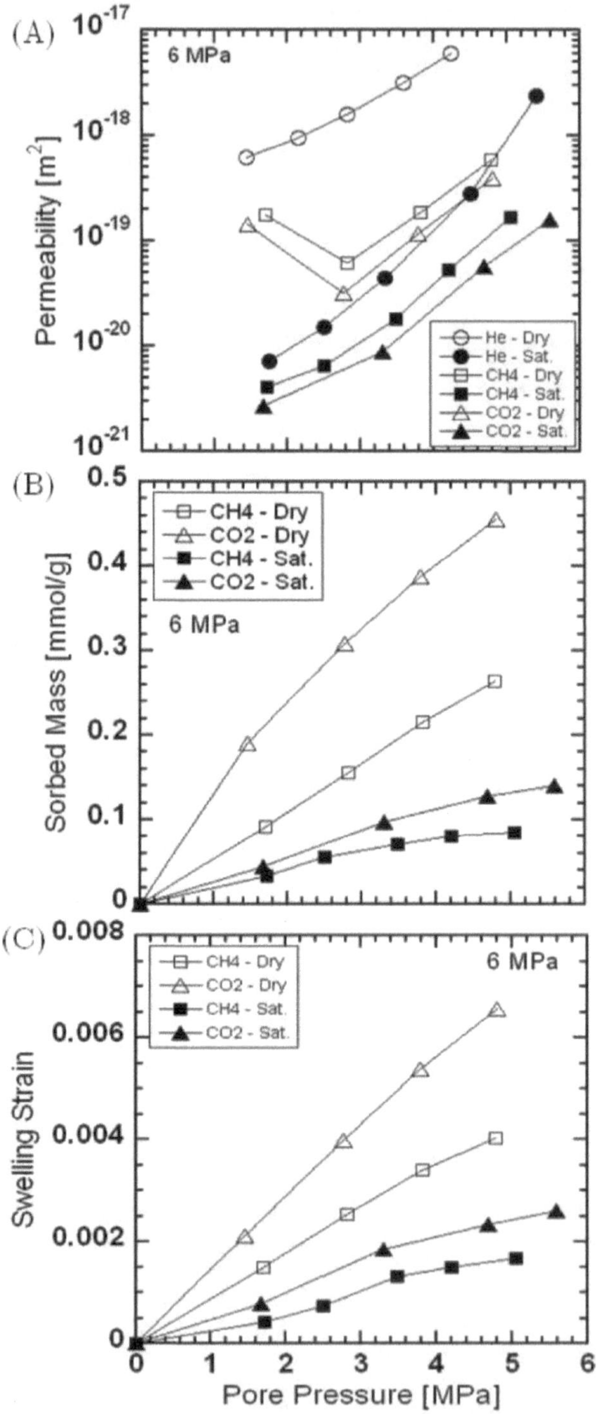

Figure 6. (A) Comparison of the evolution of permeability as a function of applied pore pressure for coal with embedded fractures under dry and water-saturated conditions for constant applied total stress at 6 MPa. (B) Comparison of the sorbed mass as a function of applied pore pressure for coal with embedded fractures under dry and water-saturated conditions for constant applied total stress at 6 MPa. (C) Comparison of the swelling strain as a function of applied pore pressure for coal with embedded fractures under dry and water-saturated conditions for constant applied total stress at 6 MPa.

ence of water reduces the permeability significantly so the sample begins with an extremely low permeability. As pore pressure increases, the amount of water in the sample is gradually displaced by the injected gas, which leads to an increase in permeability. An increment in pore pressure also enhances the permeability due to the influence of decreased effective stresses. At the same time, swelling induced by adsorption with increasing pore pressure lowers the permeability. However, the amount of gas adsorbed and induced swelling are greatly reduced by the presence of water. So the net change in permeability for wet samples depends on the overall effect of these three competitive components. In this case, the permeability enhancement due to effective stress and water displacement dominates the reduction due to coal swelling, showing that permeabilities always increase with pore pressure. This is also consistent with the observation that the rate of change in permeability to He is greater under wet conditions.

The comparison of the sorbed mass as a function of pore pressure of the ubiquitously fractured coal under dry and water-saturated conditions at 6 MPa total stress is shown in Figure 6(B). Sorbed mass is consistently higher for CO_2 versus CH_4 and is reduced by a factor of 3 to 4 when wet versus dry. The isotherms can be approximated by Langmuir-like behavior. At final pressures, the amount of gas adsorbed in the wet sample is reduced by 69% for CO_2 (0.45 mmol/g coal, dry sample; 0.14 mmol/g coal, wet sample) and 69% for CH_4 (0.26 mmol/g coal, dry sample; 0.08 mmol/g coal, wet sample). Figure 6(C) compares swelling strain as a function of pore pressure for the coal containing embedded fractures under dry and water-saturated conditions at a 6 MPa total stress. As shown before, the measured swelling under dry conditions can be reasonably described by a Langmuir type relationship.

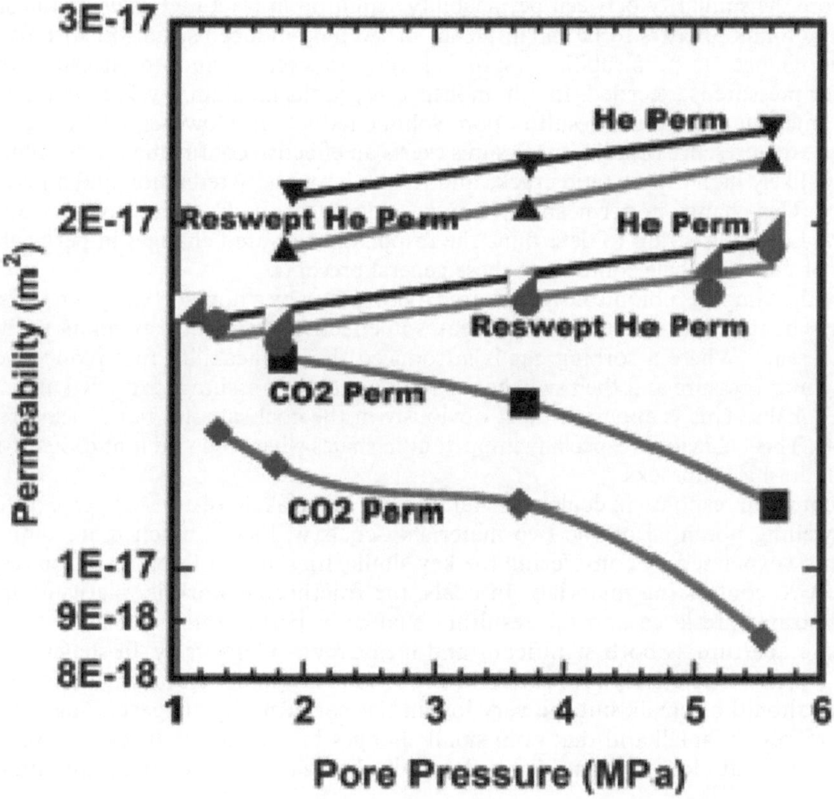

Figure 7. Variation in permeability in shale infiltrated with He then swept with CO_2 and then reswept with He. Data represent two experiments at total confining stresses of 10 MPa (filled triangles and squares) and 12 MPa (circles, diamonds and half-filled squares).

The presence of water also reduces the swelling strain by 62% for CO_2 and by 58% for CH_4 at final pressures. The presence of water also makes the initial portion of the curves largely linear and less similar to Langmuir-like behavior. However, at pore pressure above 3 MPa a significant amount of water is displaced by the injected gas and above this stress the swelling strain follows a Langmuir-like relation with pore pressure.

4 PERMEABILITY EVOLUTION IN SHALE

The observed response in coals is also apparent in shales. Figure 7 shows the evolution of permeability in a sample of porous shale where it is first infiltrated with He. As pressure is increased, the permeability increases as the effective stress concomitantly decreases. As CO_2 is introduced, the shale swells (not shown) and the reduction in permeability is proportional to the increment of gas pressure. In this particular case, the Langmuir pressure is not exceeded, and there is no turnaround in permeability evolution due to effective stress. Once the sample is reswept with He, the former response returns with an increase in permeability with incremented pressure. Apparent, however, is that the permeability is not returned to the pre-swept magnitude. In this manner, the observed response is intrinsically similar to that for coal—with the swelling of organic material dominating the response.

5 CONCLUSIONS

We explore the similarity between permeability evolution in fractured coals and in shales—the common link appears to be the influence of swelling-induced strains in sorbing media. Observed changes in permeability respond directly to swelling and are arrested when the Langmuir pressure is exceeded. In a homogeneous porous medium, swelling that accompanies gas injection should not result in pore volume reduction. However, where the medium is cracked, the presence of adjacent fissures exerts an effective confinement to swelling pores ore morel likely high aspect-ratio cracks and results in porosity reduction and a loss of permeability. This change in permeability may be indexed to the fracture geometry and magnitude of Langmuir strain to determine the response. Measured changes in permeability in samples of coal and shale conform to these general precepts.

Where the sample is ubiquitously fractured, permeation by a non-sorbing gas results in permeability changes that increase with decreases in effective stress corresponding to increases in gas pressure. Where a sorbing gas is introduced the permeability first reduces up until the Langmuir pressure and then switches as the Langmuir pressure is exceeded and swelling effectively halts. This is apparent most obviously in the coal samples but is also present in the shales. This behavior is especially important in shales where it could hint of a self-healing capability in shale caprocks.

This common response in coals and shale is surprising, as there are very large differences in the swelling potential of the two materials—coals will swell much more than shales. This can be explained by considering the key similarities and differences in the responses of these two contrasting materials. In coals, the fracture network is relatively open but induced strains are large and the resulting change in permeability, sensitive to changes in fracture aperture, is both significant and measurable. Conversely, in shales where the organic content is relatively low, the resulting strains are small. Resulting changes in permeability should be small, but the very low initial permeability suggests that initial fracture apertures are small and that even small changes in aperture induced by swelling can result in significant loss in permeability. It is this observable change that is apparent in the laboratory response.

These evaluations note the very significant changes in permeability that result in cracked media infiltrated by sorbing gases, that in turn have important ramifications for gas production from coal and shales and of storage of CO_2 in coals and beneath shale caprocks.

ACKNOWLEDGMENTS

This work is a partial result of support from NIOSH under contract 200–2008–25702, from ConocoPhillips and from KIGAM. This support is gratefully acknowledged.

REFERENCES

Ates, Y. & Barron, K., 1988. The effect of gas sorption on the strength of coal. *Mining Science and Technology* 6, 291–300.

Bae, J.-S. & Bhatia, S.K., 2006. High-Pressure Adsorption of Methane and Carbon Dioxide on Coal. *Energy & Fuels* 20, 2599–2607.

Bai, M. & Elsworth, D., 2000. Coupled Processes in Subsurface Deformation, *Flow and Transport*. ASCE Press. 336 pp.

Battistutta, E., van Hemert, P., Lutynski, M., Bruining, H. & Wolf, K.-H., 2010. Swelling and sorption experiments on methane, nitrogen and carbon dioxide on dry Selar Cornish coal. *International Journal of Coal Geology* 84, 39–48.

Busch, A., Gensterblum, Y. & Krooss, B.M., 2003. Methane and CO_2 sorption and desorption measurements on dry Argonne premium coals: pure components and mixtures. *International Journal of Coal Geology* 55, 205–224.

Busch, A., Gensterblum, Y. & Krooss, B.M., 2007. High-Pressure Sorption of Nitrogen, Carbon Dioxide, and their Mixtures on Argonne Premium Coals. *Energy & Fuels* 21, 1640–1645.

Busch, A., Gensterblum, Y., Krooss, B.M. & Siemons, N., 2006. Investigation of high-pressure selective adsorption/desorption behaviour of CO_2 and CH_4 on coals: An experimental study. *International Journal of Coal Geology* 66, 53–68.

Chikatamarla, L., Cui, X. & Bustin, R.M., 2004. Implications of volumetric swelling/shrinkage of coal in sequestration of acid gases 2004 *International Coalbed Methane Symposium Proceedings*, Tuscaloosa, Alabama (2004) paper 0435.

Clarkson, C.R. & Bustin, R.M., 1999. The effect of pore structure and gas pressure upon the transport properties of coal: a laboratory and modeling study. 1. Isotherms and pore volume distributions. *Fuel* 78, 1333–1344.

Cui, X., Bustin, R.M. & Chikatamarla, L., 2007. Adsorption-induced coal swelling and stress: Implications for methane production and acid gas sequestration into coal seams. *J. Geophys. Res.* 112, B10202.

Cui, X.J. & Bustin, R.M., 2005. Volumetric strain associated with methane desorption and its impact on coalbed gas production from deep coal seams. *AAPG Bulletin* 89(9), 1181–1202.

Czerw, K., 2011 Methane and carbon dioxide sorption/desorption on bituminous coal--Experiments on cubicoid sample cut from the primal coal lump. *International Journal of Coal Geology*, 85(1) 72–77.

Day, S., Fry, R. & Sakurovs, R., 2008. Swelling of Australian coals in supercritical CO_2. *International Journal of Coal Geology* 74, 41–52.

Durucan, S. & Edwards, J.S., 1986. The effects of stress and fracturing on permeability of coal. *Mining Science and Technology* 3, 205–216.

Elsworth, D. & Bai, M., 1992. Flow-deformation response of dual porosity media. *Journal of Geotechnical Engineering* 118, 107–124.

Goodman, A.L., Favors, R.N., Hill, M.M. & Larsen, J.W., 2005. Structure Changes in Pittsburgh No. 8 Coal Caused by Sorption of CO_2 Gas. *Energy & Fuels* 19, 1759–1760.

Gruszkiewicz, M.S., Naney, M.T., Blencoe, J.G., Cole, D.R., Pashin, J.C. & Carroll, R.E., 2009. Adsorption kinetics of CO_2, CH_4, and their equimolar mixture on coal from the Black Warrior Basin, West-Central Alabama. *International Journal of Coal Geology* 77, 23–33.

Han, F., Busch, A., van Wageningen, N., Yang, J., Liu, Z. & Krooss, B.M., 2010 Experimental study of gas and water transport processes in the inter-cleat (matrix) system of coal: Anthracite from Qinshui Basin, China. *International Journal of Coal Geology* 81, 128–138.

Harpalani, S. & Chen, G., 1997. Influence of gas production induced volumetric strain on permeability of coal. *Geotechnical and Geological Engineering* 15, 303–325.

Harpalani, S., Prusty, B.K. & Dutta, P., 2006. Methane/CO_2 Sorption Modeling for Coalbed Methane Production and CO_2 Sequestration. *Energy & Fuels* 20, 1591–1599.

Harpalani, S. & Schraufnagel, R.A., 1990. Shrinkage of coal matrix with release of gas and its impact on permeability of coal. *Fuel* 69, 551–556.

Hol, S., Peach C. & Spiers, C., 2011. Applied stress reduces the CO_2 sorption capacity of coal, *International Journal of Coal Geology*, 85(1) 128–142.

Izadi, G., Wang, S., Elsworth, D., Liu, J., Wu, Y. & Pone, D., 2011. Permeability evolution of fluid-infiltrated coal containing discrete fractures. *International Journal of Coal Geology*, 85(2), 202–211.
Karacan, C.O., 2003. Heterogeneous sorption and swelling in a confined and stressed coal during CO_2 injection. *Energy and Fuels* 17(6), 1595–1608.
Karacan, C.O., 2007. Swelling-induced volumetric strains internal to a stressed coal associated with CO_2 sorption. *International Journal of Coal Geology* 72, 209–220.
Karacan, C.O. & Okandan, E., 2001. Adsorption and gas transport in coal microstructure: investigation and evaluation by quantitative X-ray CT imaging. *Fuel* 80, 509–520.
Kelemen, S.R. & Kwiatek, L.M., 2009. Physical properties of selected block Argonne Premium bituminous coal related to CO_2, CH_4, and N_2 adsorption. *International Journal of Coal Geology* 77, 2–9.
Kiyama, T., Nishimoto, S., Fujioka, M., Xue, Z., Ishijima, Y., Pan, Z. & Connell, L.D., 2011. Coal swelling strain and permeability change with injecting liquid/supercritical CO_2 and N_2 at stress-constrained conditions. *International Journal of Coal Geology*, 85(1) 56–64.
Larsen, J.W., 2004. The effects of dissolved CO_2 on coal structure and properties. *International Journal of Coal Geology* 57, 63–70.
Levine, J.R., 1996. Model study of the influence of matrix shrinkage on absolute permeability of coalbed reservoirs. *Geological Society Special Publication* 109, 197–212.
Li, D., Liu, Q., Weniger, P., Gensterblum, Y., Busch, A. & Krooss, B.M., 2010. High-pressure sorption isotherms and sorption kinetics of CH_4 and CO_2 on coals. *Fuel* 89, 569–580.
Liu, C.J., Wang, G.X., Sang, S.X. & Rudolph, V., 2010. Changes in pore structure of anthracite coal associated with CO_2 sequestration process. *Fuel* 89, 2665–2672.
Liu, H.H. and Rutqvist, J., 2010. A New Coal-Permeability Model: Internal Swelling Stress and Fracture-Matrix Interaction. *Transport in Porous Media*, 82(1): 157–171.
Majewska, Z., Ceglarska-Stefanska, G., Majewski, S. & Zietek, J., 2009. Binary gas sorption/desorption experiments on a bituminous coal: Simultaneous measurements on sorption kinetics, volumetric strain and acoustic emission. *International Journal of Coal Geology* 77, 90–102.
Majewska, Z., Majewski, S. & Zietek, J., 2010. Swelling of coal induced by cyclic sorption/desorption of gas: Experimental observations indicating changes in coal structure due to sorption of CO_2 and CH_4. *International Journal of Coal Geology* 83, 475–483.
Mastalerz, M., Gluskoter, H. & Rupp, J., 2004. Carbon dioxide and methane sorption in high volatile bituminous coals from Indiana, USA. *International Journal of Coal Geology* 60, 43–55.
Mazumder, S. & Wolf, K.H., 2008. Differential swelling and permeability change of coal in response to CO_2 injection for ECBM. *International Journal of Coal Geology* 74, 123–138.
Ottiger, S., Pini, R., Storti, G. & Mazzotti, M., 2008. Competitive adsorption equilibria of CO_2 and CH_4 on a dry coal. *Adsorption* 14, 539–556.
Palmer, I. & Mansoori, J., 1998. How permeability depends on stress and pore pressure in coalbeds: a new model, *SPE Reservoir Evaluation and Engineering*, SPE 52607, pp. 539–544.
Palmer, I. & Mansoori, J., 1996. How permeability depends on stress and pore pressure in coalbeds: a new model. SPE-52607.
Pan, Z., Connell, L.D. & Camilleri, M., 2010. Laboratory characterisation of coal reservoir permeability for primary and enhanced coalbed methane recovery. *International Journal of Coal Geology* 82, 252–261.
Pekot, L.J. & Reeves, S.R., 2002. Modeling the effects of matrix shrinkage and differential swelling on coalbed methane recovery and carbon sequestration. *U.S. Department of Energy* DE-FC26-00 NT40924.
Pini, R., Ottiger, S., Burlini, L., Storti, G. & Mazzotti, M., 2009. Role of adsorption and swelling on the dynamics of gas injection in coal. *J. Geophys. Res.* 114, B04203.
Ranjith, P.G., Jasinge, D., Choi, S.K., Mehic, M. & Shannon, B., 2010. The effect of CO_2 saturation on mechanical properties of Australian black coal using acoustic emission. *Fuel* 89, 2110–2117.
Robertson, E.P. & Christiansen, R.L., 2005. Measurement of sorption induced strain, Paper 0532, Proceedings of the 2005 *International Coalbed Methane Symposium,* Tuscaloosa, AL.
Saghafi, A., Faiz, M. & Roberts, D., 2007. CO_2 storage and gas diffusivity properties of coals from Sydney Basin, Australia. *International Journal of Coal Geology* 70, 240–254.
Seidle, J.P. & Huitt, L.G., 1995. Experimental measurements of coal matrix shrinkage due to gas desorption and implications for cleat permeability increases. *SPE*-30010-MS.
Shi, J.-Q. & Durucan, S., 2004. A numerical simulation study of the Allison unit CO_2—ECBM pilot: the impact of matrix shrinkage and swelling on ECBM production and CO_2 injectivity. *Proceedings of 7th International Conference on Greenhouse Gas Control Technologies.*
Shi, J.-Q. & Durucan, S., 2005. A model for changes in coalbed permeability during primary and enhanced methane recovery. *SPE Reservoir Evaluation and Engineering* 8(4), 291–299.

Shimada, S., Li, H., Oshima, Y., & Adachi, K., 2005. Displacement behavior of CH_4 adsorbed on coals by injecting pure CO_2, N_2, and CO_2–N_2 mixture. *Environmental Geology* 49, 44–52.
Siemons, N. & Busch, A., 2007. Measurement and interpretation of supercritical CO_2 sorption on various coals. *International Journal of Coal Geology* 69, 229–242.
Siriwardane, H., Haljasmaa, I., McLendon, R., Irdi, G., Soong, Y. & Bromhal, G., 2009. Influence of carbon dioxide on coal permeability determined by pressure transient methods. *International Journal of Coal Geology* 77, 109–118.
Somerton, W.H., Soylemezoglu, I.M. & Dudley, R.C., 1975. Effect of stress on permeability of coal. *International Journal of Rock Mechanics and Mining Sciences & Geomechanics Abstracts* 12, 129–145.
van Bergen, F., Spiers, C., Floor, G. & Bots, P., 2009. Strain development in unconfined coals exposed to CO_2, CH_4 and Ar: Effect of moisture. *International Journal of Coal Geology* 77, 43–53.
Viete, D.R. & Ranjith, P.G., 2006. The effect of CO_2 on the geomechanical and permeability behaviour of brown coal: Implications for coal seam CO_2 sequestration. *International Journal of Coal Geology* 66, 204–216.
Wang, G.X., Massarotto, P. & Rudolph, V., 2009. An improved permeability model of coal for coalbed methane recovery and CO_2 geosequestration. *International Journal of Coal Geology* 77, 127–136.
Wang, G.X., Wei, X.R., Wang, K., Massarotto, P. & Rudolph, V., 2010. Sorption-induced swelling/shrinkage and permeability of coal under stressed adsorption/desorption conditions. *International Journal of Coal Geology* 83, 46–54.
Wang, S., Elsworth, D. & Liu, J., 2010. Evolution of Permeability in Coal to Sorbing Gases — A Preliminary Study. 44th U.S. *Rock Mechanics Symposium*. July 2010, Salt Lake City, Utah, 209.
Wang, S., Elsworth, D., & Liu, J., 2011. Permeability evolution in fractured coal: the roles of fracture geometry and water-content. *International Journal of Coal Geology*, submitted for publication, 46 pp.
Wu, Y., Liu, J., Elsworth, D., Chen, Z., Connell, L. & Pan, Z., 2010. Dual poroelastic response of a coal seam to CO_2 injection. *International Journal of Greenhouse Gas Control* 4, 668–678.
Wu, Y., Liu, J., Elsworth, D., Miao, X. & Mao, X., 2010. Development of anisotropic permeability during coalbed methane production. *Journal of Natural Gas Science and Engineering* 2, 197–210.
Zhang, H.B., Liu, J. & Elsworth, D., 2008. How sorption-induced matrix deformation affects gas flow in coal seams: a new FE model. *International Journal of Rock Mechanics and Mining Sciences* 45(8), 1226–1236.

CO_2 Storage in Carboniferous Formations and Abandoned Coal Mines –
He, Ribeiro E Sousa, Elsworth & Vargas Jr. (eds)
 ISBN 978-0-415-62079-6

A fully coupled gas flow, coal deformation and thermal transport model for the injection of carbon dioxide into coal seams

Hongyan Qu, J.S. Liu & Zhongwei Chen
School of Mechanical and Chemical Engineering, The University of Western Australia, Australia

Zhejun Pan & Luke Connell
CSIRO Earth Science and Resource Engineering, Clayton South, Australia

ABSTRACT: This paper presents solutions for non-isothermal gas flow coupled with mechanical deformation to evaluate the coupled thermo-hydro-mechanical (THM) processes in coal seams. A non-isothermal case of high temperature CO_2 injection is simulated to characterise the complex responses of coal to the injection. The results are compared with those under the isothermal conditions, indicating that temperature has complicated effect on coal properties during the CO_2 sequestration process. In particular, high temperature CO_2 injection into coal seams reduces adsorption capacity, and the coal matrix swelling caused by gas sorption enhances permeability. The main contribution of this paper lies in providing numerical solutions of challenging flow problems coupled to thermal and geo-mechanical effects in coal seams. Since this model involves the balance of thermal energy and heat transfer, it is applicable of investigating the effects of temperature variation on each term of the effective strain, which helps better understand the CO_2 injection processes.

1 INTRODUCTION

There Geological CO_2 sequestration in coal seams is one of the promising ways to reduce anthropogenic greenhouse gas emissions due to the relatively high excess sorption capacity coal exhibits and the simultaneous recovery of coal bed methane (CBM). Of all the aspects related to this problem, temperature variation in the coal seams and its effect on gas flow and coal properties during the injection and sequestration process are not well studied. However, the temperature of injected gas is often different from that of the formation, yielding significant temperature discrepancy between them, and perturbing the physical, chemical and thermal state of the subsurface formation.

Several thermal processes are involved in the temperature disequilibrium throughout the CO_2 injection. The temperature evolution starts from the vicinity of the injection well, where temperature soon reaches the same as that of the injection fluid, leading to a thermal gradient from the injection well to the formation boundary. In the meanwhile, substantial temperature drop is induced near the injection well due to the gas expansion from higher pressure to lower pressure. Furthermore, slight temperature increase is caused by the exothermic adsorption process. In consequence, heat conduction and convection are essential ways of heat transfer in balancing these temperature variations, and the heat transfer rate depends on several factors including the enthalpy of the injected CO_2, specific heat capacity and conductivity of the formation as well as the expansion of the fluid and formation.

Most early studies on the thermal effect are analytics. Elsworth (1989) presented a conceptual model describing permeability enhancement subjected to temperature change. He found that thermal diffusion between solid and fluid phases is an important phenomenon and changes in permeability induced by relatively modest temperature modifications appear significant. However, the application of this model was only limited to competent block rock. Later, a thermo-poroelastic model was developed by Zhou (1998), in which the coupled thermo-hydro-mechanical effects and material property changes were taken into account. The study demonstrated that changes in stresses, hydraulic gradient, and temperature gradient can cause permeability to decrease. Similarly, this model could be only applied to rock since no adsorption effect was considered.

Recently, a few experiments have been carried out to investigate the temperature variation in the vicinity of the injection well. The most remarkable one is the classical Joule-Thomson experiment, focusing on the cooling effect. Over 20°C temperature drop was observed due to gas expansion, which has negative effect on permeability and injectivity (Oldenburg 2007). Another experimental study was focused on coal permeability conducted under various temperatures using three different gases (Long et al., 2009). Adsorption was found to have major impacts on coal permeability. In addition, permeability decreases with the increase of temperature and stress. However, no numerical relationship between permeability and temperature was obtained.

Noticeably, the focus of most previous CO_2 sequestration study is on the development of the gas plume and the migration path of the fluid in the formation, particularly in the brine aquifers or other rocks. However, coal has different characteristics from the other formations with respect to the trapping mechanism. Coal is described as a discontinuous medium consisting of coal matrix and cleats. Gas predominantly adsorbs in the mircopores of the coal matrix, while cleats are the main pathway for gas flow. Therefore, the highlight of the effect of temperature in this paper lies in the effect on coal permeability and adsorption capacity. The variation of temperature has considerable effect on coal permeability, which decreases with thermal expansion and increases with the shrinkage of the coal matrix simultaneously as temperature rises. Consequently, the net change of permeability is uncertain due to the varying temperature during the CO_2 sequestration.

Although the effect of temperature on gas sorption and coal deformation has been widely realized, its potential impact on permeability change in coal seams has not been well understood. Especially, the numerical relationship between coal permeability and temperature has not been established. The objective of this work is to develop a non-isothermal THM model for CO_2 sequestration in coal seams, fully coupled with heat transfer, gas flow and solid deformation. To accomplish that, three steps are conducted including modeling the relationship between temperature and sorption capacity, investigating the effect of adsorption on coal permeability and the effect of thermal expansion on coal deformation.

In the following, the general conservation laws are introduced and the governing THM equations are developed, followed by the simulation of a numerical example. In the example, the simulation scenario is simplified as non-isothermal single phase flow. The main aspect of the post processing is the temperature evolution in the coal seams and the variation of coal properties with the change of temperature.

2 THERMAL-HYDRAULIC-MECHANICAL SIMULATION MODELLING

Non-isothermal consolidation of deformable porous media is the basis of modern coupled THM models. The couplings between the processes of heat transfer, fluid flow and stress/deformation in fractured rocks has become an increasingly important subject in rock mechanics and engineering design since the early 1980s (Tsang, 1987; 1991), mainly due to the modelling requirements for the design and performance assessment of underground radioactive waste repositories, and other application areas, such as reservoir simulations (Gutierrez and Makurat, 1997; Zhao et al., 1999; Yang, 2001; Sasaki and Morikawa, 1996); partially saturated porous materials (Gawin and Schrefler, 1996); advanced numerical solution techniques for coupled THM models (Wang and Schrefler, 1998; Cervera et al., 1996; Thomas et al., 1999); soil mechanics (Thomas and Missoum, 1999); simulation of expansive clays (Thomoas and Cleall, 1999); flow and mechanics of fractures (Selvadurai and Nguyen, 1999); non-Darcy flow in coupled THM processes (Nithiarasu et al., 2000); double-porosity model of porous media (Masters et al., 2000); parallel formulations of coupled hydro-mechanical and thermo-mechanical models for porous media (Zimmerman, 2000); and tunnelling in cold regions (Lai et al., 1998).

The interactions and integration of the THM processes are shown in Table 1. The three processes have large differences in characteristic time and spatial scales. The thermal process has a relatively larger time and spatial scale due to the long heating cycle, besides, thermal dispersivity smoothes out the effects of local spatial property variations. Conversely, the mechanical deformation propagates rapidly and is commonly affected by faults and fractures, although much less so by medium property variations. Finally, the hydrologic flow and transport are sensitive to the small-scale medium heterogeneity, but with a timescale corresponding to the solute transport times. These processes are generally solved numerically through methods like finite difference methods, finite element methods and discrete element methods.

CO_2 sequestration in coal seams is nothing like that in other porous medium with respect to the trapping mechanism. Unfortunately, most of previous THM models were only applied to rocks and the phenomenon of sorption was not rigorously considered. Another problem is that the current simulators like TOUGH II can simulate multiphase flow in various medium coupled with mechanical response, but thermal effect on the fluid flow and mechanical deformation was not well described and even often ignored which makes it impossible to simulate the fully coupled THM process in the coal seams. In fact, the variation of sorption capacity with changing temperature in the coal seams has significant effect on the sorption-induced coal matrix swelling/shrinkage. Furthermore, the porosity of the formation and other properties are often assumed as a constant or simply a function of pressure to reduce the number of variables. Even though this simplifies the governing equations, but may not completely reflect the specific effect of temperature and reduce the stability and efficiency of the numerical simulation since most of the fluid properties are temperature dependent. Therefore, only if a comprehensive modeling strategy for coupling the THM processes in coal seams is presented, can the effects in all aspects are better interpreted.

Figure 1 demonstrates the coupled THM processes in the coal seams. The hydro-mechanical processes are coupled via cleat aperture, coal porosity/permeability, fluid pressure and mechanical stress. The link for thermo-mechanical processes is volume expansion and thermal stress increment of the coal matrix, as well as the matrix swelling/shrinkage (indirectly the cleat aperture change) induced by adsorption/desorption due to temperature variations. The coupling in the thermo-hydraulic processes are more complex, involving variations in volume (density), viscosity and phase changes (gas, liquid and supercritical) of the fluid var-

Table 1. Interactions among the THM processes (Stephansson et al., 2004).

Geology	Rock inhomogeneity and anisotropy, & fractures (geometry, etc.) affects rock properties/stresses	Rock porosity, rock mass permeability, and water-rock interaction affects water flow	Mineralogical composition, porosity, textural and structural anisotropy effects
Stress data affecting the geo-interpretation of fracture systems and rock mass	Rock Mechanics	Spatial distribution of *in situ* stress and EDZ influences the hydro-geological regime	Rock & fracture zones subjected to higher stresses may be more thermally conductive
Hydro-geological tests and measures can affect the geo-interpretation of permeable features	Water pressure changes the effective stress	Hydro-Geology	The flow pattern will affect the temperature due to convection effects
Thermal anisotropy and measurement affecting the geo-interpretation	Change in temperature can change the local stress, possibly leading to failure	Temperature gradients and thermal expansion in rock/fractures affect the water flow	Thermal Properties

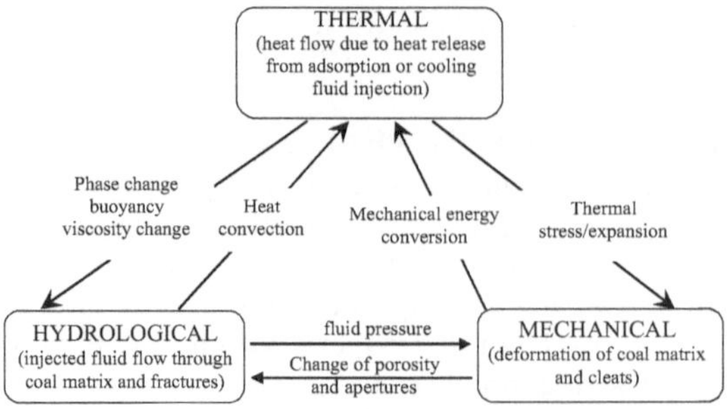

Figure 1. Coupled THM processes of fractured coal seams.

ied with temperature, as well as the conductive-convective thermal transfer through fractured coal seams by the flow of the fluids.

Three conventional individual physics (fluid flow, energy conservation and mechanical deformation) are incorporated into a multiphysics model to solve the coupled physics phenomena in coal seams simultaneously in this section, and a complete set of governing equations (a large system of nonlinear coupled partial differential equations) are derived. The main difference between this model and other ones is that the adsorption/desorption process is involved and the sorption-induced strain plays the predominant role in permeability change during the whole process. Another difference is that the properties of both coal and injected fluid are treated as functions of temperature, and heat transfer is explicitly described with energy conservation law, so that the thermal effect on gas flow and coal deformation can be well understood. The equation derivations are based on the following assumptions:

1. The target coal is linear elastic porous medium with small cleats and statistically isotropic
2. Small displacement and infinitesimal strains are assumed
3. Gas flow through the coal follows Darcy's Law
4. The medium is fully saturated and the dissolution of CO_2 is ignored
5. The fluid is supercritical and does not undergo a phase transformation.

2.1 *Coal deformation equation*

In this section, stress–strain relationship for a linear elastic porous medium are derived. The derivation makes use of an analogue between thermal expansion and matrix swelling/shrinkage associated with gas adsorption/desorption in coalbeds. Stress-strain relationships for a thermoelastic porous medium can be readily found in the literature (Nowacki, 1975; Bear and Corapcioglu, 1981).

The strain-displacement relation is expressed as

$$\varepsilon_{ij} = \frac{1}{2}\left(\mu_{i,j} + \mu_{j,i}\right) \tag{1}$$

where ε_{ij} is the component of total strain tensor and μ_i is the component of displacement.

The equilibrium equation for mechanical deformation is defined as

$$\sigma_{ij,j} + f_i = 0 \tag{2}$$

where $\sigma_{ij,j}$ denotes the component of the total stress tensor and f_i denotes the component of the body force.

According to the elasticity law, the constitutive relations of an isotropic linear poroelastic continuum representation of the coal can be expressed in the terms of the effective stress, compressive strain, sorption-induced strain and thermal strain.

$$\varepsilon_{ij} = \frac{1}{2G}\sigma_{ij} - \left(\frac{1}{6G} - \frac{1}{9K}\right)\sigma_{kk} \cdot \delta_{ij} + \frac{\alpha}{3K} p\delta_{ij} + \frac{1}{3}\varepsilon_s \delta_{ij} + \frac{1}{3}\varepsilon_T \delta_{ij}$$
$$E = R_m \cdot E_m, \quad G = \frac{E}{2(1+\nu)}, \quad K = \frac{E}{3(1-2\nu)}, \quad \alpha = 1 - \frac{K}{K_s}, \quad \sigma_{kk} = \sigma_{11} + \sigma_{22} + \sigma_{33} \tag{3}$$

K represents the bulk modulus of coal and K_s represents the bulk modulus of coal grains. G is the shear modulus of coal, ε_s is the sorption-induced strain, ε_T is the thermal strain, E_m is the Young's modulus of the matrix, R_m is the modulus reduction ratio, E is the equivalent Young's modulus of coal-fracture assemblage and v is the Poisson's ratio of coal-fracture assemblage. α represents the Biot coefficient, α_T is the coefficient of volumetric thermal expansion of coal, p is the gas pressure in pores and δ_{ij} is the Kronecker delta; 1 for $i = j$ and 0 for $i \neq j$.

In a non-isothermal body, the swelling of coal matrix with the increase of temperature due to thermal expansion leads to a decrease in the porous medium porosity. This is directly analogous to matrix swelling in coalbeds, where cleat porosity decreases as gas adsorbs during injection (Palmer and Mansoori, 1996). By combining thermal expansion/contraction and matrix swelling/shrinkage, stress–strain relationships for a non-isothermal coalbed may be written as (negative in compression).

$$\Delta\varepsilon_{ij} = \frac{1}{2G}\Delta\sigma_{ij} - \left(\frac{1}{6G} - \frac{1}{9K}\right)\Delta\sigma_{kk}\delta_{ij}$$
$$+\frac{\alpha}{3K}\Delta p\delta_{ij} + \frac{\Delta\varepsilon_s}{3}\delta_{ij} + \frac{\alpha_T}{3}\Delta T\delta_{ij} \tag{4}$$

Combining the equations above, the following governing equation for displacement of coal under a combination of applied stress, strain, gas pressure and temperature can be obtained.

$$G\mu_{i,kk} + \frac{G}{1-2v}\mu_{k,ki} - \alpha p_{,i} - K\varepsilon_{s,i} - K\varepsilon_{T,i} + f_i = 0 \tag{5}$$

From Equation (10), we obtain

$$\Delta\varepsilon_v = -\frac{1}{K}(\Delta\bar{\sigma} - \alpha\Delta p) + \Delta\varepsilon_s + \alpha_T\Delta T$$
$$\varepsilon_v = \varepsilon_{11} + \varepsilon_{22} + \varepsilon_{33}; \quad \bar{\sigma} = -\sigma_{kk}/3 \qquad 6)$$

where ε_v is the volumetric strain of coal matrix and σ is the mean compressive stress.

Considering a porous medium containing solid volume of V_s and pore volume of V_p, we assume the bulk volume and the porosity

$$V = V_p + V_s; \quad \phi = V_p / V$$

According to Eq. (2), the volumetric evolution of the porous medium with the load of σ and p can be described in terms of $\Delta V/V$ and $\Delta V_p/V_p$, the volumetric strain of coal and volumetric strain of pore space, respectively. The relations are

$$\frac{\Delta V}{V} = -\frac{1}{K}(\Delta\bar{\sigma} - \alpha\Delta p) + \Delta\varepsilon_s + \alpha_T\Delta T \tag{7}$$

$$\frac{\Delta V_p}{V_p} = -\frac{1}{K_p}(\Delta\bar{\sigma} - \beta\Delta p) + \Delta\varepsilon_s + \alpha_T \Delta T$$

$$\beta = 1 - K_p / K_s \tag{8}$$

We assume that the sorption-induced strain for the coal is the same as for the pore space. Without the gas sorption effect, the volumetric variation of the porous medium satisfies the Betti-Maxwell reciprocal theorem, (Hudson et al., 1993):

$$\left.\frac{\partial V}{\partial p}\right|_{\bar{\sigma}} = \left.\frac{\partial V_p}{\partial \bar{\sigma}}\right|_p$$

and we obtain

$$K_p = \frac{\phi}{\alpha} K \tag{9}$$

Using the definition of porosity, the following expressions can be deduced as

$$\frac{\Delta V}{V} = \frac{\Delta V_s}{V_s} + \frac{\Delta \phi}{1-\phi} \tag{10}$$

$$\frac{\Delta V_p}{V_p} = \frac{\Delta V_s}{V_s} + \frac{\Delta \phi}{\phi(1-\phi)} \tag{11}$$

Solving Eqs. (13)–(17), we obtain the relationship as

$$\Delta\phi = \phi\left(\frac{1}{K} - \frac{1}{K_p}\right)(\Delta\bar{\sigma} - \Delta p) \tag{12}$$

Substituting Equations (12), (15) into Equation (18) yields

$$\phi - \phi_0 = -(\alpha - \phi)\frac{(\Delta\bar{\sigma} - \Delta p)}{K} \tag{13}$$

Rearranging the above equation gives

$$\phi = \frac{\phi_0}{\left(1 - \frac{\Delta\bar{\sigma} - \Delta p}{K}\right)} - \frac{\alpha}{\left(1 - \frac{\Delta\bar{\sigma} - \Delta p}{K}\right)}\frac{\Delta\bar{\sigma} - \Delta p}{K} \tag{14}$$

Because generally $(\Delta\bar{\sigma} - \Delta p)/K << 1$, the above equation can be simplified into

$$\frac{\phi}{\phi_0} = 1 - \frac{\alpha}{\phi_0}\frac{\Delta\bar{\sigma} - \Delta p}{K} = 1 + \frac{\alpha}{\phi_0}\Delta\varepsilon_e \tag{15}$$

where $\Delta\varepsilon_e = -(\Delta\bar{\sigma} - \Delta p)/K$ is defined as the total effective volumetric strain.

Eq. (12) can be written as

$$\Delta\varepsilon = \Delta\varepsilon_e - \frac{\Delta p}{K_m} + \Delta\varepsilon_s + \alpha_T \Delta T \tag{16}$$

The gas sorption-induced normal strain ε_s and the gas pressure-induced strain p/K_m produce no shear strain. Its effects on all three normal components of strain are equal (Robertson, 2005). By these definitions, ε_s and p/K_m are the free-swelling volumetric strain and the coal grain compressive volumetric strain. The total effective volumetric strain is calculated by

$$\Delta\varepsilon_e = \Delta\varepsilon_v + \frac{\Delta p}{K_m} - \Delta\varepsilon_s - \alpha_T \Delta T \tag{17}$$

Only $\Delta\varepsilon_e$ is responsible for the coal permeability change.

2.2 *Permeability model*

Changes in coal permeability are determined by the redistribution of effective stresses or strains due to changed conditions such as gas injection, as shown in Figure 2. Coal fracture porosity can be determined by fracture spacing and aperture (Liu, Elsworth et al., 1999). For the two-dimensional case, the initial areal porosity for x or y directions is defined as

$$\phi_{fi0} = \frac{3 \cdot b_{j0}}{s} \cdot i \neq j \tag{18}$$

where ϕ_{fi0} is the porosity, and b_{fi0} is the fracture aperture for i direction.

It is assumed that fracture and matrix deformation are both linear and fully recoverable, and deformations in normal closure and opening are the predominant permeability alteration mode. Therefore, coal permeability changes can be defined as a function of the variation aperture in corresponding directions, where the aperture variation is realized through the elastic modulus reduction ratio, R_m.

$$R_m = \frac{E}{E_m} \tag{19}$$

E and E_m are Young's modulus of skeleton and modulus of matrixes, respectively.

The principle of modulus reduction ratio is shown in Figure 3. U_m and U_r are the displacements of coal matrix and fracture. When the coal modulus reduction ratio is unity, i.e., $R_m = 1$ then the equivalent modulus of the fractured medium is equal to that of the coal matrix. In other words the coal mass may be considered as unfractured or the fractures are infinitely small. Conversely, in the limit as $R_m = 0$ then the coal matrix is infinitely stiff and

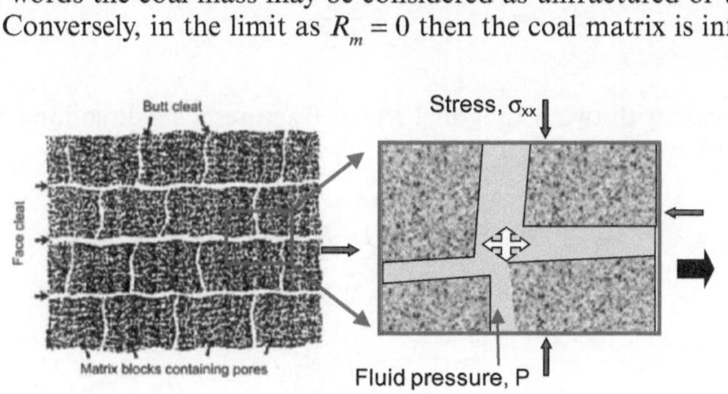

Figure 2. Micro structure of coal.

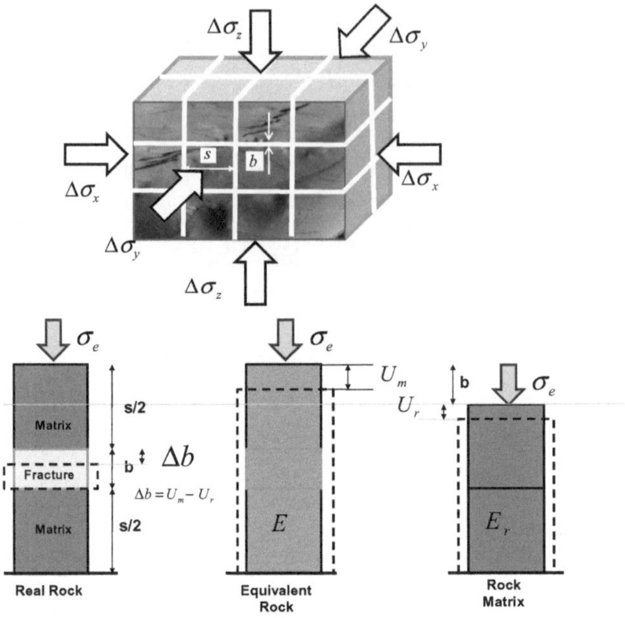

Figure 3. Schematic diagram for modulus reduction ratio.

the observed deformational response is equivalent to that of the fractures alone. Therefore the parameter $1 - R_m$ represents the ratio of the partitioned strain for the fracture system to the total equivalent strain. If $R_m = 1$, zero percent of the equivalent strain is due to coal fractures, therefore the coal permeability change is due to the deformation of the matrix only. If $R_m = 0$, 100% of the equivalent strain is due to coal fractures, the coal permeability change is due to fractures only.

Results from field and laboratory experiments indicate that coal permeability can change significantly during absorbable gas injection (e.g. CH_4 and CO_2). The injection gas pressure tends to mechanically open coal cleats and thus enhance the permeability as the initial gas pressure resides only in the fractures and any constrained change in total stress compresses the matrix blocks. If expansion of the cleat-matrix assemblage is constrained then fracture permeability reduces by narrowing and even closing cleat apertures. When the coal swelling is taken into consideration, the total effective strains in the equation (17) can be replaced by the differences between the total strain change, $\Delta\varepsilon_{vi}$, in the i direction and the free swelling strain change, $\Delta\varepsilon_s$, as follows

$$\Delta\varepsilon_{ei} = \Delta\varepsilon_{vi} - \frac{1}{3}\Delta\varepsilon_s - \frac{1}{3}\alpha_T\Delta T + \frac{\Delta p}{3K_m} \tag{20}$$

Under the 2D case with two orthogonal sets of fractures, coal directional permeability for different directions are defined as following (Liu, Elsworth et al., 1999):

$$\frac{k_{fi}}{k_{fi0}} = \left(1 + \frac{2\cdot(1-R_m)}{\phi_0}\cdot\Delta\varepsilon_{ei}\right)^3 \tag{21}$$

$$k_{fi0} = \frac{b_{j0}{}^3}{12\cdot s} \tag{22}$$

where k_{fi} is permeability of coal in i direction, k_{fi0} is the initial permeability for i direction, $\Delta\varepsilon_{ei}$ is the directional effective strain change. s is the fracture spacing.

From Eq. (13) we can see the effective strain change for the matrix system can be expressed as

$$\Delta\varepsilon_{em} = \frac{\Delta\sigma - \Delta p_m}{K_m} = R_m \frac{\Delta\sigma - \Delta p_m}{K} \tag{23}$$

Then the matrix porosity change regarding effective stress variation can be given as

$$\phi_m = \phi_{m0} + R_m \frac{\Delta\bar{\sigma} - \Delta p}{K} \tag{24}$$

From Eq. (15) we can know that fracture porosity change can be given as

$$\phi_f = \phi_{f0} + (1 - R_m)(\Delta\varepsilon_v + \frac{\Delta p}{K_m} - \alpha_T \Delta T - \Delta\varepsilon_s) \tag{25}$$

For the 2D case with two orthogonal sets of fractures, coal directional permeability, k_x and k_y were defined as

$$\frac{k_{fi}}{k_{fi0}} = \left(1 + \frac{2(1-R_m)}{\phi_{x0}} \cdot \left(\begin{array}{l} \varepsilon_{vj} - \frac{1}{3}\varepsilon_L \cdot \left(\frac{p_g}{p_g + p_L} - \frac{p_0}{p_0 + p_L} \right) \\ -\frac{1}{3}\alpha_T \cdot (T_g - T_0) + \frac{1}{3K_m}(p_g - p_0) \end{array} \right)\right)^3 \tag{26}$$

2.3 *Flow and transport analysis*

The mass balance equation for a single component gas is defined as

$$\frac{\partial m}{\partial t} + \nabla \cdot (\rho_g \cdot \upsilon) = Q_s \tag{27}$$

where ρ_g is the gas density, υ is the Darcy velocity vector and Q_s is the gas source or sink.

Assuming the effect of gravity is relatively small and can be neglected, the Darcy's velocity may be defined as

$$\mu = -\frac{k}{v} \cdot \nabla p_g \tag{28}$$

where v is the viscosity, ∇p_g is the pressure gradients, m is the gas content including both free-phase and adsorbed components w

$$m = \rho_g \cdot \phi + (1 - \phi) w \tag{29}$$

As discussed above in introduction, the modified Langmuir adsorption equation is applied here.

$$w = \frac{\rho_c}{1000} \cdot w_a = \frac{\rho_c}{1000} \cdot \left[w_0 \cdot \left(1 - \frac{\rho_g}{\rho_a}\right) \cdot \left(\frac{\rho_g}{\rho_g + \rho_L}\right) + k_a \cdot \rho_g \cdot \left(1 - \frac{\rho_g}{\rho_a}\right) \right] \tag{30}$$

where w_0 is the maximum adsorption capacity, ρ_a is the density of the adsorbed phase, ρ_L is the characteristic Langmuir density, represents the gas density at which the adsorption is half the maximum. k_a is the error collection.

2.4 *Heat conduction and convection modelling*

Energy conservation equation can be obtained on the basis of the first and second laws of thermo-dynamics, which analyze the transformations affecting all the forms of energy involved in the evolution of a system.

If the dissipation of mechanical energy is neglected, under the assumption of small perturbations, thermal equation can be written with substitution of Fourier's Law as:

$$T\left(\frac{\partial S}{\partial t} - s_f \frac{\partial m}{\partial t}\right) = \nabla \cdot (\lambda \nabla T) \tag{31}$$

where S stands for the overall density of entropy per unit of initial volume; s_f is the fluid-specific entropy, m is the unit mass, λ is the thermal conductivity, and T is the absolute temperature.

Through substituting the expression of m and the porosity ϕ, by some simple manipulations, the Energy conservation equation can be written as:

$$c_1 \frac{\partial \varepsilon_t}{\partial t} - c_2 \frac{\partial p}{\partial t} + c_3 \frac{\partial T}{\partial t} = \nabla \cdot (\lambda \nabla T) \tag{32}$$

in which, c_1, c_2 and c_3 are the coefficients for the change rate of total volume strain, pressure and temperature respectively.

Among all thermal processes, heat conduction and convection are dominant. Conduction is heat transfer by diffusion in a stationary medium, due to a temperature gradient.

The heat conductivity equation can be written as:

$$c_v \frac{\partial T}{\partial t} = \lambda \nabla^2 T \tag{33}$$

In coal seams, heat conduction happens in both coal and CO_2 fluid. An equivalent heat capacity and equivalent thermal conductivity are applied here.

$$\lambda = (1 - \phi) \cdot \lambda_s + \phi \cdot \lambda_g \tag{34}$$

$$c_p = (1 - \phi) \cdot \rho_s \cdot c_s + \phi \cdot \rho_g \cdot c_g \tag{35}$$

where λ is the equivalent thermal conductivity, c_p is the heat capacity at constant pressure. λ_g, λ_s are the thermal conductivity, and c_g, c_s are specific heat constant for gas and coal, respectively.

This model also implements a weak coupling between the flow and thermal steps. Heat convection is the way of heat transfer when heat is transported by a fluid motion, i.e., between a cold surface and a hot moving fluid.

The governing heat conduction and convection equation is defined as:

$$c_{eq}\frac{\partial T}{\partial t}+c_L v\nabla T=\lambda\nabla^2 T \tag{36}$$

where v is the flow velocity of the fluid, and c_{eq} is the equivalent heat capacity.

2.5 *Impact of thermal and pore pressure on adsorption and induced strain behavior*

This section is to investigate the effect of temperature on sorption by establishing the relationship between sorption-induced strain and temperature.

An approximately linear relationship was observed between sorption capacity and reciprocal temperature for Nitrogen (Zhou et al., 2001). Analogously, a temperature dependence of adsorption is reported with a linear curve for methane within the range of experimental temperature (Li et al., 2003; Pini et al., 2010; Bae and Bhatia, 2006), as shown in Figure 2. Consequently, a modified Langmuir volume equation is established in this paper as following to describe this relationship for CO_2.

$$V_s=V_L(T_g)\frac{p_g}{P_L+p_g}=\left(V_{L0}+A\times\Delta T_g\right)\frac{p_g}{P_L+p_g} \tag{37}$$

where V_s is the adsorbed volume. V_{L0} is the maximum sorption capacity of coal to a gas component at the initial reference temperature of T_0. V_L is the maximum sorption capacity of coal to a gas component at temperature of T_g. A is the sorption capacity change rate with temperature, which is negative as temperature increases. P_L is the Langmuir sorption pressure constant. Volumetric strain associated with gas sorption can be measured in terms of the adsorbed gas volume at standard pressure and temperature. Experiments on the volumetric strain associated with methane and carbon dioxide adsorption has shown that the sorption-induced volumetric strain is approximately proportional to the volume of adsorbed gas (Cui and Bustin, 2005; Clarkson and Bustin, 2010), which also has been confirmed by Day et al. (2010), as shown in Figure 3. The relationship between is described as

$$\varepsilon_s=\varepsilon_g V_s \tag{38}$$

where ε_s is the sorption-induced strain, and ε_g is the proportional ratio of sorption-induced volumetric strain to sorption volume.

Substituting Eq. (29) into the above one, the effect of temperature on sorption-induced strain can be obtained,

$$\varepsilon_s=\varepsilon_g\left(V_{L0}+A\times\Delta T\right)\frac{p}{P_L+p} \tag{39}$$

2.6 *Coupling governing equations*

According to the real gas law, gas density is proportional to the gas pore pressure and can be described as

$$\rho_g=\frac{M_g}{z\cdot R}\cdot\frac{p_g}{T_g} \tag{40}$$

where M_g is the molecular mass of gas, R is the universal gas constant, T_g is the gas temperature, z is the correction factor that accounts for the non-ideal behavior of the gas which changes with R and T_g.

Substitute all the equations above to Equation (27), and the governing gas flow equation with thermal gradients can be obtained.

$$\begin{aligned}&\left\{(\rho_g - w)\cdot\left[-(1-R_m)\cdot\frac{\varepsilon_L\cdot p_L}{(p_g+p_L)^2}-\frac{R_m}{K_m}\right]\right.\\&\left.+\frac{M}{R}\frac{1}{T_g}\cdot\left\{\phi+(1-\phi)\frac{\rho_c}{1000}\cdot\left[\begin{array}{l}\left(-\frac{w_0}{\rho_a}\right)\cdot\left(\frac{\rho_g}{\rho_g+\rho_L}+\frac{\rho_g\cdot\rho_L}{(\rho_g+\rho_L)^2}\right)\\+w_0\cdot\frac{\rho_L}{(\rho_g+\rho_L)^2}+k_a\cdot\left(1-2\cdot\frac{\rho_g}{\rho_a}\right)\end{array}\right]\right\}\right\}\frac{\partial p_g}{\partial t}\\&+\left\{(\rho_g-w)\cdot(1-R_m)\cdot\left(-\varepsilon_g A\frac{p_g}{P_L+p_g}-\alpha_T\right)\right.\\&\left.-\frac{M}{R}\frac{p_g}{T_g^2}\cdot\left\{\phi+(1-\phi)\frac{\rho_c}{1000}\cdot\left[\begin{array}{l}\left(-\frac{w_0}{\rho_a}\right)\cdot\left(\frac{\rho_g}{\rho_g+\rho_L}+\frac{\rho_g\cdot\rho_L}{(\rho_g+\rho_L)^2}\right)\\+w_0\cdot\frac{\rho_L}{(\rho_g+\rho_L)^2}+k_a\cdot\left(1-2\cdot\frac{\rho_g}{\rho_a}\right)\end{array}\right]\right\}\right\}\cdot\frac{\partial T_g}{\partial t}\\&+\nabla\left(-\frac{k}{\mu}\cdot\rho_g\cdot\nabla p\right)=Q_s-(\rho_g-w)\cdot(1-R_m)\cdot\frac{\partial\varepsilon_v}{\partial t}\end{aligned} \tag{41}$$

If the process is isothermal, the governing gas flow equation is described as

$$\begin{aligned}&\left\{(\rho_g - w)\cdot\left[-(1-R_m)\cdot\frac{\varepsilon_L\cdot p_L}{(p_g+p_L)^2}-\frac{R_m}{K_m}\right]\right.\\&\left.+\frac{M}{R}\frac{1}{T_g}\cdot\left\{\phi+(1-\phi)\frac{\rho_c}{1000}\cdot\left[\begin{array}{l}\left(-\frac{w_0}{\rho_a}\right)\cdot\left(\frac{\rho_g}{\rho_g+\rho_L}+\frac{\rho_g\cdot\rho_L}{(\rho_g+\rho_L)^2}\right)\\+w_0\cdot\frac{\rho_L}{(\rho_g+\rho_L)^2}+k_a\cdot\left(1-2\cdot\frac{\rho_g}{\rho_a}\right)\end{array}\right]\right\}\right\}\frac{\partial p_g}{\partial t}\\&+\nabla\left(-\frac{k}{\mu}\cdot\rho_g\cdot\nabla p\right)=Q_s-(\rho_g-w)\cdot(1-R_m)\cdot\frac{\partial\varepsilon_v}{\partial t}\end{aligned} \tag{42}$$

Substituting Eq. (39) of sorption induced strain into Eq. (5), the governing mechanical deformation equation is obtained

$$\begin{aligned}&G\nabla^2 u_i+\frac{G}{1-2\nu}e_{,i}-K\cdot\left(\alpha_T+\varepsilon_g A\frac{p}{P_L+p}\right)T_{,i}\\&-\left[\alpha+\frac{K\varepsilon_g(V_{L0}+A\times\Delta T)P_L}{(p+P_L)^2}\right]p_{,i}+f_i=0\end{aligned} \tag{43}$$

3 NUMERICAL SIMULATION MODEL

In this section, a 2D numerical model was implemented to simulate the permeability and gas adsorption response to the injection pressure and temperature variation.

3.1 *The FEM model geometry*

The model is solved with finite element method. The two-dimensional geometry domain with an area of 200 m × 200 m is shown in Figure 4. A CO_2 injection well is located in the center with a diameter of 0.1 m. Directional permeability was used in this comparison for fractured porous media case. Homogeneous properties of the coal seams in horizontal direction are assumed. Therefore, the CO_2 plume will develop radially symmetrically around the injection well and it is sufficient to model only a sector of the domain.

3.2 *Initial and boundary conditions*

An increasing pressure from the initial pressure of 0.5 MPa to the injection value pin of 15 MPa and an increasing temperature from initial 298.15 K to T_b of 323.15 K are specified on the injection well boundaries.

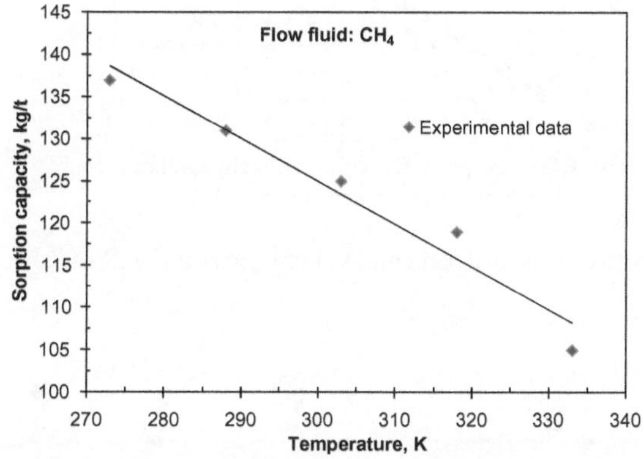

(a) Temperature dependence of coal adsorption capacity to CH_4 by charcoal (data from Li et al., 2003)

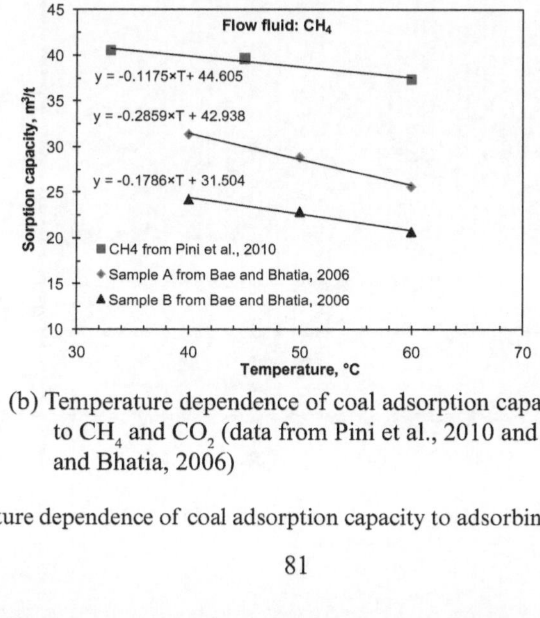

(b) Temperature dependence of coal adsorption capacity to CH_4 and CO_2 (data from Pini et al., 2010 and Bae and Bhatia, 2006)

Figure 4. Temperature dependence of coal adsorption capacity to adsorbing gas.

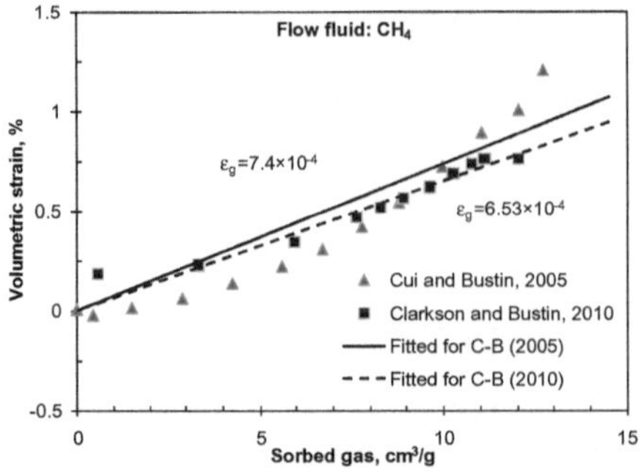

(a) CH_4 adsorption for a Mesaverde Group coal (data from Cui and Bustin, 2005) and a Wolf Mtn coal (data from Clarkson and Bustin, 2010)

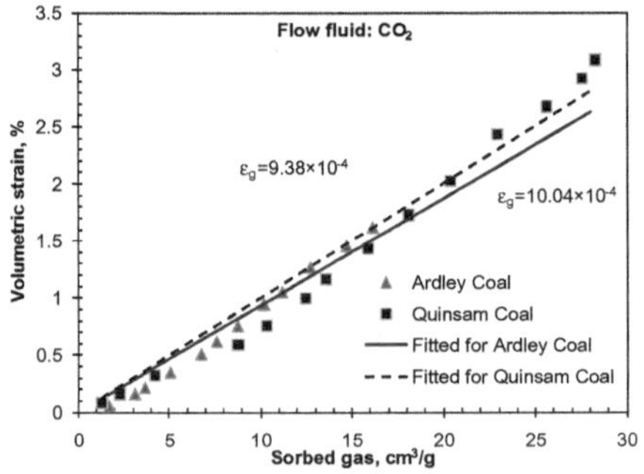

(b) CO_2 adsorption for Ardley coal and Quinsam coal (data from Clarkson and Bustin, 2010)

Figure 5. Relationship between the experimental volumetric strain and the absorbed gas volume.

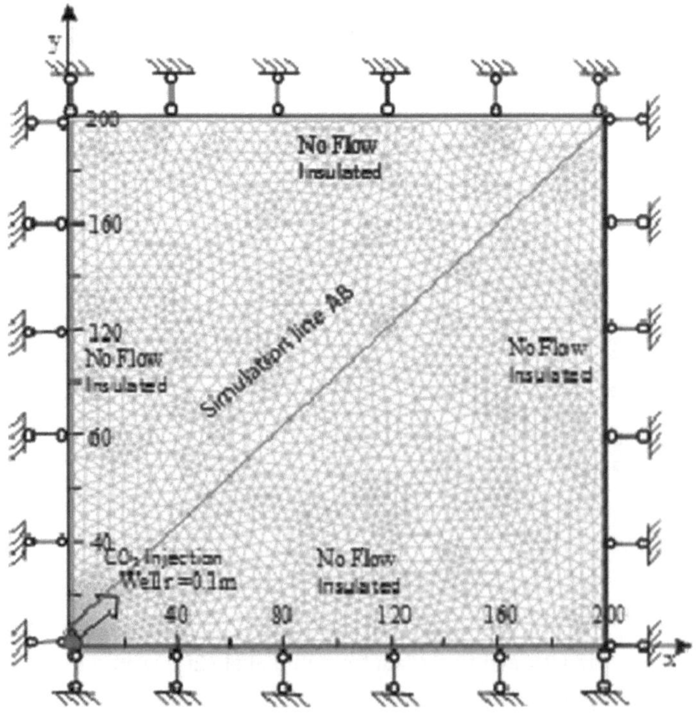

Figure 6. Geometry of the simulation model (one quarter of the model is simulated).

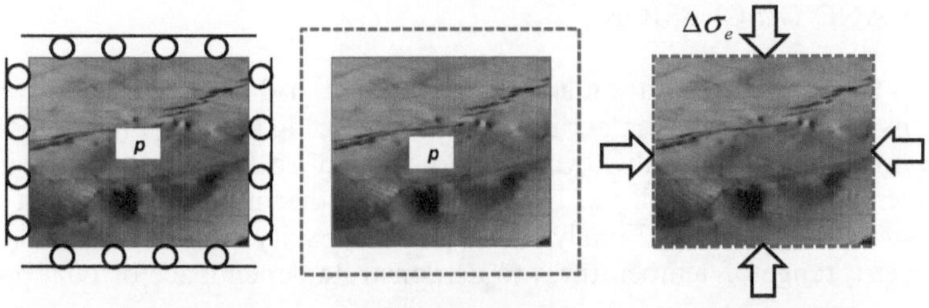

Figure 7. Illustration of free expansion plus push-back process.

Table 2. Parameters for permeability model.

Parameters	Definition	Value [units]
p_0	Initial pressure	0.5 [MPa]
p_{ir}	Injected pressure	7 [MPa]
T_0	Initial temperature	298.15 [K]
T_b	Injected temperature	323.15 [K]
ρ_a	Adsorped phase density	1000 [kg/m^3]
ρ_L	Langmuir density	50 [kg/m^3]
W_0	Maximum adsorption	80.9 [kg/t]
K_a	Correction constant	0.023 [m^3/t]
M_g	Molar mass of CO_2	44e-3 [kg/mol]
R	Gas constant	8.314472 [JK^{-1} mol^{-1}]
vis	Viscosity	1.48e-5 [Pa s]
ρ_g	Coal density	1.25e3 [kg/m^3]
α_T	Thermal expansion constant	5e-5 [K^{-1}]
R_m	Elastic modulus reduction ratio	0.5
C_s	Specific heat constant of coal	2300 [J/(kg K)]
C_g	Specific heat constant of CO_2	847 [J/(kg K)]
λ_s	Thermal conductivity of coal	0.33 [w/(m k)]
λ_g	Thermal conductivity of CO_2	0.0246 [w/(m k)]

For the gas and heat flow models, no gas or heat flow conditions are specified on boundaries. For the coal deformation model, four boundaries are constrained to keep the whole volume constant. The free expansion that coal sample experiences is illustrated in Figure 7.

In this case, four boundaries are no lateral deformations are allowed. The total strain increments are given as follows,

$$\Delta\varepsilon_{tx} = \Delta\varepsilon_{ty} = 0 \tag{44}$$

All permeability ratios are determined by the effective strain. Substituting equation (2) or (3) into (21) gives

$$\frac{k_{fi}}{k_{fi0}} = \left(1 + \frac{2(1-R_m)}{\phi_{x0}} \cdot \left(\begin{array}{l} \varepsilon_{vj} - \frac{1}{3}\varepsilon_L \cdot \left(\frac{p_g}{p_g + p_L} - \frac{p_0}{p_0 + p_L}\right) \\ -\frac{1}{3}\alpha_T \cdot (T_g - T_0) + \frac{1}{3K_m}(p_g - p_0) \end{array}\right)\right)^3 \tag{45}$$

3.3 *The main material parameters*

Detailed parameters are listed in Table 2.

4 RESULTS AND DISCUSSION

In this section, temperature evolution and distribution at grid (100,100) are displayed in the first few graphs to illustrate the temperature variation during the CO_2 injection and gas flow in the coal seams. Then permeability evolution under non-isothermal condition is compared with that under isothermal condition by analysing the thermal effect on each component of effective strain respectively. Finally, adsorption capacity and permeability isotherms are studied over a range of temperatures to demonstrate dependence of coal properties on temperature.

Temperature in the vicinity of the injection well instantly increases from the initial value of 298.15 K to the same value as that of the injected fluid. The propagation of the heat head leads to large temperature gradients from the injection well to the far away boundaries, as shown in Figure 8. However, it is noticeable that the temperature varies slightly after it reaches the equilibrium due to the heat transfer between the high temperature fluid and unheated solid surface.

The temperature evolution at the point of (60,60), (90,90) and (100,100) in 3 years are illustrated in Figure 9. The heat head reaches the point (60,60) on the 200th day, and gets equilibrium around the 250th day. However, it takes double time for the temperature to increase from the initial value to the injection value at the point (90,90), while as long as 400 days with respect to the point (100,100). Therefore, the further the area is away from the injection

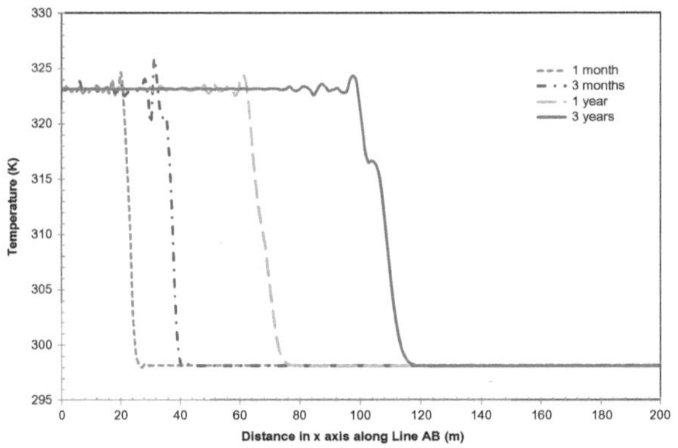

Figure 8. Comparison of temperature distribution along line AB in coal seams from 1 month to 3 years.

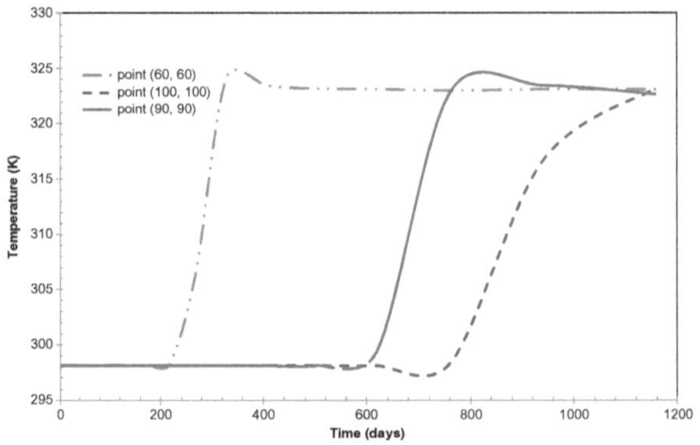

Figure 9. Comparison of temperature evolution along line AB in coal seams at points (60,60) (90,90) (100,100).

well, the longer it takes for the temperature to get equilibrium. In addition, the temperature increase rate decreases with time at each point. As the injection rate is constant, the larger area that the injected fluid passes the less heat is distributed to each area and the longer time it takes for the heating process.

The temperature distribution and evolution during the whole injection period, as shown in Figures 8 and 9, demonstrates that temperature varies considerably between the injection point and the boundary. This is due to the temperature difference between that of the injected fluid and the coal reservoir, and the role of the heat conduction and convection is to reduce the temperature gradients. Heat conduction happens within the solid medium from the higher temperature part to the lower temperature part, and the conductive speed is controlled by the thermal conductivity. Heat convection occurs between the solid phase and fluid phase through the flow of fluid and associated with flow velocity. The relationship between heat and flow is demonstrated in the thermal conservation equation and Darcy's Law.

$$c_{eq}\frac{\partial T}{\partial t}+c_L\upsilon\nabla T=\lambda\nabla^2 T \tag{46}$$

$$\upsilon=-\frac{k}{\mu}\nabla p_g \tag{47}$$

$$c_{eq}\frac{\partial T}{\partial t}-c_L\frac{k}{\mu}\nabla p_g\nabla T=\lambda\nabla^2 T \tag{48}$$

The increase of temperature mainly relies on the heat convective term associated with flow velocity since the thermal conductivity of coal matrix is quite small and it takes much longer time for temperature to get equilibrium, as shown in Figure 10. Temperature changes faster with the increase of flow velocity, as shown in the Eq. (46), and slows down as the flow velocity begins to decrease. As shown in Figure 11, velocity increases rapidly from 0 to 2.4 m/s in the first 400 days, then declines sharply to 0 till the 10000th day. As a consequence, temperature increases with a larger rate as velocity rises then with a smaller rate after the velocity begins to decrease.

However, the temperature head is always behind the pressure head, as shown in Figure 12. Pressure at the point of (100,100) starts to increase from 200 days after the injection begins, about 2000 days earlier than the increase of temperature, which remains on the initial value and does not change in the first 3000 days. This retardation of temperature increase is determined by the velocity of the fluid phase, which is calculated through Darcy's Law and proportion to the pressure gradients, as shown in Eq. (47). Furthermore, the discrepancy of pressure and temperature evolution has an impact on the evolution of effective strain, affecting the permeability as well.

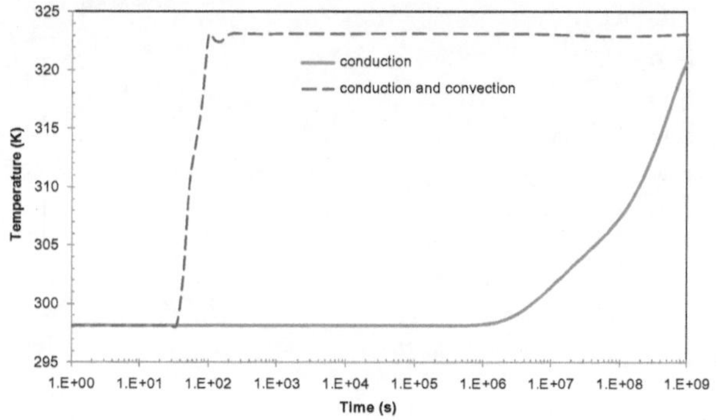

Figure 10. Temperature evolutions under heat conduction and convection.

Permeability is the other factor besides the pressure gradients that controls the rate of fluid flux and hence affects the heat convection term, which can be seen in Eq. (47) and (48). As shown in Figure 13, permeability ratio decreases consistently over 90% in the coal seams in 100000 days under both isothermal condition and non-isothermal condition.

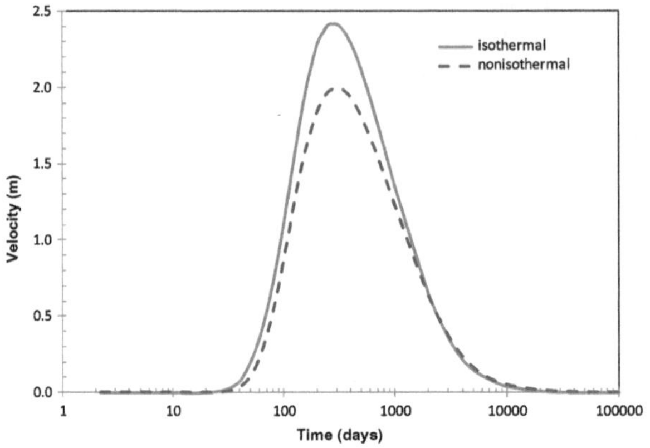

Figure 11. Comparison of velocity evolution for isothermal and non-isothermal cases at point (100,100).

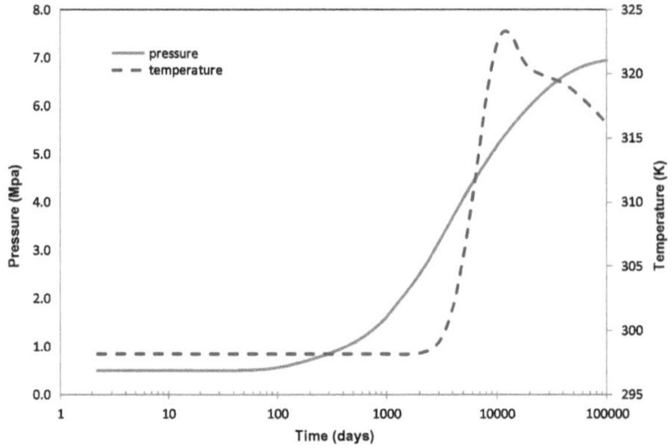

Figure 12. Pressure and temperature evolution under non-isothermal condition at point (100,100).

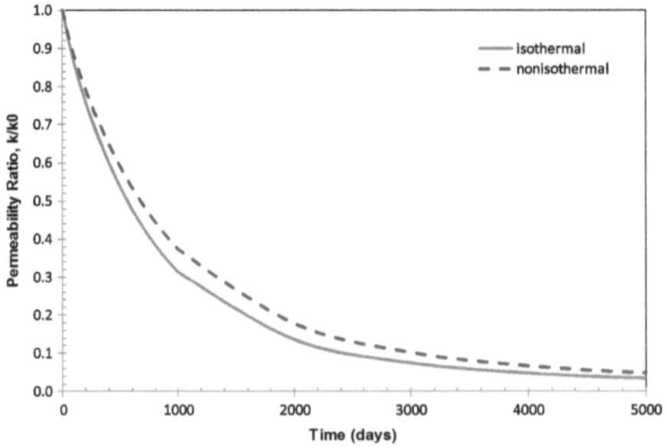

Figure 13. Comparison of permeability ratio evolution for isothermal and non-isothermal cases at point (100,100).

The permeability ratio difference between the isothermal and non-isothermal conditions is only 1% in the first 100 days, but the ratio decreases steeply to 0.3 on the 1000th day, as much as 10% more than the permeability decrease under the isothermal condition. Then it keeps declining with almost the same rate for both cases until the 2000th day when the permeability difference reduces to 5%. After 3000 days, permeability ratio keeps decreasing and stables under 0.1 for both cases.

Permeability distributions on the 1000th day for both isothermal and non-isothermal cases are compared in Figure 14. The lowest permeability ratio on these two lines lies near the injection well, which is under 0.05 for both cases, suggesting that permeability starts to decrease from this area. The permeability difference begins to increase about 10 m away from the injection well. The permeability at the point (100,100) under non-isothermal condition is 0.37, 7% higher than that under isothermal condition. Similarly, permeability on the boundary under non-isothermal condition is about 0.45, 10% higher than that under isothermal condition. The permeability on the boundary is about 30% to 40% higher than that in the vicinity of the injection well under both cases, indicating that permeability near the boundary decreases much slower than the other part of the other area.

The decrease of permeability and the difference between the isothermal and non-isothermal conditions are caused by effective strain, as suggested in Eq. (27), combined the effect of total volume strain, sorption-induced strain and thermal strain. Sorption-induced strain is determined by both pressure and temperature. Thermal strain is caused by thermal expansion and controlled by thermal expansion coefficient and temperature difference. Permeability is determined by pressure and temperature with different weight factors at different stages according to the discrepancy of their evolution time.

The components of the effective strain under two thermal cases are compared in Figure 15. Sorption-induced strain increases substantially from 0 to 0.5% in the first 6000 days. Thus the decrease of the permeability in the first 6000 days is mainly caused by the sorption-induced strain due to the increase of pressure. However, the sorption-induced strain under the non-isothermal condition is smaller compared with that under the isothermal case since the sorption capacity decreases with increasing temperature, leading to a less decrease of permeability. After 6000 days, sorption-induced strain keeps increasing with a smaller rate to 0.7%, and thermal strain begins to increase and rises to 0.05%, as shown in Figure 15, increasing the effective strain and causing permeability to decrease more under non-isothermal condition.

The effect of temperature on permeability for fully constrained model can be better understood by analyzing the displacement evolution since permeability is determined by volumetric strain via displacement. As shown in Figure 16, the displacement increases steeply from 0 to 0.075 m in the first 3000 days for the isothermal case, then decreases consistently to 0 again. The displacement evolution for the non-isothermal case shows similar trend in the

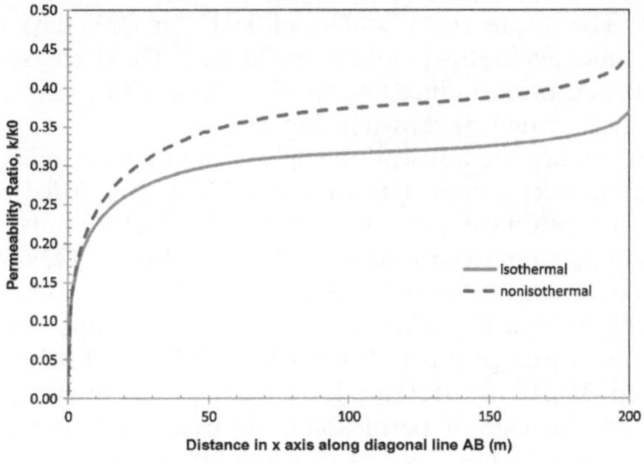

Figure 14. Comparison of permeability ratio distribution along line AB in coal seams for isothermal and non-isothermal cases on the 1000th day.

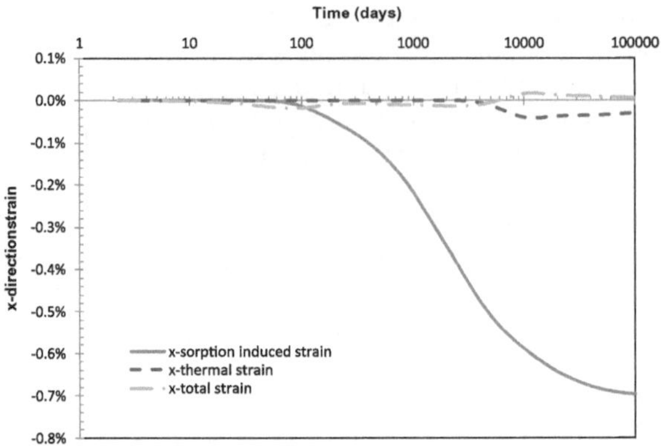

a. Total strain, sorption-induced strain and thermal strain evolution for non-isothermal case

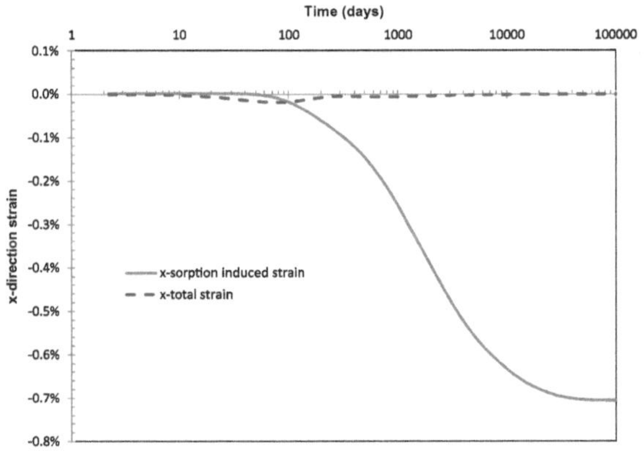

b. Total strain and sorption-induced strain evolution for isothermal case

Figure 15. Comparison of strain evolution at point (100,100) for isothermal and non-isothermal cases.

first period, but the maximum value reaches 0.068 m 200 days later than that of the isothermal case. In addition, the decrease is not consistent but with different decreasing rate at different time period. The displacement stables at 0.032 for 2000 days after the first sharp decrease, and then declines again slowly with a smaller rate. The decrease of the displacement is retarded because the sorption-induced strain decreases with the increasing temperature and then reaches the steady state, as shown in Figure 15.

The last 2 graphs compare the adsorption capacity and permeability distribution under the isothermal conditions over a range temperature of 298.15 K, 328.15 K and 358.15 K on the 1000th day. The adsorption capacity at the point of (100,100) under 3 temperatures are 37 kg/m^3, 34 kg/m^3 and 32 kg/m^3 respectively, as shown in Figure 17, suggesting a 2.5 kg/m^3 decrease of adsorption amount for every 30 K increase of temperature. The results validate the theory that adsorption capacity decreases with increasing temperature.

Permeability ratio decreases gradually from 0.34 to 0.325 and 0.30 respectively, as temperature decreases from 358.15 to 298.15 on the 1000th day, as shown in Figure 18, suggesting an average 2% more decrease of permeability for every 30 K decrease of temperature, which demonstrates that permeability decreases less as temperature increases. This is because sorption-induced strain is the predominant reason for permeability change for isothermal cases. Even though permeability decreases with the increase of pressure, adsorption capacity

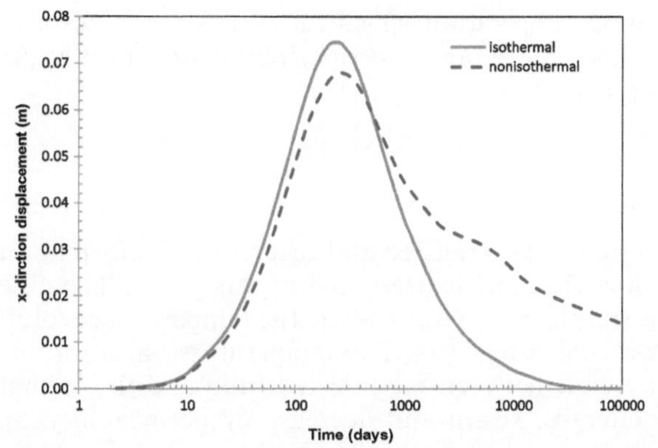

Figure 16. Comparison of displacement evolution in coal seams at point (100,100) for isothermal and non-isothermal cases.

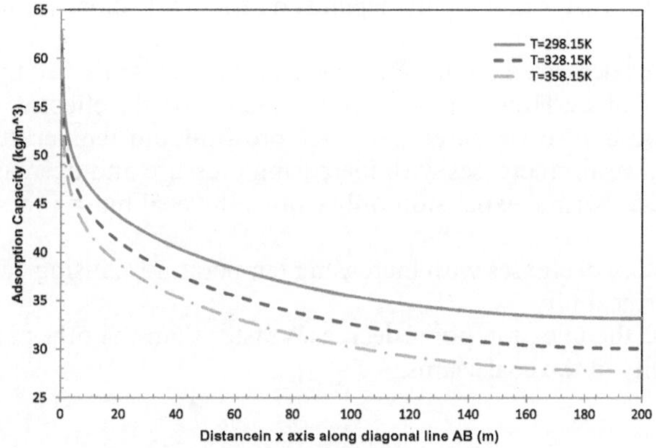

Figure 17. Comparison of adsorption capacity along line AB in coal seams under isothermal condition over a range of temperature of 298.15 K, 323.15 K and 358.15 K.

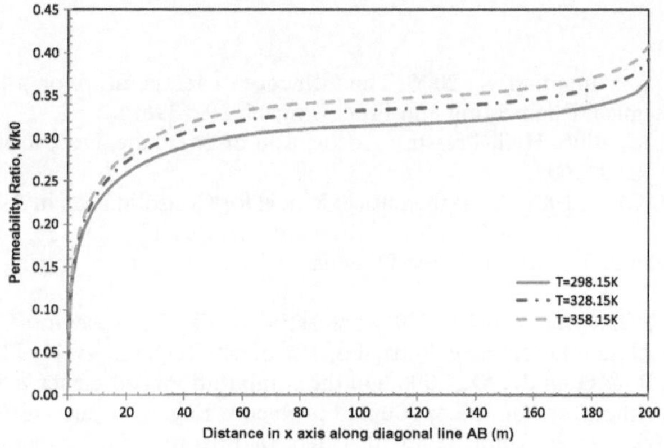

Figure 18. Comparison of permeability ratio along line AB in coal seams under isothermal condition over a range of temperature of 298.15 K, 323.15 K and 358.15 K.

decreases with increasing temperature, which causes less gas to adsorb on the coal surface, and coal matrix to shrink, enhancing the permeability. Consequently, permeability at higher temperature decreases less.

5 CONCLUSIONS

A fully coupled gas flow, heat transfer and mechanical deformation model applied to CO_2 sequestration in coal seams is presented in this paper. The case of high temperature CO_2 injection is simulated to characterise the temperature evolution and investigate the thermal effect on coal properties. The temperature variation in coal seams due to the temperature difference between the injected fluid and the formation leads to a significant change of effective strain and decrease of permeability. Specific conclusions include:

1. Temperature in the vicinity of the injection well reaches the injected gas temperature instantly. The further the area away from the injection well, the smaller rate temperature increases with and the longer time it takes for temperature to get equilibrium.
2. Temperature difference between the high temperature injected gas and the colder solid surface as well as the heat release from the adsorption process is the main reason for heat transfer. The heat transfer rate highly depends on the fluid velocity and coal permeability that controls the heat convective term.
3. Permeability is well described by the effective strain, which is affected by pressure and temperature simultaneously. However, their weight factors on the effective strain are different in different period due to the discrepancy of pressure and temperature evolution time. Sorption-induced strain increases with increasing pressure and decreases with increasing temperature, while thermal expansion only causes the coal matrix to swell and decreases permeability.
4. Adsorption capacity decreases with increasing temperature, causing coal matrix to shrink and enhances permeability.
5. Permeability near the injection point decreases faster than the other areas, leading to the lowest permeability in the coal seams.

ACKNOWLEDGEMENTS

This work was partially funded by the Chinese Scholarship Council and CSIRO National Flagship. These supports are gratefully acknowledged.

REFERENCES

Azmi, A.S., Yusup, S. & Muhamad, S., 2006. The influence of temperature on adsorption capacity of Malaysian coal. Chemical Engineering and Processing 45, 392–396.

Bae, J.-S. & Bhatia, S.K., 2006. High-Pressure Adsorption of Methane and Carbon Dioxide on Coal. Energy & Fuels 20, 2599–2607.

Bear, J. & Corapcioglu, M.Y., 1981. A Mathematical Model for Consolidation in A Thermoelastic Aquifer Due to Hot Water Injection or Pumping. Water Resour. Res. 17, 723–736.

Biot, M.A., 1941. General Theory of Three-Dimensional Consolidation. Journal of Applied Physics 12, 155–164.

Boade, R.R., Chin, L.Y. & Siemers, W.T., 1989. Forecasting of Ekofisk Reservoir Compaction and Subsidence by Numerical Simulation. SPE Journal of Petroleum Technology 41, 723–728.

Cervera, M., Codina, R. & Galindo, M., 1996. On the computational efficiency and implementation of block-iterative algorithms for non-linear coupled problems. Eng. Comput., 13(6), 4–30.

Charrière, D., Pokryszka, Z. & Behra, P., 2010. Effect of pressure and temperature on diffusion of CO_2 and CH_4 into coal from the Lorraine basin (France). International Journal of Coal Geology 81, 373–380.

Chen, Z., Liu, J., Elsworth, D., Connell, L.D. & Pan, Z., 2010. Impact of CO_2 injection and differential deformation on CO_2 injectivity under in-situ stress conditions. International Journal of Coal Geology 81, 97–108.

Chilingar, G.V., 1964. Relationship Between Porosity, Permeability, and Grain-Size Distribution of Sands and Sandstones, in: Straaten, L.M.J.U.v. (Ed.), Developments in Sedimentology. Elsevier, pp. 71–75.

Clarkson, C.R. & Bustin, M., 2010. Coalbed Methane: Current Evaluation Methods, Future Technical Challenges, SPE Unconventional Gas Conference. Society of Petroleum Engineers, Pittsburgh, Pennsylvania, USA.

Connell, L.D., 2009. Coupled flow and geomechanical processes during gas production from coal seams. International Journal of Coal Geology 79, 18–28.

Connell, L.D. & Detournay, C., 2009. Coupled flow and geomechanical processes during enhanced coal seam methane recovery through CO_2 sequestration. International Journal of Coal Geology 77, 222–233.

Connell, L.D., Lu, M. & Pan, Z., 2010. An analytical coal permeability model for tri-axial strain and stress conditions. International Journal of Coal Geology doi:10.1016/j.coal.2010.08.011.

Crosdale, P.J., Moore, T.A. & Mares, T.E., 2008. Influence of moisture content and temperature on methane adsorption isotherm analysis for coals from a low-rank, biogenically-sourced gas reservoir. International Journal of Coal Geology 76, 166–174.

Cui, X. & Bustin, R.M., 2005. Volumetric strain associated with methane desorption and its impact on coalbed gas production from deep coal seams. AAPG Bulletin 89, 1181–1202.

Day, S., Fry, R., Sakurovs, R. & Weir, S., 2010. Swelling of Coals by Supercritical Gases and Its Relationship to Sorption. Energy & Fuels 24, 2777–2783.

Gawin, D. & Schrefler, B.A., 1996. Thermo-hydro-mechanical analysis of partially saturated materials. Eng. Comput., 13(7), 113–143.

Gilman, A. & Beckie, R., 2000. Flow of Coal-Bed Methane to a Gallery. Transport in Porous Media 41, 1–16.

Gutierrez, M. & Makurat, A., 1997. Coupled THM modelling of cold water injection in fractured hydrocarbon reservoirs. Int. J. Rock Mech. Min. Sci., 34(3–4), Paper No. 113.

Gray, I., 1987. Reservoir Engineering in Coal Seams: Part 1-The Physical Process of Gas Storage and Movement in Coal Seams. SPE Reservoir Engineering 2, 28–34.

Gu, F. & Chalaturnyk, J.J., 2005. Analysis of Coalbed Methane Production by Reservoir and Geomechanical Coupling Simulation, 44.

Harpalani, S. & Chen, G., 1997. Influence of gas production induced volumetric strain on permeability of coal. Geotechnical and Geological Engineering 15, 303–325.

Harpalani, S. & Schraufnagel, R.A., 1990. Shrinkage of coal matrix with release of gas and its impact on permeability of coal. Fuel 69, 551–556.

Hettema, M.H.H., Schutjens, P.M.T.M., Verboom, B.J.M. & Gussinklo, H.J., 2000. Production-Induced Compaction of a Sandstone Reservoir: The Strong Influence of Stress Path. SPE Reservoir Evaluation & Engineering 3, 342–347.

Hudson, J.A. & Fairhurst, C., et al., 1993. Comprehensive rock engineering: principles, practice, and projects, Vol. II, Analysis and Design Method. Pergamon, Oxford.

Jaeger, J.C., Cook, N.G.W. & Zimmerman, R.W., 2007. Fundamentals of rock mechanics, Fourth ed. Blackwell Publishing, Victoria, Australia.

Lai, Y.M., Wu, Z.W., Zhu, Y.L. & Zhu, L.N., 1998. Nonlinear analysis for the coupled problem of temperature, seepage and stress fields in cold-region tunnels. Tunnelling Underground Space Technol., 13(4), 435–440.

Laxminarayana, C. & Crosdale, P.J., 1999. Role of coal type and rank on methane sorption characteristics of Bowen Basin, Australia coals. International Journal of Coal Geology 40, 309–325.

Li, M., Gu, A.-Z., Lu, X.-S. & Wang, R.-S., 2003. Supercritical Methane Adsorption Equilibrium Data on Activated Carbon with Prediction by the Adsorption Potential Theory. Journal of Chemical & Engineering Data 49, 73–76.

Liu, J. & Elsworth, D. et al. 1999. "Linking stress-dependent effective porosity and hydraulic conductivity fields to RMR." International Journal of Rock Mechanics and Mining Sciences 36(5): 581–596.

Liu, H.-H. & Rutqvist, J., 2010. A New Coal-Permeability Model: Internal Swelling Stress and Fracture–Matrix Interaction. Transport in Porous Media 82, 157–171.

Liu, J., Chen, Z., Elsworth, D., Miao, X. & Mao, X., 2010. Evaluation of stress-controlled coal swelling processes. International Journal of Coal Geology 83, 446–455.

Liu, J., Chen, Z., Elsworth, D., Miao, X. & Mao, X., 2010b. Linking gas-sorption induced changes in coal permeability to directional strains through a modulus reduction ratio. International Journal of Coal Geology 83, 21–30.

Long, Q.-M., Wen, G.-C., Zou, Y.-H. & Zhao, X.-S., 2009. Experimental study on gas permeability by adsorption under 3D-stress. Journal of Coal Science and Engineering (China) 15, 148–151.

Lu, M. & Connell, L.D., 2008. Non-isothermal flow of carbon dioxide in injection wells during geological storage. International Journal of Greenhouse Gas Control 2, 248–258.

Macrae, J.C. & Ryder, C., 1955. Thermal Expansion of Coal. Nature 176, 265–265.

Masters, I., Pao, W.K.S. & Lewis, R.W., 2000. Coupling temperature to a double-porosity model of deformable porous media. Int. J. Numer. Anal. Methods Geomech., 49, 421–438.

Myer, L.R., 2003. Geomechanical Risks in Coal Bed Carbon Dioxide Sequestration. Lawrence Berkeley National Laboratory. LBNL Paper LBNL–53250.

Nithiarasu, P., Sujatha, K.S., Ravindran, K., Sundararajan, T. & Seetharamu, K.N., 2000. Non-Darcy natural convection in a hydrodynamically and thermally anisotropic porous medium. Comput. Methods Appl. Mech. Eng., 188, 413–430.

Nowacki, W., 1995. Dynamic Problems of Thermoelasticity. Noordhoff, Leiden, The Netherlands.

Oldenburg, C.M. (2007). "Joule-Thomson cooling due to CO_2 injection into natural gas reservoirs." Energy Conversion and Management 48(2007): 1808–1815.

Palmer, I., 2009. Permeability changes in coal: Analytical modeling. International Journal of Coal Geology 77, 119–126.

Palmer, I. Mansoori, J., 1996. How Permeability Depends on Stress and Pore Pressure in Coalbeds: A New Model, SPE Annual Technical Conference and Exhibition. 1996 Copyright 1996, Society of Petroleum Engineers, Inc., Denver, Colorado.

Pan, Z., Connell, L.D. Camilleri, M., 2010. Laboratory characterisation of coal reservoir permeability for primary and enhanced coalbed methane recovery. International Journal of Coal Geology 82, 252–261.

Pekot, L.J. Reeves, S.R., 2002. Modeling the effects of matrix shrinkage and differential swelling on coalbed methane recovery and carbon sequestration. U.S. Department of Energy, DE-FC26-00 NT40924.

Pini, R., Ottiger, S., Burlini, L., Storti, G. & Mazzotti, M., 2009. Role of adsorption and swelling on the dynamics of gas injection in coal. J. Geophys. Res. 114, B04203.

Pini, R., Ottiger, S., Burlini, L., Storti, G. & Mazzotti, M., 2010. Sorption of carbon dioxide, methane and nitrogen in dry coals at high pressure and moderate temperature. International Journal of Greenhouse Gas Control 4, 90–101.

Reeves, S., Taillefert, A., Pekot, L. & Clarkson, C., 2003. The Allison unit CO_2-ECBM pilot: a reservoir modelling study. U.S. Department of Energy. Topical Report DE-FC26-0 NT40924. February.

Robertson, E.P., 2005. Modeling Permeability in Coal Using Sorption-Induced Strain Data, SPE Annual Technical Conference and Exhibition. Society of Petroleum Engineers, Dallas, Texas.

Robertson, E.P. & Christiansen, R.L., 2007. Modeling Laboratory Permeability in Coal Using Sorption-Induced Strain Data. SPE Reservoir Evaluation & Engineering 10, 260–269.

Saghafi, A., Faiz, M. & Roberts, D., 2007. CO_2 storage and gas diffusivity properties of coals from Sydney Basin, Australia. International Journal of Coal Geology 70, 240–254.

Sakurovs, R., Day, S., Weir, S. & Duffy, G., 2008. Temperature dependence of sorption of gases by coals and charcoals. International Journal of Coal Geology 73, 250–258.

Seidle, J.R. & Huitt, L.G., 1995. Experimental Measurement of Coal Matrix Shrinkage Due to Gas Desorption and Implications for Cleat Permeability Increases, International Meeting on Petroleum Engineering. 1995 Copyright 1995, Society of Petroleum Engineers, Inc., Beijing, China.

Selvadurai A.P.S., Nguyen T.S., 1999. Mechanics and fluid transport in a degradable discontinuity. Eng. Geo. 53, 243–249.

Settari, A., 2002. Reservoir Compaction. SPE Journal of Petroleum Technology 54(8), 62–69.

Shi, J.Q. & Durucan, S., 2004. Drawdown Induced Changes in Permeability of Coalbeds: A New Interpretation of the Reservoir Response to Primary Recovery. Transport in Porous Media 56, 1–16.

Siriwardane, H., Haljasmaa, I., McLendon, R., Irdi, G., Soong, Y. & Bromhal, G., 2009. Influence of carbon dioxide on coal permeability determined by pressure transient methods. International Journal of Coal Geology 77, 109–118.

Thomas, H.R., Yang, H.T. & He, Y., 1999. A sub-structure based parallel solution of coupled thermo-hydro-mechanical modelling of unsaturated soil. Eng. Comput. 16(4), 428–442.

Thomas, H.R. & Cleall, P.J., 1999. Inclusion of expansive clay behaviour in coupled thermo hydraulic mechanical models. Eng. Geol. 54, 93–108.

Thomas, H.R. & Missoum, H., 1999. Three-dimensional coupled heat, moisture and air transfer in a deformable unsaturated soil. Int. J. Numer. Methods Eng. 44, 919–943.

Tsang, C.-F., editor. 1987. Coupled processes associated with nuclear waste repositories. New York: Academic Press.

Tsang, C.-F., 1991. Coupled thermomechanical and hydrochemical processes in rock fractures. Rev. Geophys. 29, 537–548.

Wang, G.X., Massarotto, P. & Rudolph, V., 2009. An improved permeability model of coal for coalbed methane recovery and CO_2 geosequestration. International Journal of Coal Geology 77, 127–136.

Wang, X. & Schrefler, B.A., 1998. A multi-frontal parallel algorithm for coupled thermo-hydro-mechanical analysis of deformable porous media. Int. J. Numer. Methods Eng. 43, 1069–1083.

Wong, S., Law, D., Deng, X., Robinson, J., Kadatz, B., Gunter, W.D., Jianping, Y., Sanli, F. & Zhiqiang, F., 2007. Enhanced coalbed methane and CO_2 storage in anthracitic coals--Micro-pilot test at South Qinshui, Shanxi, China. International Journal of Greenhouse Gas Control 1, 215–222.

Zhao, C., Hobbs, B.E., Baxter, K., Mühlhaus, H.B. & Ord, A., 1999. A numerical study of pore-fluid, thermal and mass flow in fluid-saturated porous rock basins. Eng. Comput. 16(2), 202–214.

Zhang, H., Liu, J. & Elsworth, D., 2008. How sorption-induced matrix deformation affects gas flow in coal seams: A new FE model. International Journal of Rock Mechanics and Mining Sciences 45, 1226–1236.

Zhou, L., Zhou, Y., Bai, S., Lü, C. & Yang, B., 2001. Determination of the Adsorbed Phase Volume and Its Application in Isotherm Modeling for the Adsorption of Supercritical Nitrogen on Activated Carbon. Journal of Colloid and Interface Science 239, 33–38.

Zhou, Y., Rajapakse, R.K.N.D. & Graham, J., 1998. A coupled thermoporoelastic model with thermo-osmosis and thermal-filtration. International Journal of Solids and Structures 35, 4659–4683.

Zimmerman, R.W., 2000. Coupling in poroelasticity and thermoelasticity. Int. J. Rock Mech. Min. Sci, 37(1), 79–87.

CO_2 Storage in Carboniferous Formations and Abandoned Coal Mines – He, Ribeiro E Sousa, Elsworth & Vargas Jr. (eds)
© 2012 Taylor & Francis Group, London, ISBN 978-0-415-62079-6

Micro-scale modeling of gas-coal interaction in coalbed seam—heterogeneity effect

J.G. Wang & J.S. Liu
School of Mechanical and Chemical Engineering, The University of Western Australia, Crawley, Australia

ABSTRACT: Carbon-dioxide geological storage has an important issue—the interaction of gas-coal interaction in micro-scale. In this report, micro-scaling modeling is implemented for the gas-coal interaction in coalbed seam. First, the mathematical model for the gas-coal interaction is formulated. This model considers the interaction of gas flow, density change of gas due to pressure and deformation of coalbed seam. The absorption /desorption of gas on coal matrix is also incorporated into the framework. Then, the evolution of porosity is developed based on the gas absorption and compaction of coal skeleton. Third, an input method is developed to describe the local heterogeneity through local porosity. Fourth, two examples are numerically simulated to investigate the effect of fracture-induced heterogeneity on gas flow in coalbed seam. Finally, an idealized micro-structure model is investigated to consider the swelling-induced permeability change in coalbed seam.

1 INTRODUCTION

Coal is generally characterized as a heterogeneous material such as a dual porosity system model (Wu et al., 2010). The micro-pores in coal serve as storage spaces for over 95% of the gas in absorption form and have very low permeability. However, macro-fracture network isolates the matrix into individual blocks. How such a heterogeneous system affects the methane production as well as the underground storage of carbon dioxide (CO_2) is an interesting and important issue for carbon capture and storage scheme.

Heterogeneity of coal seam and other stones has been investigated by many researchers. For example, Zhang et al. (2007) investigated the geological CO_2 storage in gas reservoirs through a system-level modeling for economic evaluation. Perrin and Benson (2010) experimentally studied the influence of sub-core scale heterogeneity on CO_2 distribution in reservoir rocks. The CO_2 saturation in steady flow is measured by X-ray CT image techniques. During transient flow such as methane displacing carbon dioxide process, the saturation evolution is measured by Seo (2004) through porosity distribution and fluid density contrast. Wang et al. (2010) numerically simulated the impact of rock microstructures on the supercritical CO_2 enhanced gas recovery for both Weibull heterogeneity and potential preferential paths input by X-Ray CT images. All of these experimental observations can directly measure the heterogeneity of porosity. Recently, the heterogeneity distribution of permeability was experimentally measured if the Darcy's law is true locally (Uh and Watson, 2010). How the heterogeneity of porous medium affect the storage and flow rate of CO_2 in coalbed seam has not been investigated from micro-scale experimental observations and numerical simulations.

In this report, the interaction of gas and coal in micro-scale is numerically investigated. Particularly, the fracture network effect on the flow patterns is investigated. The fracture opening/closing under coal matrix shrinkage/swelling is studied through a simple micro-scale fracture model. Some understandings may be helpful for the carbon dioxide geological storage, particularly in the coalbed seam.

2 MATHEMATICAL MODEL

2.1 *Basic assumptions*

In this research, the coal as porous medium is treated based on the following assumptions: (1) the porous medium is continuous, homogeneous, isotropic and elastic. (2) The deformation and gas flow processes are all pseudo-static. (3) The deformation is infinitesimal. (4) The system is isothermal. Temperature does not change in temporal and spatial domains. (5) Real gases are stored in pore spaces and absorbed on the surface of coal particles. The gas viscosity is constant under isothermal conditions. (6) The gas flow follows the Darcy's law. (7) Gas absorption/adsorption is described by Langmuir law.

2.2 *Deformation of coal porous medium*

Because the deformation is pseudo-static, the inertial force is ignorable. For a gas-saturated coal element, its stress state satisfies the equilibrium equation as

$$\frac{\partial \sigma_{ij}}{\partial x_j} + f_i = 0 \tag{1}$$

where σ_{ij} = total stress components of coal elements; f_i = body force in the ith direction; x_j = the jth coordinates.

The total stress is shared by both skeleton stress or effective stress (σ'_{ij}) and pore fluid pressure (p). This effective stress principle can be expressed as

$$\sigma_{ij} = \sigma'_{ij} + \alpha p \delta_{ij} \tag{2}$$

Where δ_{ij} is the Kronecker delta which is 1 for $i = j$ and 0 for other cases. α is called as Biot's coefficient which is related to the bulk modulus of the porous skeleton K and the bulk modulus K_s of individual grain as

$$\alpha = 1 - \frac{K}{K_s} \tag{3}$$

Combining Eq. (1) and (2) gets the following form of equilibrium equation:

$$\frac{\partial \sigma'_{ij}}{\partial x_j} + \frac{\partial (\alpha p)}{\partial x_i} + f_i = 0 \tag{4}$$

For linear elasticity, the constitutive model is expressed by the Hooke's law as

$$\sigma'_{ij} = G(u_{i,j} + u_{j,i}) + \lambda u_{k,k}\delta_{ij} - K\varepsilon_s \delta_{ij} \tag{5}$$

where G = shear modulus; λ = Lame constant; u = displacement of skeleton; ε_s = gas absorption/adsorption induced volumetric strain (swelling-induced strain). Further, $u_{i,j} = \partial u_i / \partial x_j$ and $\varepsilon_v = u_{k,k} = \partial u_1 / \partial x_1 + \partial u_2 / \partial x_2 + \partial u_3 / \partial x_3$.

Combining all above equation gets the Navier equation as

$$G u_{i,jj} + \frac{G}{1-2\nu} u_{k,kj} = K\varepsilon_{s,i} - \alpha p_{,i} - f_i \tag{6}$$

This equation shows that the body forces have three sources: gravity-induced body force, pore fluid flow induced friction force (drag force) and swelling induced body forces. It is noted that the drag force depends on the Biot's coefficient and the swelling-induced body force is related to the bulk modulus of skeleton.

2.3 *Gas absorption/adsorption on coal*

The gas absorption is assumed to observe the Langmuir law as

$$\varepsilon_s = \frac{\varepsilon_L p}{p_L + p} \tag{7}$$

Where ε_L= Langmuir strain of gas; p_L= Langmuir pressure.

2.4 *Gas flow equation*

The mass conservation law for each gas component can be expressed as

$$\frac{\partial m}{\partial t} + \nabla g\left(\rho_g \vec{v}_g\right) = Q_s \tag{8}$$

The gas mass content m contains free-phase gas and absorbed gas; ρ_g= gas density at the in-situ conditions; $\vec{v}_g$= Darcy's velocity vector; Q_s= gas source; t = time. For coal, the mass content is expressed by

$$m = \rho_g \phi + \rho_{ga} \rho_c \frac{V_L p}{p + p_L} \tag{9}$$

where ρ_{ga}= gas density at the standard atmospheric conditions. ρ_c = density of rock; V_L = Langmuir volume constant.

If the gas is ideal gas, its density can be calculated by

$$\rho_g = \frac{M_g}{RT} p \tag{10}$$

where M_g= molecular weight of gas; R = universal gas constant; T = temperature.

The Darcy's law can be expressed as follows if the gravity effect is ignorable:

$$\vec{v}_g = -\frac{k}{\mu} \nabla p \tag{11}$$

where k = coal permeability; μ = dynamic viscosity of free gas.

3 EVOLUTION OF COAL POROSITY AND PERMEABILITY

3.1 *Evolution of coal porosity*

A general porosity model was proposed by Zhang et al. (2008) as

$$\phi = \frac{1}{1+S}[(1+S_0)\phi_0 + \alpha(S - S_0)] \tag{12}$$

where the variables are defined as

$$S = \varepsilon_v + \frac{p}{K_s} - \varepsilon_s, \quad S_0 = \varepsilon_{v0} + \frac{p_0}{K_s} - \varepsilon_{s0} \tag{13}$$

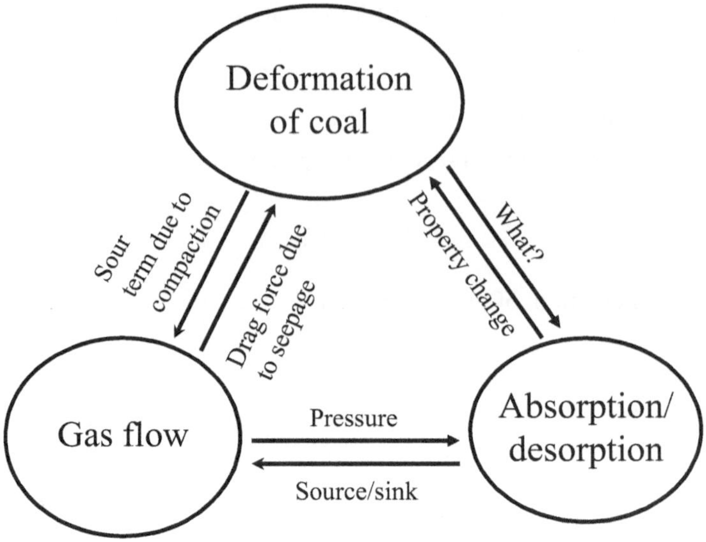

Figure 1. Interactions among multi-physical processes in coalbed seam.

Where ε_v= the current volumetric strain; ε_{v0} = the initial volumetric strain; ε_{s0} = the initial absorption volumetric strain.

3.2 *Cubic relationship between porosity and permeability*

The evolutions of porosity and permeability is described by a cubic law as

$$\frac{k}{k_0}=\left(\frac{\phi}{\phi_0}\right)^3=\left\{\frac{1}{1+S}\left[\left(1+S_0\right)+\frac{\alpha}{\phi_0}\left(S-S_0\right)\right]\right\}^3 \tag{14}$$

Where k_0 = initial permeability; ϕ_0 = initial porosity.

The final form of gas flow equation is then obtained as

$$\left[\phi+\frac{\alpha-\phi}{1+S}\frac{p}{K_s}+\frac{\rho_c p_a V_L P_L}{\left(p+P_L\right)^2}-\frac{\alpha-\phi}{1+S}\frac{\varepsilon_L P_L p}{\left(p+P_L\right)^2}\right]\frac{\partial p}{\partial t}-\nabla g\left(\frac{k}{\mu}p\nabla p\right)=Q_s-\frac{(\alpha-\phi)p}{1+S}\frac{\partial \varepsilon_V}{\partial t} \tag{15}$$

This equation and Eq. (6) together describe the interactions among the deformation process of coal, the free gas flow in coalbed and the gas absorption/desorption from coalbed. Figure 1 shows the details of their couplings.

4 LOCAL POROSITY AND HETEOGENEITY INPUT

4.1 *Concept of local porosity*

Porous medium has three fundamental parameters: porosity, percolation, and anisotropy of permeability. The porosity is defined as the ratio of the volume of pore space to the total volume of the porous medium. For diffusion and convection of fluids in porous medium, local porosity proposed by Hilfer (1992) is a good index for the presentation of the pore microstructure. Local porosity is a comprehensive index for the spatial distributions of both pore space and solid particles. On computational efficiency, the local porosity is still within the framework of continuum and thus be easily incorporated into the mathematical formulation stated above through the initial local porosity and the initial permeability.

4.2 *Input of heterogeneity through images*

Usually, the pore microstructures of coal can be obtained through a series of images such as X-ray CT images and other images. These XY images can be arranged according to the z coordinates in increase or decrease order. For each pixel in an image, a digital number is assigned through its gray scale:

$$D(i,j) = \left(image(n,o,1) + image(n,o,2) + image(n,o,3)\right)/(3*256) \tag{16}$$

where image (:,:,1), image (:,:,2), image (:,:,3) refer to the values of R,G, B colors, respectively. Such a color is taken a value between 0 and 255, thus the length is 256. Through the above conversion, each pixel in the X-Y image has only one value between 0 and 1.0. This value is called as pixel label. All pixel labels in the X-Y images form a pixel label database.

Another important step is to transform this pixel label database into porosity/permeability database or distribution. This transformation can be completed through some conversion techniques such as widely used image threshold method in image processing (Holden et al., 2009). That is,

$$q(i,j) = \begin{cases} q_1 & D(i,j) \le T_1 \\ q_2 & T_1 < D(i,j) \le T_2 \\ \vdots & \\ q_n & T_{n-1} < D(i,j) \le T_n \end{cases} \tag{17}$$

where the threshold values T_1, T_2, …, T_n are prescribed for different applications. $q(i,j)$ = local porosity at the point (i, j).

5 NUMERICAL EXAMPLES

5.1 *Effect of image-based heterogeneity on flow patterns*

Example 1: Multi-fracture flow model

The model is shown in Figure 2 and the details of fractured zone are shown in Figure 3. This model has two zones: matrix zone (zone 1) and fractured zone (zone 2). The matrix zone has only coal matrix but the fractured zone has both fracture and matrix. The fracture networks are input through a photo of fractures. In the matrix, the permeability is 0.1 mD and the porosity is 1.0%. In the fractured zone, the matrix has the same properties as zone 1 but fractures have the permeability of 10 mD and porosity of 15%.

The pressure history curves at points P1 and P2 are shown in Figure 4. These two points are all in the matrix but P1 is in the fractured zone and near the wellbore. The point P2 is in the zone 1 (matrix) but far away from the wellbore. This figure shows that the pressure at point P1 dissipates faster. Further, Figure 4 shows the pressure dissipation profiles (different times) along a horizontal section from the wellbore. This section intercrosses two main fractures. Their effect on pressure dissipation profile is observed and indicated with two vertical lines.

Example 2: Single-fracture flow model

Only one fracture is in the zone 2. The pressure history curves at four points are compared in Figure 5. The points P1 and P3 are geometrically symmetric on wellbore and point pair P2 and P4 are also symmetric if there is no fracture. Due to existence of one fracture at the right hand side, the pressure history is different.

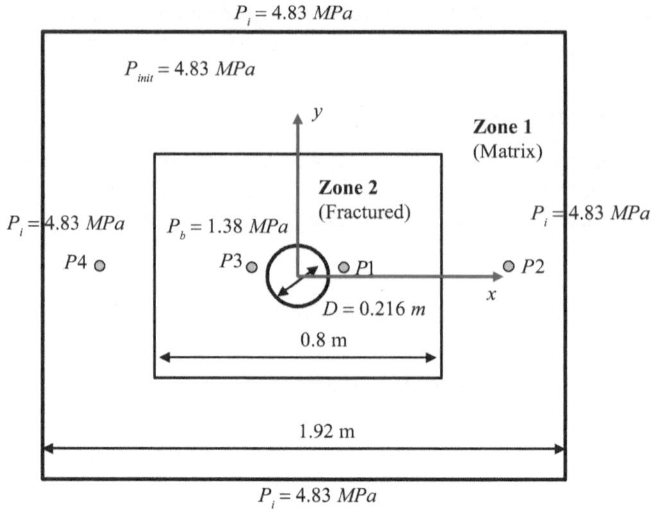

Figure 2. Computational model for fractured microstructures.

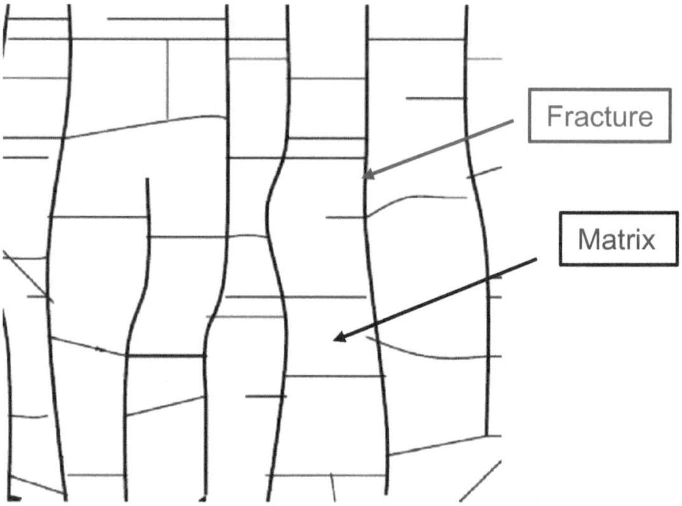

Figure 3. Microstructures in zone 2.

5.2 *A micro-scale fracture model*

Conceptually, the gas flow in coal seam are mainly in fracture and the matrix supplies and store gas through absorption/desorption. On this sense, the absorption induced matrix swelling may change the opening/closing of the fracture on micro-scale. In this section, a micro-scale fracture model is shown in Figure 7. In this model, the fracture is assumed to be empty (no filling) and the matrix is still porous medium. The fracture permeability is usually calculated by

$$k_f = \frac{a^3}{12s} \tag{18}$$

where a = fracture opening or aperture; s = fracture spacing.

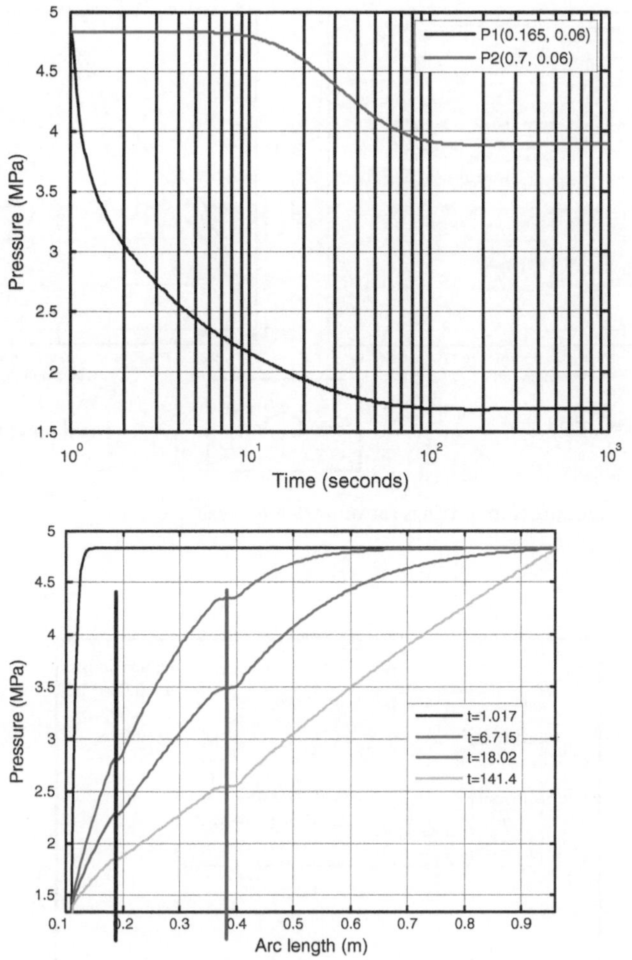

Figure 4. Pressure dissipation history (up) and profiles (down) for multi-fracture case.

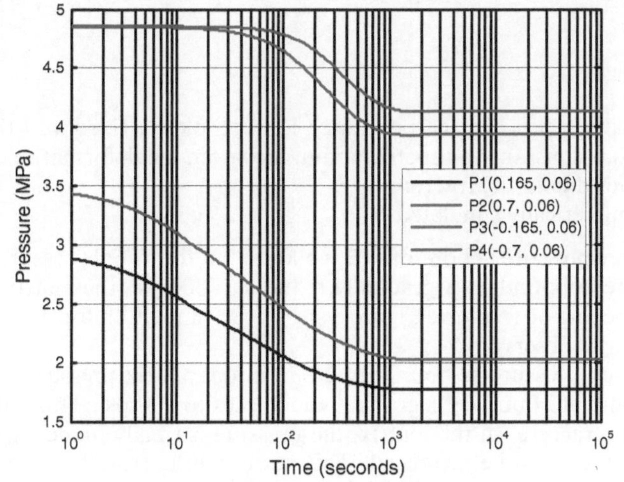

Figure 5. Comparison of pressure dissipation history for one-fracture case.

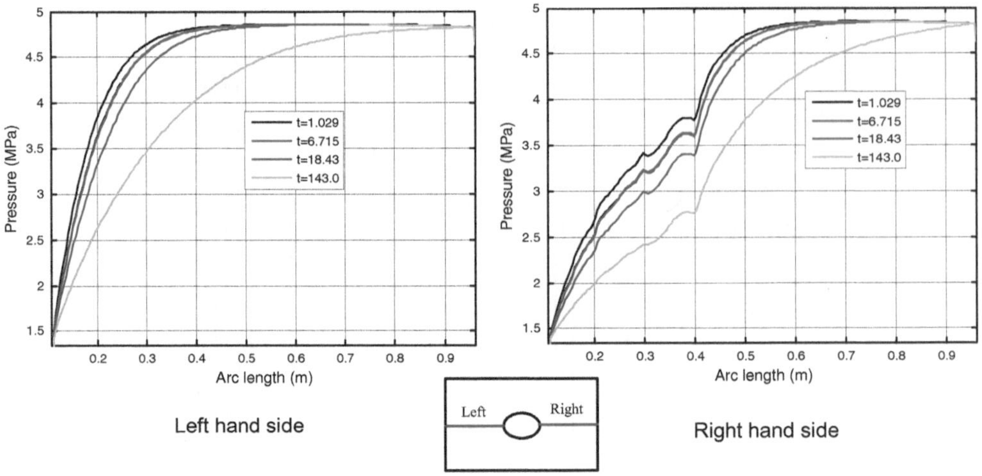

Figure 6. Left and right pressure profiles for one-fracture case.

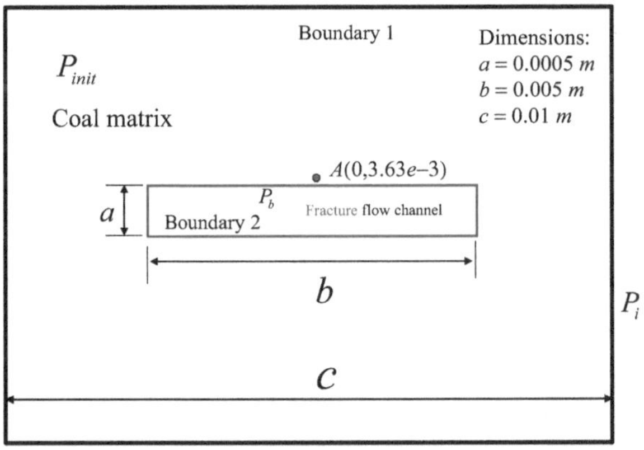

Figure 7. Micro-scale model for swelling-induced fracture opening study.

For the deformation process, the boundary 1 is completely fixed and the boundary 2 is free. This sample has constant volumetric strain. However, the deformation within the model is not homogeneous due to the fracture.

Two cases are numerically simulated.

Case 1: The fracture pressure at boundary 2 is 500 psi but the initial pressure is 700 psi. When the matrix pressure at boundary 1 gradually reduces to 500 psi, the matrix will shrink. This may induce the opening of fracture. Figure 8 shows the simulation for this case. The increase of fracture opening is observed.

Case 2: The fracture pressure at boundary 2 and initial matrix pressure are all 500 psi but the external pressure at boundary 1 gradually increases to 700 psi. This will produce a flow from matrix to the fracture. In the matrix, the pressure is finally higher than 500 psi. Thus, swelling-induced strain will be produced. This strain will narrow the opening as shown in Figure 9, thus reducing the permeability of the fracture.

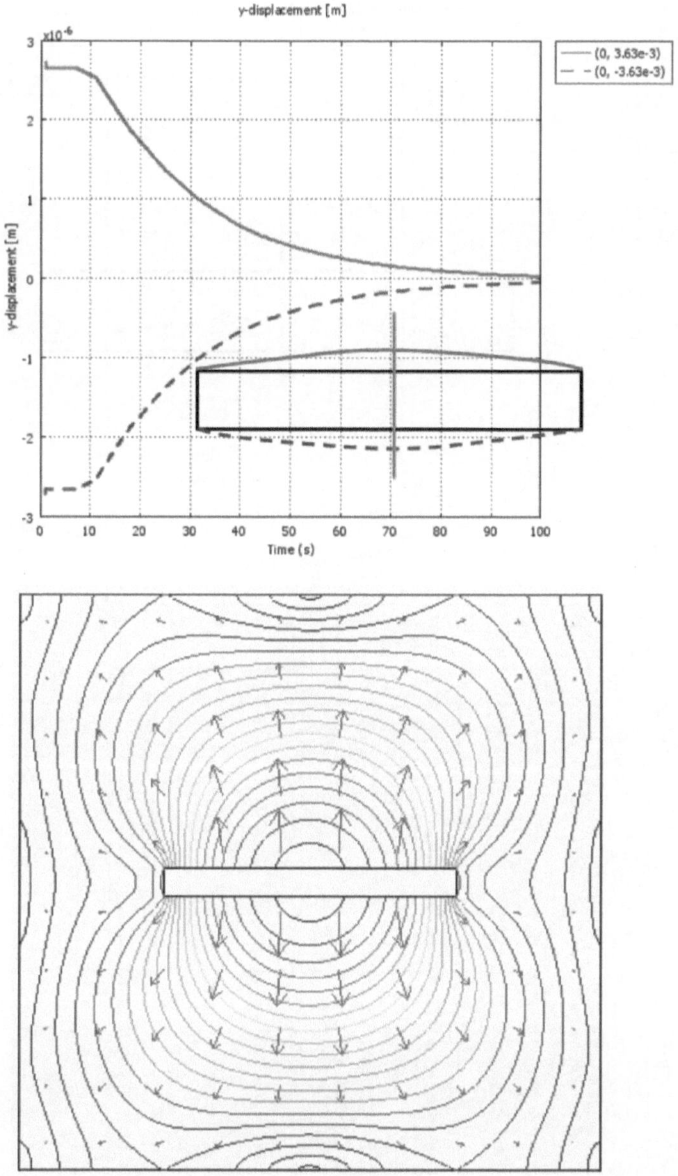

Figure 8. Fracture opening due to matrix shrinkage.

6 CONCLUSIONS

These simulations observed the following phenomena: First, gas flow has different sources and flow channels at different flow stages. The gas flow is mainly in fractures particularly at the initial stage. The coal matrix supplies gas for fracture channel flow at the later stages. This supply determines the flow rate at the steady-state flow rate. Second, the gas flow is generally heterogeneous on micro-scale observation. This heterogeneity depends on the fracture density and orientation. The gas-coal interaction induced swelling may significantly change the permeability of fractures. This change is determined by both swelling properties of coal and external conditions such as constraint responses to the swelling strain.

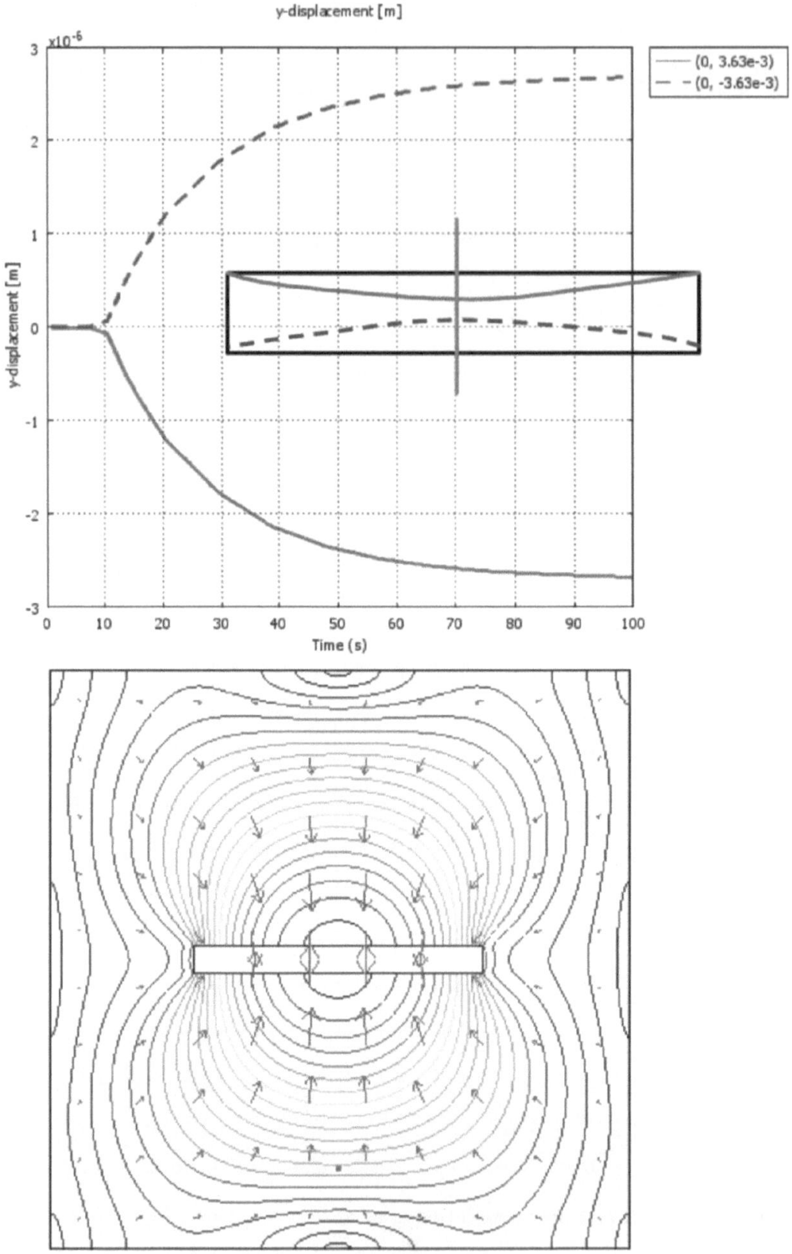

Figure 9. Coal matrix swelling-induced closing of fracture.

REFERENCES

Hilfer R, 1992. Local porosity theory for flow in porous medium. Physical Review B, 4(13):7115–7121.

Holden EJ, Moss S, Russell JK & Dentith MC, 2009. An image analysis method to determine crystal size distribution of olivine in kimberlite. Computers & Geosciences, 13:255–268.

Perrin JC & Benson S, 2010. An experimental study on the influence of sub-core scale heterogeneities on CO_2 distribution in reservoir rocks. Transport in Porous Media, 82:93–109.

Seo JG, 2004. Experimental and simulation studies of sequestration of supercritical CO_2 in depleted gas reservoirs. PhD Thesis, Texas A&M University.

Uh J & Watson AT, 2010. Determining spatial distributions of permeability. Transport in Porous Media, DOI: 10.1007/s11242–010–9627–3.
Wang JG, Liu JX, Liu JS & Chen ZW, 2010. Impact of rock microstructures on the supercritical CO_2 enhanced gas recovery. SPE 131759, CPS/SPE International Oil & Gas Conference and Exhibition in China held in Beijing, China, 8–10 June 2010.
Wu Y, Liu JS, Elsworth D, Chen ZW, Connel L & Pan ZJ, 2010. Dual poroelastic response of a coal seam to CO_2 injection. International Journal of Greenhouse Gas Control, 4(4):668–678.
Zhang HB, Liu J & Elsworth D, 2008. How sorption-induced matrix deformation affects gas flow in coal seams: A new FE model. International Joirnal of Rock Mechanics and Mining Sciences, 45(8):1226–1236.
Zhang Y, Oldenburg CM, Finsterle S & Bodvarsson GS, 2007. System-level modeling for economic evaluation of geological CO_2 storage in gas reservoirs. Energy Conversion and Management, 48:1827–1833.

CO_2 Storage in Carboniferous Formations and Abandoned Coal Mines –
He, Ribeiro E Sousa, Elsworth & Vargas Jr. (eds)
 ISBN 978-0-415-62079-6

Considerations on the numerical modelling of injection processes of CO_2 in geological formations with emphasis on carboniferous formations and abandoned coal mines

E.A. Vargas Jr.
Department of Civil Engineering, Catholic University of Rio de Janeiro (PUC Rio), Rio de Janeiro, Brazil
Department of Geology, Federal University of Rio de Janeiro, Rio de Janeiro, Brazil

R.Q. Velloso, W.N. Ribeiro & A.L. Muller
Department of Civil Engineering, Catholic University of Rio de Janeiro (PUC Rio), Rio de Janeiro, Brazil

L.E. Vaz
Department of Civil Engineering, Fluminense Federal University, Niteroi, Brazil

ABSTRACT: Injection of anthropogenic CO_2 in geological formations is now considered to be a promising proposition in order to diminish the effects of global warming. However, this proposition is not without risks. In order to ascertain and to study alternatives in order to diminish these risks, it is necessary to predict the short and long term behaviour of the created CO_2 reservoirs. Numerical techniques are the obvious choices for these purposes. However, complex processes are to be simulated both in the large scale and on the micro-scale. The present paper makes considerations on the available techniques. In particular the paper concentrates on the possible alternative of injecting CO_2 in carboniferous formations and abandoned coal mines

1 INTRODUCTION

The long and short term predictions of injection of CO_2 in geological formations require careful and detailed analysis of the created reservoirs. A number of escape routes may exist (IPCC, 2005) and should be analyzed. These escape routes depend on the chosen geological formation for those purposes, existing/abandoned wells and injection procedures adopted. The CO_2 injected must remain in the geological formations for hundreds of years and therefore predictive tools should be able to simulate these processes both for short and long term situations. CO_2 reservoirs occupy large areas, similarly to oil and gas reservoirs. Analytical tools should therefore be able to model these large areas/volumes. Furthermore, one would need to study in detail CO_2 interactions with geological media at the microstructure. This type of study is very important in order to derive constitutive relationships which could be used in larger scale simulations. Numerical techniques today (finite differences, finite volumes, finite elements, boundary elements, etc) are of common use today for the prediction of behaviour of engineering structures as CO_2 reservoirs could be classified. However, injection of CO_2 in geological formations present some challenging and not completely resolved questions at both large and micro-scale. This paper raises some of these questions and presents some of the available and presently under development techniques. The paper considers separately modelling at the large and at the small scale.

One of the great challenges related to the geological disposal of carbon dioxide is the identification of possible sites that present the necessary safety requirements. Decision taking in relation to the definition of possible storage sites requires detailed analysis of the risks involved. On the other hand, in order to carry out risk analysis studies it is mandatory that realistic predictions of the reservoir behaviour along its life time. Numerical analysis is the usual tool used for carrying out these predictive studies. Establishment of related costs can benefit from the use of predictive capabilities. Flow conditions in the reservoirs and the injection depths may increase considerably the injection costs. In particular, the evaluation of risks in the disposal of CO_2 in geological media concerns possible leakages reaching the atmosphere, which may pose threats to human and animal life. Besides, eventual leakages may cause substantial modifications of the water chemistry, impact on ecosystems and modifications in the chemical equilibrium of soils (changes in vegetables and the microbial ecology). Risk analyses may be carried out based on the effects over humans and the biosphere. There may be risks in the CO_2 injection at the well scale and at the reservoir scale. External factors such as seismic activities, well perforations, mining activities, terrorism, which may pose threats to the long term performance of the created reservoirs. According to Hepple and Benson (2003), one should consider the time span of 10000 years as the duration of the CO_2 storage in the geological reservoirs.

According to IPCC (2005), risks related to reservoir fracturing and CO_2 leakages to the surface should be carried out in different time scales, short, medium and long term predictions. For short term predictions, analyses should be carried out for periods just before the injection operations, for medium term predictions, analyses should be carried out for the injection operations. Finally, for the long term predictions, analyses must be carried out after injection activities stopped. These analyses should be carried out at the well (small) scale and at the reservoir (large) scale.

Numerical analysis is generally the preferred tool able to simulate (predict) the varied scenarios related to the geological disposal of CO_2. Is is also able to perform the execution of parametric and sensitivity analyses. For those purposes, commercial programs and in-house developed codes are used. Commercial software is in constant development and upgrade and is today able to simulate most of the physical, chemical and biological processes involved in the CO_2 injection in geological media. There are frequently however situations or processes, sometimes coupled processes that require the development of special, in-house developed computer codes. In-house developed codes are in general more flexible and can be modified in order to contemplate additional processes or conditions.

Independent from the type of numerical analysis code used, it is fundamental to identify and to describe the various involved processes. These comprise physical, chemical and biological processes with the addition of their coupling processes. The above mentioned processes may have strong and/or weak couplings which have to be identified. Furthermore these couplings can go in one way or two ways. Some of them are described below:

- **Fluid-mechanical coupling:** flow processes may influence geomechanical processes through the dependency of the former regarding fluid pressure/saturation changes. Geomechanical processes may influence flow processes through the dependency of the former in relation to porosity/permeability changes.
- **Fluid-thermal coupling:** flow processes influence thermal processes through the dependency of heat/energy transfer in the fluid velocities field. Thermal processes influence flow processes through density (convection) driven flows.
- **Fluid-biogeochemical coupling**: flow processes cause changes in biogeochemical processes through the dependency of the former in fluid velocities.
- **Transport-thermal coupling:** transport processes influence thermal processes through possibilities of volatilization and dissolution. Thermal processes influence transport indirectly through density driven density.
- **Transport-biogeochemical coupling:** transport processes influence biogeochemical processes through their dependency on solute concentrations. Biogeochemical processes influence transport processes through speciation/degradation mechanisms.
- **Thermo-biogeochemical coupling:** thermal processes influence biogeochemical processes through the temperature dependency of chemical reactions. Bio-geochemical processes

influence thermal processes through the eventual type of chemical reactions involved (exothermic, endothermic).

- **Thermo-mechanical coupling:** thermal processes influence geomechanical processes through the influence of temperature in deformations arising from thermal gradients. Geomechanical processes influence thermal processes through the changes in thermal properties arising from mechanical effects.

The above list mentions two by two coupled processes. It may be that more than two processes may be coupled. In the more general case of CO_2 injection in geological media, TFMBC (thermal, fluid, mechanical, chemical/biogeochemical) couplings can occur but until this moment the most relevant processes have not yet been fully established.

An important question to be evaluated in the geological disposal of CO_2 concerns the risk of leakages reaching the atmosphere, a fact that can affect human and animal health. Besides, eventual leakages may cause the dissolution of CO_2 in the groundwater, chemically modifying it, making considerable impacts in the ecosystems and modifications of the chemical equilibrium of soils producing changes in the vegetal physiology and microbiological activities. Therefore, numerical modelling is an important component of risk analysis activities (Manchao et al., 2011).

Amongst the candidate geological formations for CO_2 injection, three have been the main focus of attention for this purpose: saline aquifers (in this case salt content is equal or greater to levels found in oceans), coal seams at depth (unfeasible for commercial exploration) and already explored oil and gas fields. However other possibilities have been proposed such as evaporitic deposits, pelitic rocks and storage in spaces left underground by already explored mines, coal in particular. In general, risks in the geological disposal of CO_2 may be related to the following: (i) cap (sealing) rock heterogeneity (continuity of the sealing formation, connectivity between formations, stratigraphical heterogeneities, existing faults and fracture systems); (ii) biochemical alterations (mineral dissolution through acidification of the water, dissolution through microbiological activity, among others); (iii) geomechanical effects through the generation of new fractures and opening of existing fractures/faults and (iv) hydrogeological effects (diffusion of CO_2 due to special differences in concentration, capilar breaches in pores and fractures and advection through faults and fractures which are preferential pathways). One may state that the possible of CO_2 is very fast through (conducting) faults, fast in fractures and slow through the intact cap rock.

Each material has its own, particular response to the injection of CO_2. In the case of storage of CO_2 in coal seams some additional effects come into play and should be considered. The so-called macerals, basic constituents of coal are responsible for its expansion as CO_2 is injected. Coal has a particular double porosity structure composed of a porous matrix surrounded by regular fractures/fissures, the so-called cleats. Therefore, the injection of CO_2 can be considered one of advection through cleats and diffusion through the porous matrix. Considerable changes in the cleat system permeability may occur due to the expansion properties mentioned above. Concerning the use of space existing in abandoned coal mines (Piessens and Dusar, 2003a, Piessens and Dusar, 2003b) for CO_2 injection, one could state that numerical modelling has an important place in feasibility studies and risk analyses. In this case besides the processes mentioned above one would have to consider as well the interaction between water flowing into the excavations and the pressurization of CO_2 remaining inside. The coupling of these processes will pose additional challenges in the modelling stages.

The present Chapter focuses on the modelling of pertinent processes in two scales. Initially, an advanced numerical formulation is presented for two-phase flow coupled with geomechanics. As presented above this just a part of the possible coupling processes involved in CO_2 injection in geological media. However it is believed that amongst the coupled processes, the fluid-geomechanical coupling is a basic and crucial one. The presented formulation is pertinent to simulation of problems at the meso and macro scale. In practical terms, it may be used for modelling of problems at the scale of a well and the reservoir scale. Subsequently, a numerical formulation is described for the simulation of problems at the micro-scale. In this case, the numerical simulations will provide a phenomenological understanding of processes at the micro-scale which in turn contributes for the establishment of constitutive relationships to be used at the macro-scale. A final word should be said in relation of parameters. Although numerical modelling has advanced considerably, gaps remais regarding the definition of appropriate/realistic models and respective parameters for the representation of the coupling processes described above (Gomes, 2010). These activities will most certainly be the focus of attention in the years to come if CO_2 injection in geological media will indeed take place.

2 CONTINUUM BASED NUMERICAL METHODS

The present section presents a formulation for accurate consideration of the numerical analysis of fluid-mechanical coupling which as mentioned in the previous section constitutes a backbone of the various coupling processes occurring when injecting CO_2 in geological media. Main ingredients of the formulation are: use of finite elements, considerations of elastic and elasto-plastic media and two-phase flow (Muller et al., 2009; Ribeiro, 2011). Examples are presented at the end of the section to demonstrate the applicability of the developed codes to simulation of CO_2 injection in geological media.

2.1 *Processes and equations*

The mechanical problem can be represented mathematically by the equilibrium statement:

$$\int_{\Omega} \delta\boldsymbol{\varepsilon}^{\mathrm{T}}\dot{\boldsymbol{\sigma}}d\Omega - \int_{\Omega} \delta\mathbf{u}^{\mathrm{T}}\dot{\mathbf{b}}d\Omega - \int_{\Gamma} \delta\mathbf{u}^{\mathrm{T}}\dot{\mathbf{t}}d\Gamma = 0 \tag{1}$$

Equation 1 relates the rates of the static real quantities, as the total stress rate vector $\dot{\boldsymbol{\sigma}}$, the body forces rate vector $\dot{\mathbf{b}}$ and the surface forces rate vector $\dot{\mathbf{t}}$ to virtual kinematic quantities as virtual strains $\delta\boldsymbol{\varepsilon}$ and virtual displacements $\delta\mathbf{u}$.

Applying the effective stress principle and considering small strains relationships, the relations 2, and 3 may be applied in equation 1.

$$\dot{\boldsymbol{\sigma}} = \mathbf{D_T}\dot{\boldsymbol{\varepsilon}} + \mathbf{D_T}\mathbf{m}\frac{\dot{p}}{3K_s} + \dot{\boldsymbol{\sigma}}_0' - \mathbf{m}\dot{p} \tag{2}$$

$$\dot{\boldsymbol{\varepsilon}}_{\mathbf{ij}} = \frac{1}{2}(\dot{\mathbf{u}}_{i,j} + \dot{\mathbf{u}}_{j,i}) \tag{3}$$

In the above equations, $\dot{p}$ are the pore pressure rates, $\dot{\boldsymbol{\varepsilon}}$ the skeleton total strain rates, $\mathbf{D}_{\mathrm{T}}$ the tangent constitutive tensor, $\dot{\boldsymbol{\sigma}}_0'$ the effective initial stress rates, $\mathbf{m}\dot{p}/3K_s$ the volumetric strain rates caused by uniform compression of the grains, $\mathbf{K}_s$ the bulk modulus of solid grains and $\mathbf{m} = \{1\ 1\ 1\ 0\ 0\ 0\}$.

The two-phase flow problem must have equations for each phase separately and various forms can be used for that purpose (Aziz and Settari, 1959). For the present work, two equations considering respectively the averaged pressure p and the saturation of the wetting phase S_w as primary variables are considered (Ribeiro, 2011):

$$\phi c_t \frac{\partial p}{\partial t} = \nabla\cdot(\mathbf{k}\lambda_t \nabla p) + \mathbf{kg}\nabla\cdot(\lambda_w\rho_w + \lambda_n\rho_n) + Q_t \tag{4}$$

$$\phi\frac{\partial S_w}{\partial t} = -\nabla\cdot(f_w(\mathbf{v}_t + \mathbf{kg}(\lambda_{nw}\rho_{nw} - \lambda_w\rho_w))) + Q_w \tag{5}$$

In these equations, the subscripts w and nw represents respectively wetting and non wetting fluid type, ρ represents the density of the fluid, ∇ the differentiation operator, ϕ the porosity of the medium. **k** the absolute permeability of the porous medium, **g** the gravity,

Q_t is the total fluid volume source or sink given by $Q_t = Q_w + Q_n$, f_w represents the wetting fractional flow function, $f_w = \lambda_w / \lambda_t$, the total mobility λ_t and the wetting and non-wetting phase mobilities λ_w and λ_{nw} are given by $\lambda_t = \lambda_w + \lambda_{nw}$, $\lambda_w = k_{rw}/\mu_w$, $\lambda_{nw} = k_{rnw}/\mu_{nw}$, k_{rw} and k_{rnw} the relative permeability of the wetting and non-wetting phases, the total velocity is given by $v_t = v_w + v_{nw}$., where v_w and v_{nw} are wetting and non-wetting velocities.

c_t in Equation 4 is the total compressibility of the fluids and the porous media, it is a relationship $c_t=(1\backslash\phi) \cdot d\phi/dp$, where $d\phi$ can be given by (Pao et al., 2001):

$$d\phi = \mathbf{m}^T\boldsymbol{\varepsilon} - \mathbf{m}^T\mathbf{D_T}\frac{\boldsymbol{\varepsilon}}{3K_s} - \mathbf{m}^T\mathbf{D_T}\mathbf{m}\frac{dp}{9K_s^2} + \frac{dp}{K_s} \tag{6}$$

Note that in equations 4 and 5, capillary pressures are not taken into account. This situation constitutes an extreme case from the numerical point of view as the corresponding equations are hyperbolic and requires appropriate numerical procedures for their solution. Two important ingredients, as proposed by Hoteit et al. (2008), Raviart-Thomas elements have been used for accurate determination and continuity of normal velocities across element boundaries. This is essential for the numerical simulation of problems in heterogeneous media as it will probably be the case when injecting CO_2 in geological formations. Also, discontinuous Galerkin formulations (Hoteit et al., 2008) were used for solution of the (hyperbolic) saturation equations. The pressure equations are solved using standard finite elements techniques.

The corresponding finite element equations are:

$$\int_\Omega \phi c_t \mathbf{N}_p^T \mathbf{N}_p \frac{d\mathbf{p}}{dt} d\Omega - \int_\Omega (\nabla \mathbf{N}_p)^T \lambda_t (\nabla \mathbf{N}_p)\mathbf{p} d\Omega + \int_\Omega \mathbf{k}(\lambda_{nw}\rho_{nw} + \lambda_w\rho_w)\mathbf{g}\mathbf{N}_p d\Omega - \int_\Omega \mathbf{N}_p^T \mathbf{q} d\Omega = 0 \tag{7}$$

A residual vector from Equation 7 at an iteration i can be represented by:

$${}^{t+\Delta t}\mathbf{F}_p^i = {}^{t+\Delta t}\left| \begin{array}{l} \int_\Omega \mathbf{N}_p^T \mathbf{q} d\Omega - \int_\Omega \mathbf{k}(\lambda_{nw}\rho_{nw} + \lambda_w\rho_w)\mathbf{g}\mathbf{N}_p d\Omega \\ - \int_\Omega \phi c_t \mathbf{N}_p^T \mathbf{N}_p d\Omega \frac{\mathbf{p}}{\Delta t} + \int_\Omega (\nabla \mathbf{N}_p)^T \lambda_t (\nabla \mathbf{N}_p) d\Omega \mathbf{p} \end{array} \right|^i \tag{8}$$

These equations can be represented, using incremental pressures, in a compact form as:

$$\left[\Delta t\theta \mathbf{H} + \mathbf{G}\right]\left\{\delta^{t+\Delta t}\mathbf{p}^{i+1}\right\} = \left\{\Delta t^{t+\Delta t}\mathbf{F}_p^i\,(\mathbf{p},t)\right\} \tag{9}$$

where

$$\mathbf{H} = \int_\Omega (\nabla \mathbf{N}_p)^{\mathrm{T}} \lambda_t \nabla \mathbf{N}_p d\Omega \tag{10}$$

$$\mathbf{G} = \int_\Omega \mathbf{N}_p^T [\phi c_t] \mathbf{N}_p d\Omega \tag{11}$$

H and **G** are the permeability and compressibility matrix, respectively.

$$\frac{d\mathbf{S}}{dt}\int_\Omega \phi \mathbf{N}_p^T \mathbf{N}_p d\Omega = f_{w,\Omega}\int_\Omega (\mathbf{N}_p)^T \mathbf{v}_t.(\nabla \mathbf{N}_p) d\Omega - f_{w,\Gamma}\int_\Gamma \mathbf{N}_p^T \mathbf{N}_p \mathbf{v}_t \cdot \mathbf{n} d\Omega + \int_\Omega \mathbf{N}_p^T \mathbf{q}_w d\Omega = 0 \tag{12}$$

$$\frac{d\mathbf{S}_w}{dt} = L(\mathbf{S}_w) \tag{13}$$

where $L(\mathbf{S}_W)$ represents the RHS of the equation 12 multiplied by the inverted mass matrix. In these formulations, an IMPES temporal discretization is used for the description of the above equations in finite elements, N_p being the nodal interpolation of pressure.

2.2 *Examples*

2.2.1.1 *The Buckley Leverett Problem*

In this problem we consider the classical Buckley-Leverett problem, it is a 1-D dimensional case of two-phase flow. It consists of a horizontal column initially filled by oil and a water injection boundary.

The relevant parameters are presented in table 1, this data was also used by Durlorsfy (1993). The total velocity is equal to one, $v_t = 1$, the boundary condition at $x = 0$, is $S_w = 1$ and the domain is considered homogeneous and initially oil saturated, the units are introduced in a consistent system of units.

The analyzed domain is rectangular, 2D domain with dimension (4,1), and discretized using a structured mesh with 320 quadrilaterals elements.

Figure 1 shows the results by the DFEM and analytical solutions at step times, t = 0.1s, 0.2s, 0.3s, 0.4s e 0.5s, there is good capture of saturation front by the DFEM.

2.2.2 *Two-phase flow—five spot problem in a heterogeneous media*

Another classical problem analyzed in petroleum industry is the five spot problems. In this example, a random heterogeneous media to the intrinsic permeabilities, K_x e K_y is taken into account. Figure 2 shows a schematic representation of the problem and the intrinsic permeability field employed for K_x. This problem is similar to the Buckley-Leverett problem, the media is initially oil saturated and there is a water injection well and four oil production wells. At the water injection well, the boundary conditions are $S_w = 1$ and $Q_w = 1$, at the oil production well, are $S_o = 1$ and $Q_o = -1$.

This problem is analyzed considering bi-axial symmetry and the employed parameters are presented in table 2. A structured mesh with 32 × 32 quadrilateral elements is used to model the domain (1,1).

The results found by DFEM are presented in Figure 3, it shows the wetting-phase saturation field at several time steps.

2.2.3 *Two-phase flow in a layered reservoir*

In this example, extracted from Hoteit et al. (2008), is studied a two-phase flow condition in a two-dimensional layered reservoir. The domain is constituted by heterogeneous permeabilities

Table 1. Relevant parameters employed in Buckley-Leverett problem.

$\rho_{nw} = \rho_w$	K_{nw}	K_w	ϕ	$S_{rnw} = S_{rw}$	$\mu_{nw} = \mu_w$	K_x
0.0	$(S_{rw})^2$	$(1 - S_{rw})^2$	0.2	0.0	1.0	1.0

where: ρ = fluid density; K = intrinsic permeability; ϕ = porosity; Sr = residual saturation; μ = fluid viscosity.

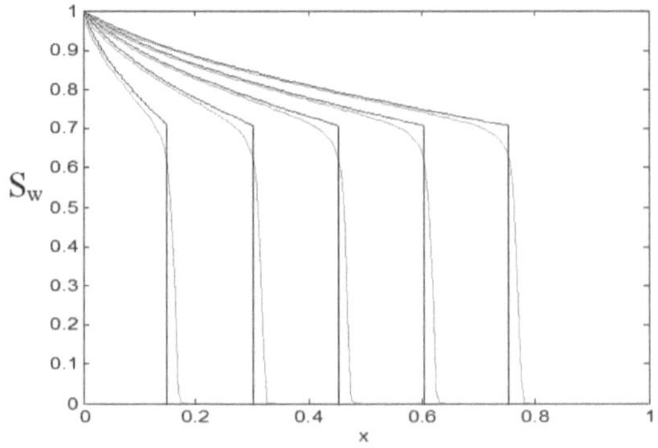

Figure 1. Results of the Buckley-Leverett problem at t = 0.1s, 0.2s, 0.3s, 0.4s e 0.5s and dimensionless length x by DFEM, grey, and analytical solution, black.

Table 2. Parameters used for the five spot problem.

Q_w	Q_o	K_w	K_{nw}	ϕ	$S_{rnw} = S_{rw}$	$\mu_{nw} = \mu_w$
1.0	−1.0	$(S_{rw})^2$	$(1 - S_{rw})^2$	0.2	0.0	1.0

where Q_w and Q_o are the injection water and extraction oil flow rates, respectively, the others parameters were defined in before sections.

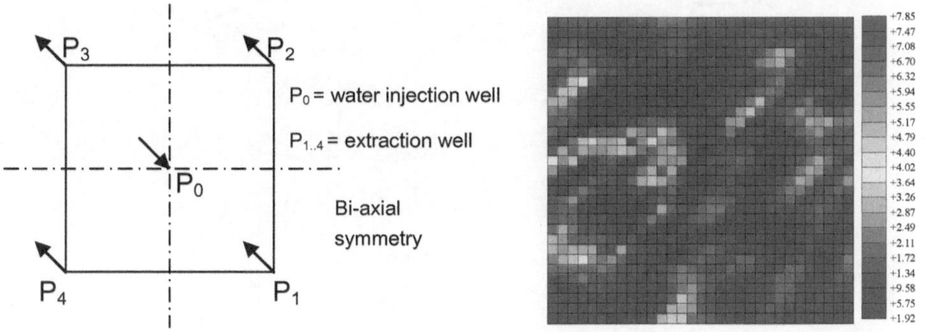

Figure 2. Representation of the five spot problem and random permeability field for Kx.

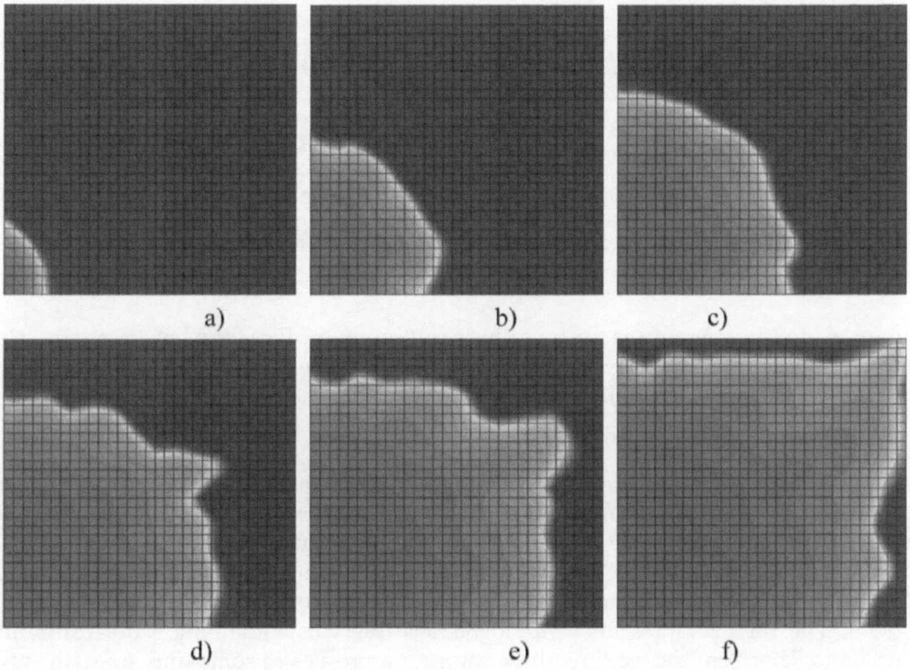

Figure 3. Wetting-phase saturation field for the five spot problem in a random permeability case. a) t = 0.7s, b) t = 4.2s, c) t = 7.7s, d) t = 11.2s, e) t = 14.7s, f) t = 19.6s.

layers, the figure 4 presents the representation of the problem. The domain is initially oil saturation and a water injection boundary is applied at x = 0. The parameters used in this analysis are presented in table 3.

The structured mesh employed is showed in Figure 5, the domain is rectangular (500, 270) and the mesh has 4500 quadrilateral elements with 4641 nodes.

Table 3. Parameters to the layered reservoir problem.

Q_w	$\rho_{nw} = \rho_w$	K_w	K_{nw}	ϕ	$S_{rnw} = S_{rw}$	$\mu_{nw} = \mu_w$	K_1	K_2
1.0	1.0	$(S_{rw})^2$	$(1 - S_{rw})^2$	0.2	0.0	1.0	1.0	10^{-5}

Figure 4. Representation of the layered reservoir problem.

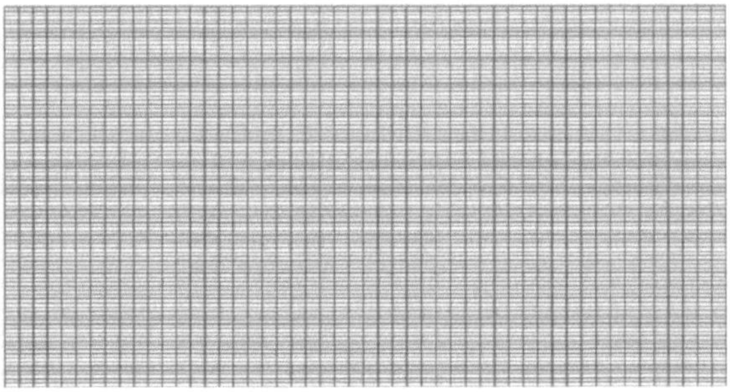

Figure 5. Finite element mesh used in the layered reservoir problem.

Figures 6a to 6e represent the wetting-phase saturation profiles for a few time steps, showing the saturation front in the layer with permeability K_1. These results are similar to the ones presented by Hoteit et al. (2008).

2.2.4 *Two-phase flow in faults*

To exemplify the application of the DFEM to faulted media, a simplified problem is elaborated considering a horizontal fault in a reservoir. In this example the fault is modeled by quadrilaterals elements with different intrinsic permeability with a width equal to 6. The same data, domain and mesh of the layered reservoir problem are applied in this case, except the fault and the reservoir media have different intrinsic permeability, the data are presented in table 3. The Figure 7 presents a homogeneous reservoir containing a horizontal fault. Figures 7a to 7e present the wetting-phase saturation profiles to some time steps, the results are similar to the layered reservoir problem, this faulted reservoir is a particular case of the layered problem presented in before section.

3 PORE SCALE MODELLING

In this section, the basics of LBM (Lattice-Boltzmann methods) are presented (Velloso, 2010). This method is suited for the simulation of flow and transport phenomena at the micro-scale (porous/fissure) as shown subsequently through the examples.

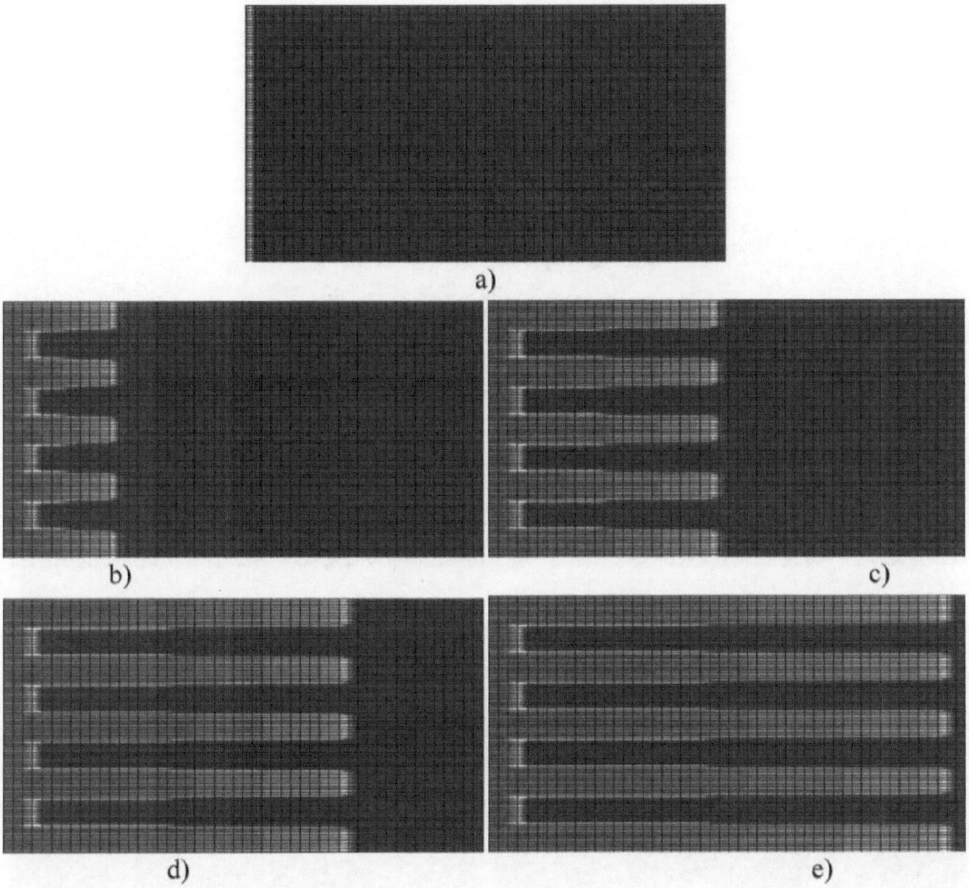

Figure 6. Wetting-phase saturation fields for a layered reservoir problem. a) t = 0s, b) t = 1s, c) t = 2s, d) t = 3s, e) t = 4s.

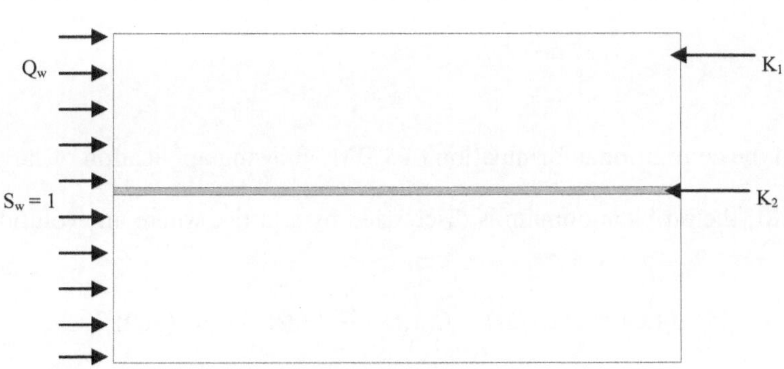

Figure 7. Representation of the faulted reservoir problem.

3.1 *One phase flow*

One phase flow occurring in the pores of a porous medium can be described by Navier-Stokes equations which represent momentum conservation of a fluid. LBM is a technique for the solution of these equations. In the present work the formulation proposed by He & Luo (1997) is used. This formulation reduces considerably the errors due to compressibility

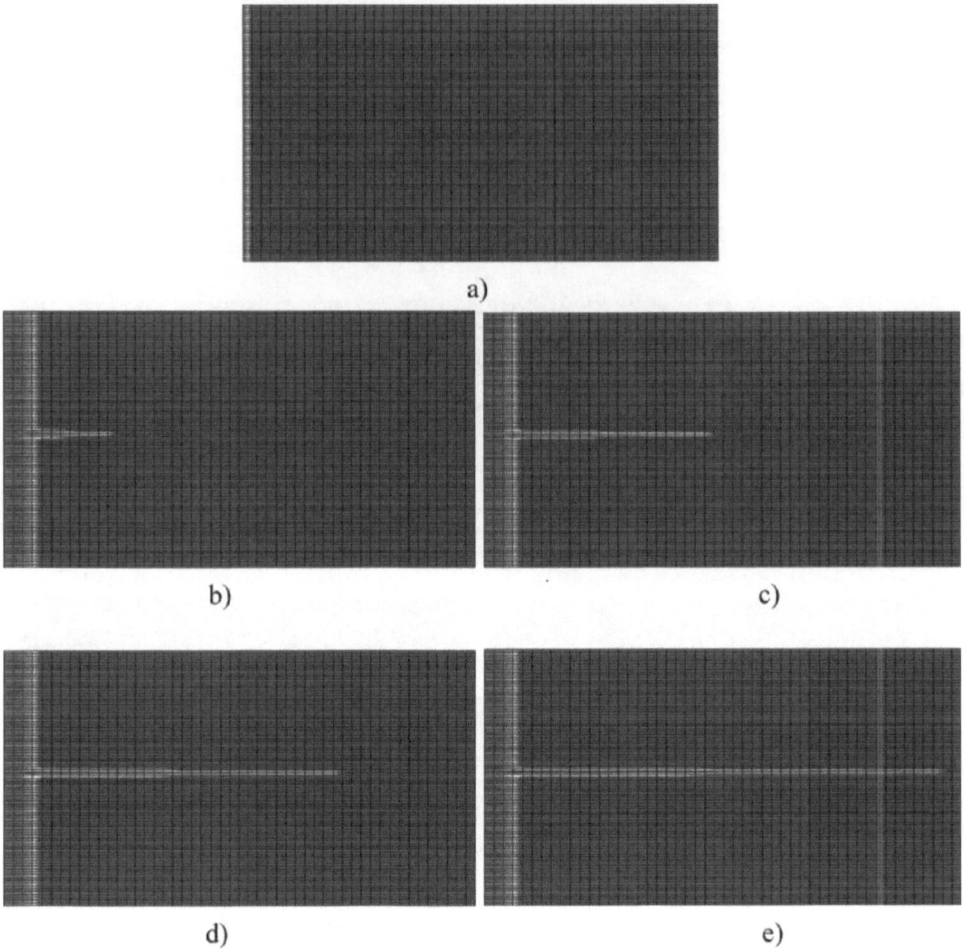

Figure 8. Wetting-phase saturation fields for the faulted reservoir problem. a) t = 0s, b) t = 1s, c) t = 2s, d) t = 3s, e) t = 4s.

in relation to the conventional formulation of LBM, allowing application of larger pressure gradients.

In the LBM, the problem domain is discretized by a lattice where an evolution equation is given by:

$$f_i(\mathbf{x}+\mathbf{e}_i\Delta t, t+\Delta t) = f_i(\mathbf{x},t) - \frac{\Delta t}{\tau}\left(f_i(\mathbf{x},t) - f_i^{eq}(\mathbf{x},t)\right) \tag{14}$$

where τ is the collision relaxation time, Δt is time interval, fi is the local pressure distribution function, ei represents the lattice where lattice velocity directions and i = 1, …, b, b being the number of discrete velocities of the lattice. In the present work, D2Q9 (two dimensional, 9 velocities) and D3Q19 (three dimensional, 19 velocities) lattices are used. The equilibrium distribution function is given by (He & Luo, 1997):

$$f_i^{eq}(\mathbf{x},t) = w_i\left\{p + p_0\left[3\frac{(\mathbf{e}_i\cdot\mathbf{u})}{c^2} + \frac{9}{2}\frac{(\mathbf{e}_i\cdot\mathbf{u})^2}{c^4} - \frac{3}{2}\frac{\mathbf{u}^2}{c^2}\right]\right\} \tag{15}$$

where p_0 is the initial (constant) pressure and $p = p_0 + \delta p$, δp being the pressure fluctuation, **u** is the fluid velocity, wi are coefficients that depend on the lattice and c is magnitude of the lattice velocity given by:

$$c = \frac{\Delta x}{\Delta t} \tag{16}$$

Δx being the lattice spacing. After imposing initial and boundary conditions, LBM equations are solved and the macroscopic variables of flow can be obtained through the following expressions:

$$p(\mathbf{x},t) = \sum_{i=1}^{b} f_i(\mathbf{x},t) \tag{17}$$

$$\mathbf{u}(\mathbf{x},t) = \frac{1}{p_0}\sum_{i=1}^{b} \mathbf{e}_i f_i(\mathbf{x},t) \tag{18}$$

Viscosity and fluid density are given by:

$$\nu = \frac{c^2}{3}\Delta t\left(\frac{\tau}{\Delta t} - \frac{1}{2}\right) \tag{19}$$

$$\rho(\mathbf{x},t) = \frac{p(\mathbf{x},t)}{c_s^2} \tag{20}$$

where c_s is the sound velocity in the fluid and for D2Q9 and D3Q19 lattices is given by:

$$c_s = \sqrt{\frac{c^2}{3}} \tag{21}$$

3.2 *Fluid-mechanical coupling*

Coupling of flow processes simulated with LBM and particle movement was initially proposed by Ladd (1994) with the objective of the simulating the particular case of suspended particles. Contact between particles was not modeled in that work. Cook & Noble (2004) implemented the coupling between DEM and LBM for the solution of 2D fluidmechanical problems using the immersed moving boundary scheme as proposed by Noble & Torczynsky (1998).

Coupling between fluid flow and particle movement involves two stages:

1. Definition of the boundary condition imposed to the fluid by the solid particles in motion.
2. Calculation and transfer of the drag forces from the fluid flow to the solid particles.

In order to simulate the fluid-mechanical interactions between solid particles and fluid flow, LBM must be modified in order to incorporate the boundary condition of the solids in motion. The boundary condition at the solid surface is one of no slip, that is, the solid is impermeable and the fluid adjacent to the solid surface moves with the same velocity of the solid. The immersed moving boundary condition, proposed by Noble & Torczynsky (1998), the one adopted in the present work, modifies LB equation (Eq. (14)) in order to impose the no-slip condition in the nodes of the lattice covered by the particles. The modified LB equation is then given by:

$$f_i(\mathbf{x}+\mathbf{e}_i\Delta t, t+\Delta t) = f_i(\mathbf{x},t) - \frac{\Delta t}{\tau}(1 - B(\mathbf{x},t))(f_i(\mathbf{x},t) - f_i^{eq}(\mathbf{x},t)) + B(\mathbf{x},t)\Omega_i^s(\mathbf{x},t) \tag{22}$$

where B is a weighing function given by:

$$B(\mathbf{x},t) = \frac{\varepsilon(\mathbf{x},t)(\tau/\Delta t - 0.5)}{(1-\varepsilon(\mathbf{x},t)) + (\tau/\Delta t - 0.5)} \tag{23}$$

and ε is the solid fraction at the site, defined by fraction of the volume at the lattice site occupied by the solid particle. In Eq. (22), Ωs is the additional collision term that modifies pressure distribution functions at the lattice nodes covered by solid particles in a way that the no-slip condition is imposed. This term is given by:

$$\Omega_i^s(\mathbf{x},t) = f_{-i}(\mathbf{x},t) - f_i(\mathbf{x},t) + f_i^{eq}(\rho,\mathbf{u}_s) - f_{-i}^{eq}(\mathbf{x},t) \tag{24}$$

where u_s is particle velocity at position **x**, in time t and –i represents opposite direction to direction i.

In LBM, forces that are transferred to the solid particles can be determined directly from the summation of momentum transfer from the fluid sites to the solid surface. In the immersed moving boundary scheme, the fluid force in a particle is given by:

$$\mathbf{F}_f = \frac{\Delta x^3}{c_s^2 \Delta t} \sum_n B_n \sum_i \Omega_i^s \mathbf{e}_i \tag{25}$$

n being the sites covered by the particle. Torque regarding the center of mass of the particle, x_p, is given by:

$$\mathbf{T}_f = \frac{\Delta x^3}{c_s^2 \Delta t} \sum_n (\mathbf{x} - \mathbf{x}_p) \times \left(B_n \sum_i \Omega_i^s \mathbf{e}_i \right) \tag{26}$$

3.3 *Two-phase flow*

The formulation for LBM used in the present work for the case of two phase fluid flow follows the work of Gunstensen & Rothman (1991) and Rothman & Zaleski (2004). In that formulation, at each lattice site, two types of fluid particles can coexist, the red (white grey in print version) and blue (white grey in print version) particles. One defines Ri (x,t) as the red population at site x, and time t, velocity in direction i. Bi(x,t) can be defined likewise for the blue fluid in a way that Ni(x,t) = Ri(x,t) + Bi(x,t). The main idea is to separate the evolution of the population in four stages:

i. Collision

$$N_i'(\mathbf{x},t) = N_i(\mathbf{x},t) - \frac{\Delta t}{\tau}[N_i(\mathbf{x},t) - N_i^{eq}(\mathbf{x},t)] \tag{27}$$

being:

$$N_i^{eq}(\mathbf{x},t) = w_i N(\mathbf{x},t)\left[1 + 3\frac{(\mathbf{e}_i \cdot \mathbf{u})}{c^2} + \frac{9}{2}\frac{(\mathbf{e}_i \cdot \mathbf{u})^2}{c^4} - \frac{3}{2}\frac{\mathbf{u}^2}{c^2}\right] \tag{28}$$

$$N(\mathbf{x},t) = \sum_{i=1}^{b} N_i(\mathbf{x},t) \tag{29}$$

ii. Generation of surface tension

Initially, the color gradient is determined:

$$g(\mathbf{x},t) = \sum_i \mathbf{e}_i \sum_j \left\lfloor R_j(\mathbf{x} + \mathbf{e}_i \Delta t, t) - B_j(\mathbf{x} + \mathbf{e}_i \Delta t, t) \right\rfloor \tag{30}$$

and with this value a perturbation is added to the populations which create a surface tension:

$$N_i''(\mathbf{x},t) = N_i'(\mathbf{x},t) + A|\mathbf{g}(\mathbf{x},t)|\left[\frac{(\mathbf{e}_i \cdot \mathbf{g})^2}{\mathbf{g} \cdot \mathbf{g}} - \frac{1}{2}\right] \tag{31}$$

A is a parameter chosen to set the magnitude of the surface tension.

iii. Color redistribution in order to minimize the diffusion from one color to the other
At this stage, the method proposed by Latva-Kokko & Rothman (2004) was adopted:

$$R_i''(\mathbf{x},t) = \frac{R(\mathbf{x},t)}{R(\mathbf{x},t) + B(\mathbf{x},t)} N_i''(\mathbf{x},t) + \Delta_i \tag{32a}$$

$$B_i''(\mathbf{x},t) = \frac{B(\mathbf{x},t)}{R(\mathbf{x},t) + B(\mathbf{x},t)} N_i''(\mathbf{x},t) - \Delta_i \tag{32b}$$

$$\Delta_i = \beta \frac{R(\mathbf{x},t)B(\mathbf{x},t)}{\left(R(\mathbf{x},t) + B(\mathbf{x},t)\right)^2} N_i^{eq} \cos\varphi_i \tag{32c}$$

being R and B the total number of red and blue particles at one specific site, β is the parameter that provides the tendency of the two fluids to separate, ϕ_i is the angle between the color gradient g and direction e_i, and N_i^{eq} is given by Eq. (28) assuming that velocity **u** is zero.

iv. Propagation of populations to adjacent sites

$$\begin{aligned} R_i(\mathbf{x} + \mathbf{e}_i\Delta t, t + \Delta t) &= R_i''(\mathbf{x},t) \\ B_i(\mathbf{x} + \mathbf{e}_i\Delta t, t + \Delta t) &= B_i''(\mathbf{x},t) \end{aligned} \tag{33}$$

Wettability of a fluid in the solid phase is controlled by only one parameter $p \geq 0$, one that measures the red color fraction at the solid sites. At the beginning of the simulation, color at the solid sites is attributed with the objective of calculating color gradients. If the red fluid is totally wetting then $p = 1$, if it is partially wetting then $1 < p < 0$, and if both fluids have the same wettability then $p = 0$.

3.4 *Examples*

This section presents a few examples of applications of LBM to flow problems in geometries pertinent to CO_2 injection in geological media.

3.5 *Verification*

Two examples are initially shown in order to demonstrate the ability of the incompressible formulation of LBM in order to correctly determine the force transferred from the fluid to the solid particle. A 2D situation simulates a still particle while a 3D situation simulates a moving particle.

The first example consists of a fixed disk located between two parallel walls where one phase flow is imposed as shown in Figure 9. This verification is particularly relevant as it is representative of drag forces transmitted to a particle is high solids concentrations (in this case the walls located near to the particle). The parameters used in the simulations are: Reynolds number = 0.0002; Δx = 5.0E-5 m; Δt = 1.0E-4 s; υ = 2.0E-6 m²/s; ρ = 1000 kg/m³; disk radius, a = 10 Δx; domain length, L = 20a. The purpose of the exercise was to calculate the force transmitted by the fluid to the particle.

The results obtained with LBM were compared with the results obtained by Richou et al. (2004) who simulated the same flow problem using finite differences. Figure 10 shows a comparison between the two procedures and a good agreement was obtained.

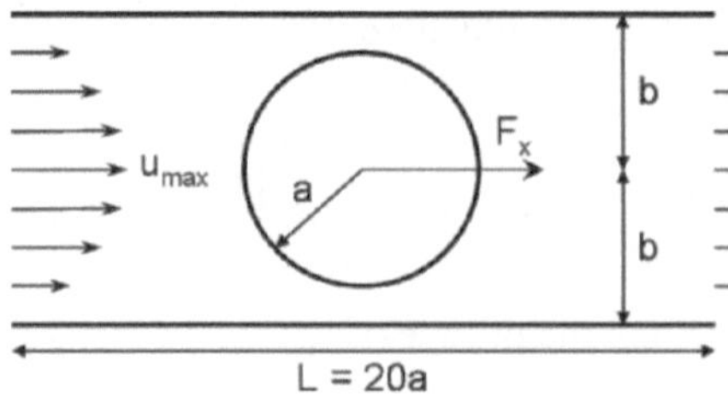

Figure 9. Geometry and boundary conditions of 2D example for the verification of the drag force determination.

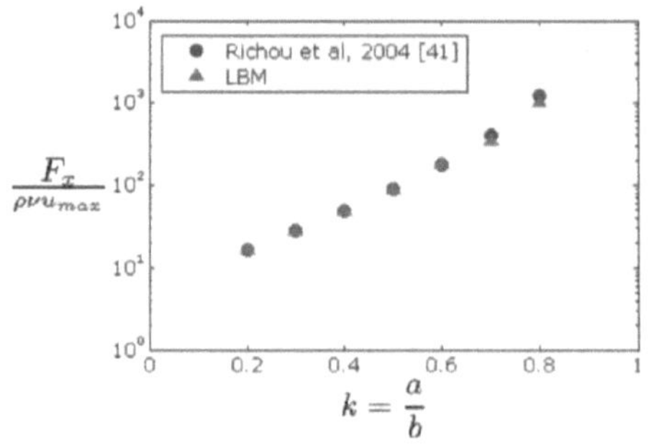

Figure 10. Comparison of the results obtained with incompressible LBM and results obtained by Richou et al. (2004).

In order to evaluate the motion of solid particles in the lattice and the evaluation of forces transferred from the fluid to the particles, a sphere immersed into an initially still fluid is considered. A constant velocity is imposed for the particle and a periodical flow condition is imposed. For the case of a sphere moving with constant velocity in a fluid, considering low Reynolds numbers, Stokes law provides the drag force:

$$F_d = 6\pi\rho\upsilon R(-u_p) \tag{34}$$

where R is the sphere radius and up is the sphere velocity. Simulation data, in non dimensional values are: up = 0.005; R = 5Δx; υ = 1/6; ρ = 1.0; Δx = 1.0; Δt = 1.0. For these values Reynolds number is 0.15 and the drag force is 0.0785. The lattice dimension is 60 × 60 ×60. An evolution of the drag force is presented in Figure 11 where a satisfactory agreement is obtained between LBM and the analytical solution, an error of approximately 5% was found.

A further verification example involves the determination of capillary forces. The results try to demonstrate the ability of LBM in simulating correctly the capillary force that appears in a meniscus between solid particles. The example consists in two fixed solid spheres surrounded by a non-wetting fluid and having an amount of a wetting fluid at their contact. Simulation using LBM proceeds until a permanent and stable condition is reached (Figure 12). The capillary forces determined with two phase LBM were compared with an analytical solution presented by Gili and Alonso (2002) for the capillary force generated by the meniscus between the two spheres (Figure 13). One notices that the analytical solution is valid for wetting angles up to approximately 50°. Figure 14 shows meniscus evolution when a constant force is applied upwards on the upper sphere. This force has a magnitude larger than the capillary force generated at the meniscus shown.

3.5.1 *LBM simulation of injection processes of CO_2 in coal formation cleats*

The simulation of CO_2 into the complex cleat structure of coal can be carried out through continuum based, homegeneized models (Wang & Liu, 2011) or through discrete models that focus

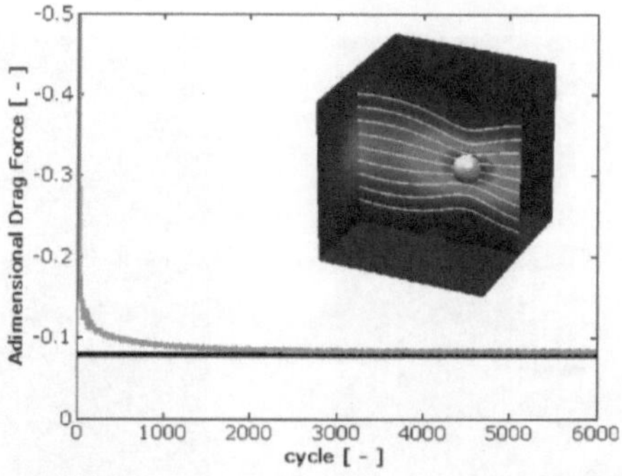

Figure 11. Comparison between the analytical solution (black line) and the numerical solution (grey line) for the drag force. The figure on the side shows the velocities and flow lines obtained in the numerical simulation.

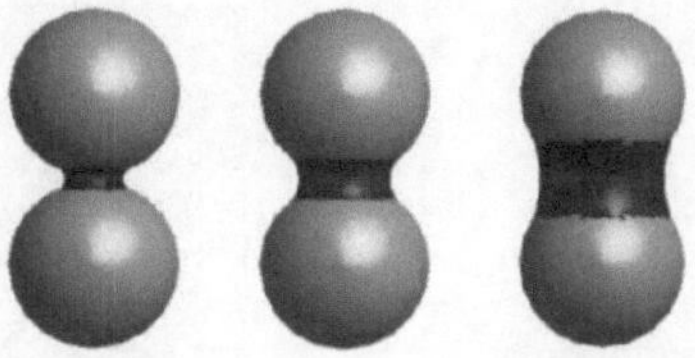

Figure 12. LBM simulation results for equilibrium configurations of meniscus between two spheres.

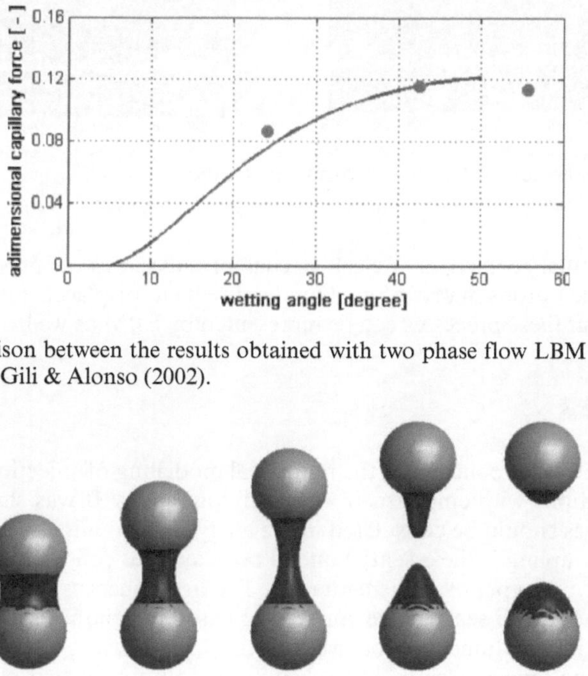

Figure 13. Comparison between the results obtained with two phase flow LBM simulations and the solution presented in Gili & Alonso (2002).

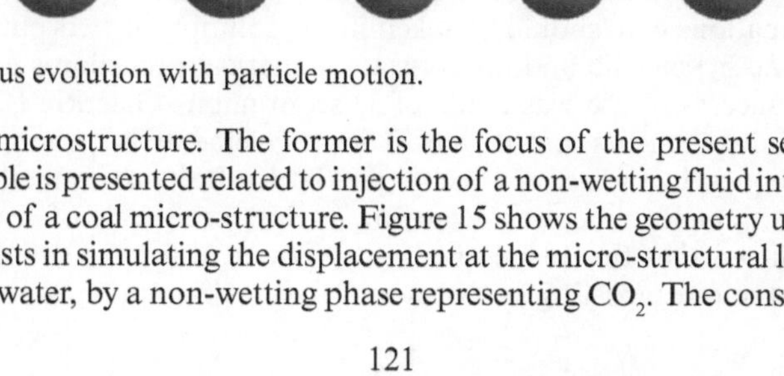

Figure 14. Meniscus evolution with particle motion.

on details of the microstructure. The former is the focus of the present section using LBM method. An example is presented related to injection of a non-wetting fluid into a fissured geometry representative of a coal micro-structure. Figure 15 shows the geometry used in the analysis. The example consists in simulating the displacement at the micro-structural level, of an existing phase, in this case water, by a non-wetting phase representing CO_2. The considered geometry is

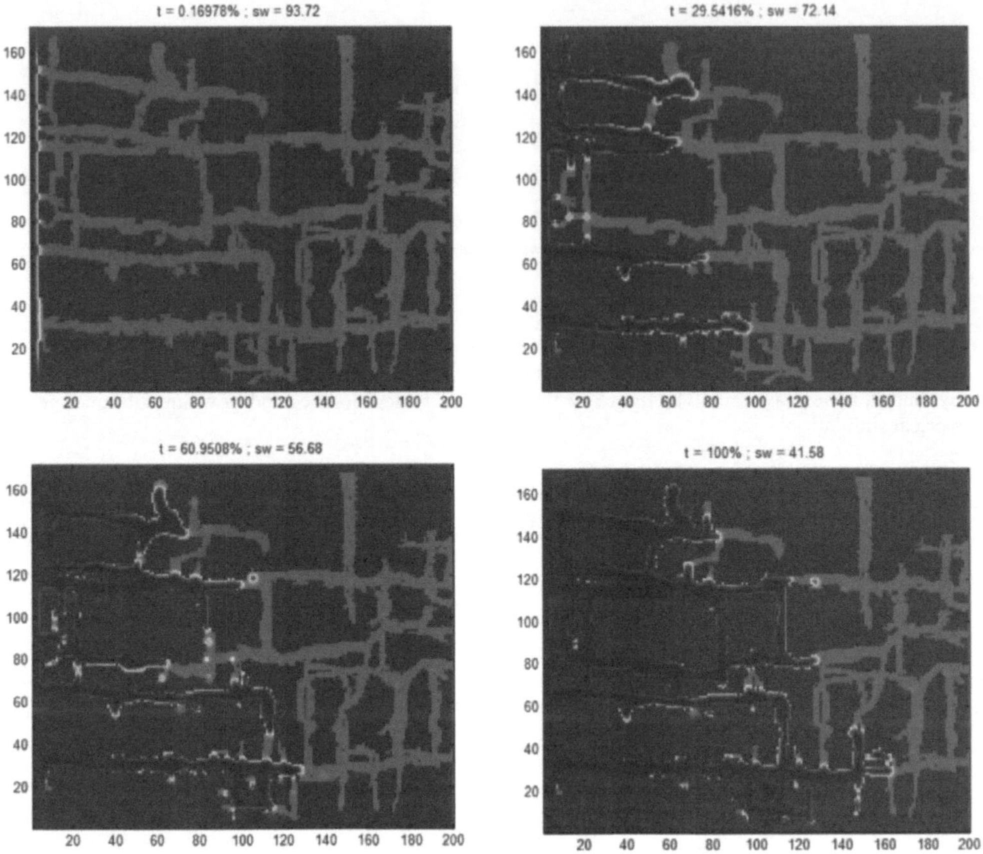

Figure 15. Injection process of CO_2 in coal formation cleats.

representative of actual geometries of cleats in coal formations. In the present work, the diffusion of CO_2 into the porous matrix of coal (and consequent displacement of methane) is not being considered but these processes can be represented by LBM as well.

4 CONCLUSIONS

The paper analyzed issues concerning the numerical modelling of injection processes of CO_2 in geological fomations with emphasis in carboniferous rocks. It was shown that a number of coupled processes should be considered in the analysis although it is possible to state that fluid-mechanical coupling is the essential one to be taken into consideration. The paper presented examples of two types of implementation. The first concerns simulations to be carried out at the meso and macro scale where continuum based techniques such as finite elements, finite differences, finite volumes can be used. In this case, it was shown that for analysis of heterogeneous formations care should be taken in the solution of two-phase flow equations as they may become hyperbolic and appropriate numerical procedures should be used. In the present work, successful use was made of Discontinuous Galerkin (DG) techniques in association with finite elements and Raviart-Thomas elements for accurate pot processing of velocities. Subsequently, the use if a technique (lattice Boltzmann methods) was demonstrated in order to simulate processes occurring at the micro-scale of the rock formations involved. It was shown that this technique is very powerful for the simulation of complex geometries as are the ones existing in pores and fissures/fractures. A connection between micro and macro-scale is envisaged as well as the implementation of additional modules to simulate other coupled processes.

REFERENCES

Aziz, K. & Settari, A. 1959. Petroleum reservoir simulation. Applied Science Publishers, Londres.

Cook B.K., D.R. Noble. 2004. A direct simulation method for particle-fluid systems. Eng. Comp. 21: 151–168.

Durlofsky, L.J. 1993. A triangle based mixed finite-element-finite volume technique for modelling two phase flow through porous media. Journal of Computational Physics,105:252–66.

Gili, J.A. & Alonso, E.E. 2002. Microstructural deformation mechanisms of unsaturated granular soils. Int. J. Numer. Ana. Meth. Geomech. 26:433–468.

Gomes, A. 2010. CO_2 injection processes in carboniferous formations (in Portuguese). MSc thesis, Porto University, Portugal, 116p.

Gunstensen, A.K. & Rothman, D.H. 1991. Lattice Boltzmann model for immiscicle fluids. Phys. Rev. A 43(8): 4320–4327.

He, X. & L.-S. Luo. 1997. Lattice Boltzmann Model for Incompressible Navier-Stokes Equation. J. Stat. Physics, 88(3/4): 927–944.

Hepple, R.P. & Benson, S. 2003. Implications of surface seepage on the effectiveness of Geologic Storage of Carbon Dioxide as a Climate Change Mitigation Strategy. In: J. Gale and Y. Kaya (eds.), Greenhouse Gas Control Technologies, pp. 261–266, Elsevier Science, Ltd., Amsterdam, The Netherlands.

Hoteit, H. & Firoozabadi, A. 2008. Numerical modelling of two-phase flow in heterogeneous permeable media with different capillarity pressures. Advances in Water Resources, 31:56–73.

IPCC (Intergovernmental Panel on Climate Change). 2005. Carbon Dioxide Capture and Storage. Cambridge University Press.

Ladd, A.J.C. 1994. Numerical simulations of particle suspensions via a discretized Boltzmann equation: Part I. Theoretical foundation. J. Fluid Mech. 271: 285–309.

Latva-Kokko, M. & Rothman, D.H. 2005. Diffusion properties of gradient-based lattice Boltzmann models of immiscible fluids. Phys. Rev. E 71: 056702.

Manchao, H., Sousa, R., Sousa, L., Gomes, A., Vargas Jr, A. & Na, Z. 2011. Risk assessment of CO_2 injection processes and storage in carboniferous formations: a review. Journal of Rock Mechanics and Geotechnical Engineering, 3(1) pp 39–56.

Muller, A.L., Vargas Jr., E.A., Vaz, L.E. & Gonçalves, C.J. 2009. Three dimensional analysis of boreholes considering spatial variability of properties and poroelastoplasticity. Journal of Petroleum Science and Engineering, 68:268–276.

Noble, D.R. & Torczynsky, J.R. 1998. A lattice-Boltzmann method for partially saturated computational cells. Int. J. Modern Phys. C 9(8): 1189–1201.

Pao. W.K.S., Lewis, R.W. & Masters, I. 2001. A fully coupled hyfro-thermo-poro-mechanicl model for black oil reservoir simulation. Int. J. Numer. Anal. Mth. Geomech, 25:1229–1256.

Piessens, K. & Dusar, M. 2003a. CO_2 Sequestration in abandoned coal mines. Royal Belgium Institute for Natural Sciences. Geological Survey of Belgium.

Piessens, K. & Dusar, M. 2003b. The vertical reservoir Simulator CO_2 —VR. Royal Belgium Institute for Natural Sciences. Geological Survey of Belgium.

Ribeiro, W.N. 2011. Evaluation of numerical solutions for analysis of coupled two-phase flow with geomechanical behavior in porous media. Ph.D. Thesis, Pontifical Catholic University of Rio de Janeiro.

Richou, A.B., Ambari, A. & Nacin, J.K. 2004. Drag force on a circular cylinder midway between two parallel plates at very low Reynolds numbers—Part 1: Poiseuille flow (numerical). Chem. Eng. Sci. 59: 3215–3222.

Rothman, D.H. & Zaleski, S. 2004. Lattice-Gas cellular Automata. 1st paperback ed. Cambridge. Cambrige University Press.

Velloso, R.Q. 2010. Numerical analysis of fluid mechanical coupling in porous media using the discrete element method. Ph.D. Thesis, Pontifical Catholic University of Rio de Janeiro.

Wang, J.G. & Liu, J.S. 2011. Micro-scale modeling of gas-coal interaction in coalbed-seam - Heterogeneity effect. Int. Workshop on CO_2 Storage in Carboniferous Formations and Abandoned Coal Mines, Taylor & Francis, Ed. He, Sousa, Elsworth & Vargas, Beijing, pp. 95–103.

CO_2 Storage in Carboniferous Formations and Abandoned Coal Mines –
He, Ribeiro E Sousa, Elsworth & Vargas Jr. (eds)

Methodologies for risk analysis and decision making

R. Leal e Sousa
Department of Civil and Environmental Engineering, Massachusetts Institute of Technology, Cambridge, USA

ABSTRACT: There are a number of models available for data analysis and representation, including event trees, rule-based systems, fuzzy-rule based systems, artificial neural networks, and Bayesian networks. In this paper several models for data analysis and representation are described, and common techniques for risk assessment are presented. The fundamentals of Bayesian Networks and Decision graphs are introduced, since these will be used in following chapters. Applications of Bayesian Networks and decision graphs to risk analysis problems are presented.

1 INTRODUCTION

There are a number of models available for data analysis and representation, including event trees, rule-based systems, fuzzy-rule based systems, artificial neural networks, and Bayesian networks. There are also several techniques for data analysis such as classification, density estimation, regression and clustering.

Knowledge representation systems (or knowledge based systems) and decision analysis techniques were both developed to facilitate and improve the decision making process. Knowledge representation systems use various computational techniques of AI (artificial intelligence) for representation of human knowledge and inference. Decision Analysis uses decision theory principles supplemented by judgment psychology (Henrion, 1991). Both emerged from research done in the 1940's regarding development of techniques for problem solving and decision making. John von Neumann and Oscar Morgensten, who introduced game theory in "Games and Economic Behavior" (1944), had a tremendous impact on research in decision theory.

Although the two fields have common roots, since then they have taken different paths. More recently there as been a resurgence of interest by many AI researchers in the application of probability theory, decision theory and analysis to several problems in AI, resulting in the development of Bayesian Networks and Influence diagrams, an extension of Bayesian Networks designed to include decision variables and utilities.

There are several advantages that Bayesian Networks have over other methods. In this paper some of the most common methods available for knowledge representation, risk and decision making are briefly presented. Their main advantages and shortcomings are discussed.

2 RISK RELATED DEFINITIONS

In this section some of the most important risk related definitions are introduced. They are Definitions are based on Technical Committee on Risk Assessment and Management: Glossary of Risk Assessment Terms from ISSMGE (International Society for Soil mechanics and Geotechnical Engineering), and are presented below:

Hazard: Probability that a particular danger (threat) occurs within a given period of time.
Risk: Measure of the probability and severity of an adverse effect to life, health, property, or the environment.

Risk = Hazard × Potential Worth of Loss.
Consequence: Result of a hazard being realized.
Damage: Another way of expressing detrimental consequences.
Vulnerability:

- Degree of loss to a given element or set of elements within the area affected by a hazard.
- Expresses the fact that even if a threat materializes, it is not necessarily 100% certain that the consequences materialize.
- Can be formulated as a conditional probability.

3 METHODOLOGIES FOR RISK ANALYSIS AND DECISION MAKING

3.1 *Rule based systems*

Ruled Based Systems are computer models of experts of a certain domain. The building blocks for modeling the experts are called production rules. A production rule is of the form:

If A **then** B

Where A (premise) is an assertion, and B (conclusion) can be either an action or another assertion. A rule based system consists of a library of such rules. These rules reflect essential relationships within the domain, or rather: they reflect ways to reason about the domain. When specific information about the domain comes in, the rules are used to draw conclusions and to point out appropriate actions.

A rule based system (or expert system) consists of a knowledge base and an inference engine. The knowledge base is the set of production rules and the inference engine combines rules and observations to come up with conclusions on the state of the world and on what actions to take.

One of the major problems of rule based systems is how to treat uncertainty. A way to incorporate uncertainty in rule based systems is to have production rules of this type:

If condition with certainty x
then fact with certainty f (x)

where f is a function.

There are many schemes for treating uncertainty in rule based systems. The most common are fuzzy logic, certainty factors and (adaptations of) Dempster—Shafer belief functions (Dempster, 1968; Russell and Norvig, 2004). However, it is not easy to capture reasoning under uncertainty with inference rules for production rules. The reason for this is that in all the schemes for treating uncertainty, mentioned above, the uncertainty is treated locally. More specifically, it is difficult to combine (un)certainties from different rules.

Despite their shortcomings, Rule Based Systems have been used in many applications in different domains, such as Medicine, for diagnosis and assisting in the selection of antibiotics (MYCIN, Stanford University, in 1976 see Shortliffe, 1976 and Melle et al., 1981); Banking, to detect fraud in use of credit cards (FRAUDWATCH, Touche Ross, UK, 1992); Aerospace Engineering for scheduling operations for the recycling Space Shuttle flights (GPSS, NASA, USA, 1993) and Civil Engineering for recommendation system in the maintenance and repairing of tunnels (MATUF, Silva, C, 2001) and a recommendation systems for repairing bridges (Sousa, R. 2000), among others (Darlington, 2000). More recently, these types of systems have been substituted by other techniques that allow one to better and more efficiently incorporate uncertainty.

3.2 *Fuzzy-rule approach*

As mentioned previously there are many schemes for treating uncertainty in rule based systems. Fuzzy logic is one way of introducing uncertainty into rule based systems. It is a superset of conventional logic that has been extended to handle the concept of "partial truth", i.e. a value between (completely) true and (completely) false (Zadeh, 1965 and 1999). Based on fuzzy logic, fuzzy rule expert systems were created. They use a collection of fuzzy membership functions and rules drawn-out from the experts (Figure 1). The rules to evaluate the fuzzy "truth" T of a sentence are presented in (Equation 1):

$$\begin{aligned} T(A \wedge B) &= \min(T(A), T(B)) \\ T(A \vee B) &= \max(T(A), T(B)) \\ T(\neg A) &= 1 - T(A) \end{aligned} \tag{1}$$

where T is the fuzzy "truth" and A and B are variables or complex sentences. The AND ($\wedge$), OR ($\vee$), and NOT ($\neg$) operators of Boolean logic exist in fuzzy logic; usually define the minimum, maximum, and complement. For example if A represents Low Pressure of the value p* then $T(A) = \mu_p$. Imagine that B represents High Temperature, of the value t* and $T(B) = \mu_t$. The result of Low Pressure (p*) and High Temperature (t*), i.e. $T(A \wedge B)$ would be the $\min(T(A), T(B)) = \min(\mu_p, \mu_t)$. The way this process of fuzzification and defuzzification works will be demonstrated through an example, presented next:

Imagine the rule about deciding whether or not a liquid is potable. The factors to consider are toxicity, measured in parts per million, and the alcohol content, measured in percent (Figure 2).

The rule is

IF	nontoxic	
AND	low alcohol	(2)
THEN	potable	

Imagine that we have a situation in which the toxicity of a liquid Z is equal to 210 ppm and the fuzzy membership function is presented in Figure 3. The liquid Z is nontoxic with membership 0.6.

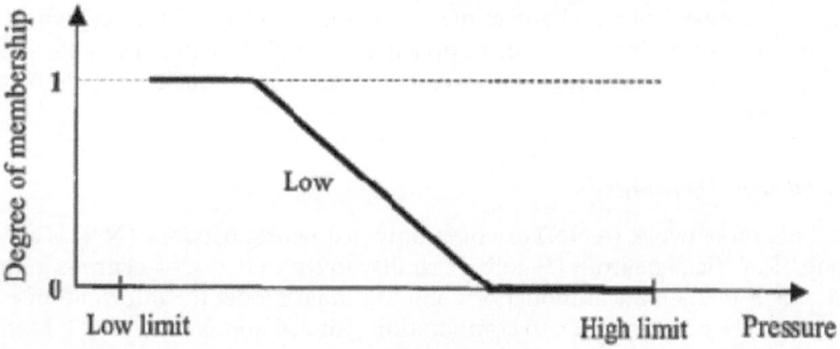

Figure 1. Fuzzy membership function (for low pressure).

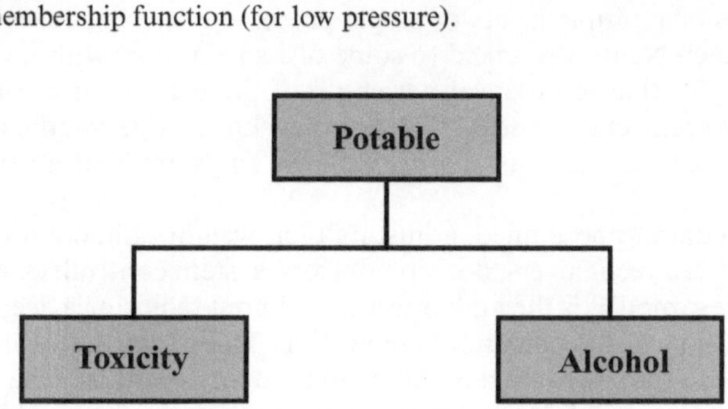

Figure 2. Factors in deciding whether a liquid is potable or not.

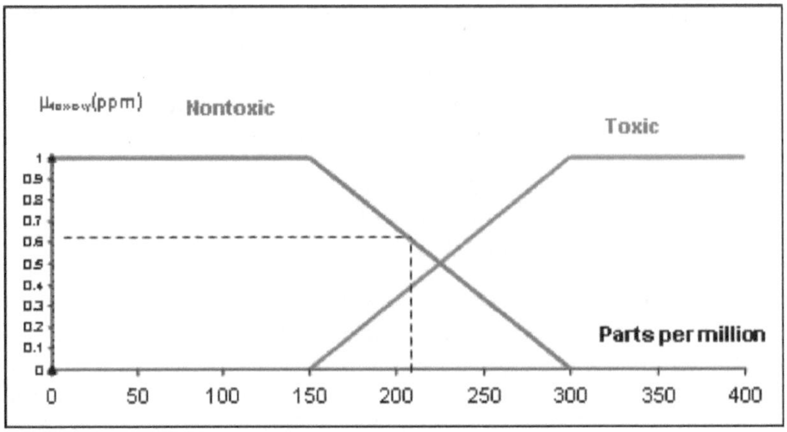

Figure 3. Membership function for toxicity.

The alcohol content of the liquid is 20%, resulting in low alcohol content with membership 0.75 (see Figure 4)

Applying the rule in Equation 2, one will obtain that the truth value of the sentence liquid Z is potable is 0.6 (Equation 3):

IF	nontoxic (Z)		
AND	low alcohol (Z)		(3)
THEN	potable (Z)	(=0.75)	
	Min (0.6, 0.75) = 0.6		

The inference mechanisms of these rules have some weaknesses; they have a weak theoretical foundation, inconsistency and sometimes oversimplification of the real world. Despite their shortcomings, fuzzy logic has been applied to several domains. In geotechnical engineering an application of fuzzy logic is use of Fuzzy set rules in rock mass characterization (Sonmez et al., 2003).

3.3 *Artificial neural networks*

An artificial neural network (ANN) or commonly just neural network (NN) is an interconnected group of artificial neurons (Figure 5), similar to the network of neurons in the human brain, that uses a mathematical model or computational model for information processing based on a connectionist approach to computation (Russell and Norvig, 2003; Mehrotra, K. et al., 1997).

An ANN consists of multiple layers of single processing elements called neurons and of their connections. Each Neuron is linked to some of its neighbors with a varying coefficient of connectivity (weight) that represent the strength of these connections. This is stored as a weight value on each connection. The ANN learns new knowledge by adjusting these weights and the connections between neurons. Figure 5 shows an example of a neural network with one hidden layer.

The ANN rely on data to be trained, adjusting their weights and connections to optimize their behavior as pattern recognizers, decision makers, system controllers, predictors, etc.

The strength of these models is their adaptiveness, without requiring a deep knowledge about the complex relationships of the domain of application. This adaptiveness allows the system to perform well even when the system that is being modeled, or controlled, changes over time.

The objective of using an ANN is to make predictions in the future. Although, an ANN network could provide almost perfect answers to the set of data with which it was trained, it

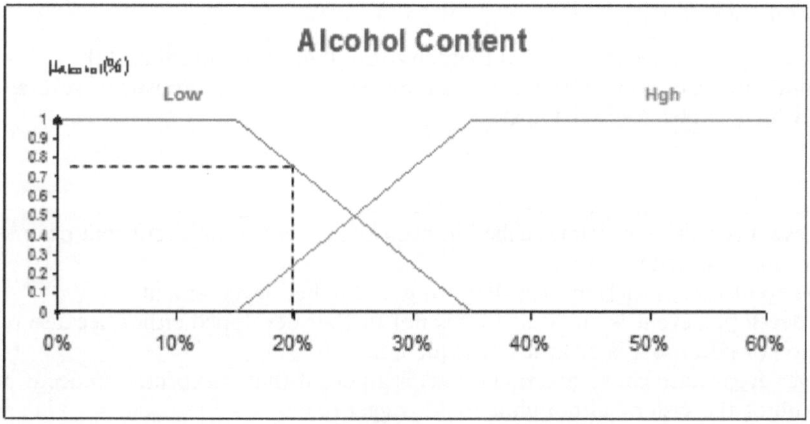

Figure 4. Membership function for alcohol content.

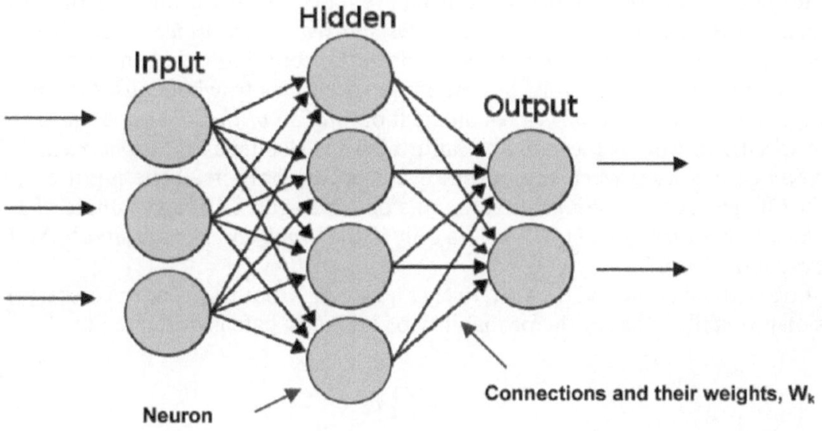

Figure 5. Neural network with one hidden layer.

may fail to produce an adequate answer when "new" data surfaces. This is a result of "over-fitting" (Suwansawat, 2002; Suwansawat and Einstein, 2006). In order to perform adequately and produce good results, these systems require a large number of sample data in order to be trained. Also, since there is not a complete understanding of the learning process, the analysis of the results may be difficult. Thus, this is not the right approach in cases in which one needs to have a complete understanding of the problem domain and relationship among variables of the domain.

3.4 *Classical decision analysis*

Decision Analysis is a logical procedure for the balancing of the factors that influence a decision. The procedure incorporates uncertainty, values, and preferences in a structure that models decision (Howard, 1966 and 1984). A classical tool used to model decisions and incorporate in a formal manner the relevant components of decision analysis is the decision tree. Prior to decision analysis, Fault trees and event trees can be used to model on one hand the different ways an event can occur (fault tree) and on the other hand, systematically identify the possible sequence of events and their consequences (event tree).

3.5 *Fault trees*

Fault tree analysis is a technique used to analyze an undesirable event and the different ways that the undesirable event can be caused. A typical fault tree is composed of several different symbols, which will be described next.

Events

The commonly used symbols for events are represented in Figure 7.

A top event (or also sometimes called intermediate event) is an event that occurs because of one or more antecedent causes.

A basic event is an initiating event requiring no further development.

An undeveloped event is an event that is not further developed either because of lack of information or because it is of little consequence.

A trigger event (also called external event) is an event that is expected to occur but is not itself a fault of the system, although it could trigger one.

Gates

There are two basic types of fault tree gates, the OR-gate and the AND-gate. The symbols are shown in Figure 6.

The OR- gate is used to show that the output event occurs only if one or more of the input events occur. In the example the "*Failure of the sub-sea tunnel project*" can occur only if a "*technical failure*" or an "*economical failure*", or both occur. Note that the inputs to an OR-gate are restatements of the output but are more specific as to what causes them, i.e. in the case of Figure 6 *Technical failure* is a restatement of "*failure of the sub-sea tunnel project*", but it is more specific to what is the cause of failure. This is also true for "*economical failure*".

The AND-gate is used when the output event occurs only if all the input events occur. Unlike the OR-gate, causes can be direct inputs of AND-gates. In the example of Figure 6 a "*total collapse, seawater fills tunnel*" occurs only if the "*rock cover is too small*" AND "*investigations are insufficient*".

A fault tree can be evaluated quantitatively and often is, but this is not necessary. Based on the rules of probability theory the probability of an AND gate is evaluated by

$$P = \prod_{i=1}^{n} p_i \tag{4}$$

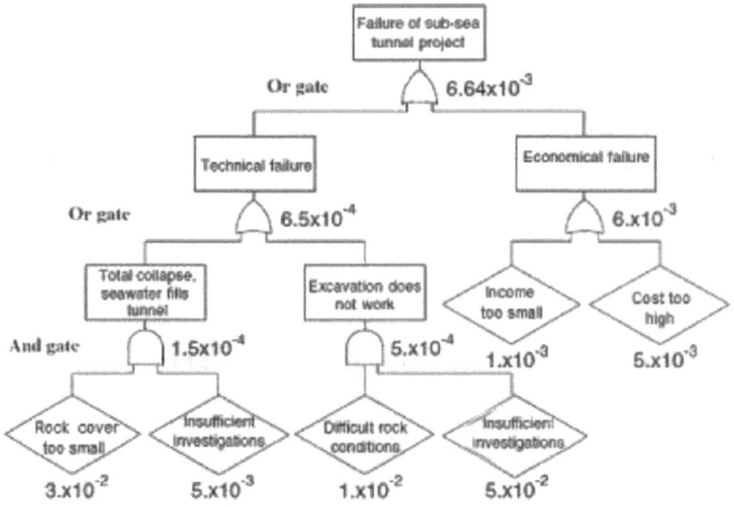

Figure 6. Example of a fault tree for evaluation of failure on sub-sea tunnel project (Eskesen, 2004).

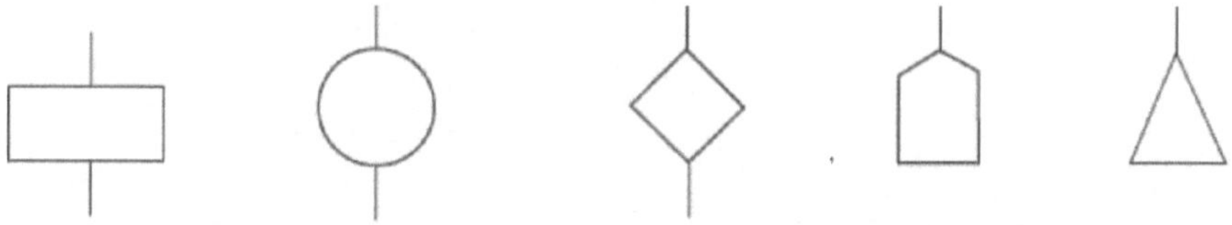

Figure 7. Symbols commonly used for events in fault tree.

And an OR-gate by

$$P = 1 - \prod_{i=1}^{n} (1 - p_i) \tag{5}$$

Where n is the number of ingoing events to the gate. p_i are the probabilities of failure of the ingoing events and it is assumed that the ingoing events are independent.

In Figure 6 the undesirable event being analyzed is the *Failure of a sub-sea tunnel*. According to the model failure can occur only if a *Technical Failure* OR an *Economical Failure* (or both) occur. A *Technical Failure* can occur if a *total collapse occurs* OR *excavation construction does not work* (or both). According to the model a total collapse can only occur if the *rock cover of the tunnel is insufficient* AND the *geotechnical investigations are insufficient*. On the other hand, the *Excavation may not work* if both *difficult rock conditions* are encountered AND *geotechnical investigations are insufficient*.

3.5.1 *Event trees*

An event tree is a representation of the logical order of events leading to consequences. In contrast to the fault tree it starts from a basic initiating event and develops from there in time until all possible states with consequences (adverse or not) have been reached. A typical graphical representation of an event tree is shown in Figure 8. This is an example regarding the non-destructive testing of a reinforced concrete structure for corrosion. The inspection may or not detect the corrosion. The event CI denotes that corrosion is present, and the event I that the corrosion is found by the inspection. The bars over the events represent the complementary events. Based on this tree, one can evaluate the probability that corrosion is in fact present given that the inspection says so.

Event trees can become very complex to analyze rather quickly. For a tree with n two-state components the total number of paths is 2^n. If each component has m states the total number of branches is m^n.

Fault trees and event trees (or decision trees) can be combined. The top event of a fault tree, in example of Figure 6, Failure of the tunnel, can be used as an initiating event for an event tree to assess the risk associated with that particular event. The combined fault tree and event tree is illustrated in Figure 9, which shows how fault trees can model an initiating event for the event tree. Note that the same fault tree can be combined with a decision where one can assess whether or not it would be worth taking measures to avoid or mitigate damage.

3.5.2 *Decision trees*

A decision tree is a formal representation of the various components of a decision problem. It consists of a sequence of decisions, namely a list of possible alternatives; the possible outcomes associated with each alternative; the corresponding probability assigments; monetary consequences and utilities (Ang and Tang, 1975). The typical configuration of a simple decision tree is shown in Figure 10. There are three types of nodes in a decision tree. The decision nodes, which are squared, represent different decisions or actions. The chance nodes, which are circular, are nodes that identify an event in a decision tree where a degree of uncertainty

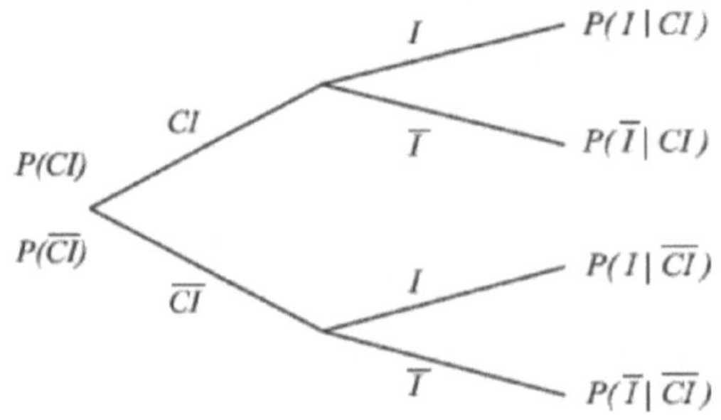

Figure 8. Typical event tree (Faber, 2005).

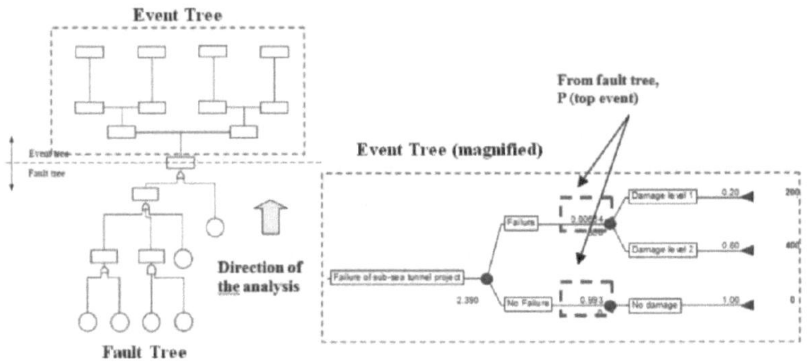

Figure 9. Combination of a fault tree and an event tree.

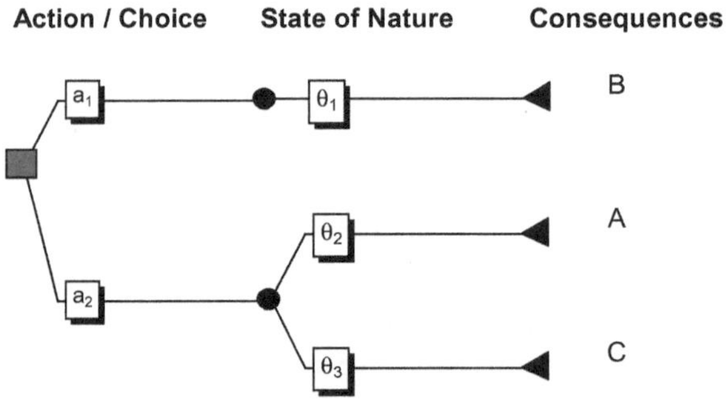

Figure 10. Typical decision tree (Faber, 2005).

exists. The utility nodes, which are triangular, are nodes that terminate a branch path and represent the utilities associated with the path.

Figure 10 models a case where the decision maker is faced with two decisions/actions, a_1 and a_2. The consequence of action a_1 is with certainty B. However the consequence of decision a_2 depends on the state of nature. Before the true state of nature is known the optimal decision depends upon the likelihood of the various states of nature θ_i and of the consequences *A*, *B* and *C*.

The decision maker will choose action a_1 over a_2 if the expected utility associated with action a_1 is greater than that of a_2.

$$E[u(a_1)] > E[u(a_2)]$$
$$u(B) > pu(A) + (1-p)u(C)$$

where
u (A), u (B)—utility of consequence A and B, respectively
p—probability of state θ_2
(1 – p)—probability of state θ_3

The valuation of an outcome, or the utility of an outcome, translates the relative preference of the decision maker towards different outcomes. The utilities are commonly based on monetary values, but they can also be based on other dimensions such as time or environmental effects. Multiattribute theory provides a way to combine all different measures of preference to come out with one single scalar utility to represent the relative preference of any outcome.

4 BAYESIAN NETWORKS

4.1 *Background and probability theory*

A Bayesian network, also known as belief network, is a graphical representation of knowledge for reasoning under uncertainty. Over the last decade, Bayesian networks have become a popular model for encoding uncertain expert knowledge in expert systems (Heckerman et al., 1995). Bayesian networks can be used at any stage of a risk analysis, and may substitute both fault trees and event trees in logical tree analysis. While common cause or more general dependency phenomena pose significant complications in classical fault tree analysis, this is not the case with Bayesian networks. They are in fact designed to facilitate the modeling of such dependencies. Because of what has been stated, Bayesian networks provide a good tool for decision analysis, including prior analysis, posterior analysis and pre-posterior analysis. Furthermore, they can be extended to influence diagrams, including decision and utility nodes in order to explicitly model a decision problem.

The concepts of Bayes' theorem, essential for Bayesian networks is presented below. For the basic concepts of probability theory (such as event, random variable, probability function, among others) necessary to understand the methodology of Bayesian networks, please refer to Ang & Tang, 1975.

Bayes' Theorem

$$P(A \mid B) = \frac{P(B \mid A)P(A)}{P(B)} \tag{6}$$

where the $P(B) = \Sigma_{i=1}^{n} P(A_i)P(B \mid A_i)$

Bayes Theorem has a many uses. Many times it is much easier to estimate the probabilities on the right side of Equation 6 than the one on the left side. A good example is the case where one want to estimate the probability of the disease given a certain symptom, P (A|B), being A = disease and B = symptom.

In order to estimate P (A|B) one would have to go through the population and then find people that had the symptom (B) and from these find out how many of these had the disease (A). Counting these cases maybe very hard especially if the disease is very rare; one may have to look at millions and millions of people. However, finding the probability of the symptom given the disease, P (B|A) is much easier. One just has to check hospital records and find people that had the disease and count how many of them had the symptom. Then one will also have to find the probability of the symptom and the probability of the disease, these also easier to get than P (A|B).

For random variables the Bayes' theorem can be written as follows:

$$P_X(x \mid Y = x) = \frac{P_Y(y \mid X = x)P_X(x)}{P_Y(y)}, \quad \text{where } P_Y(y) = \sum_{x \in \chi} P_Y(y \mid X = x)P_X(x)$$

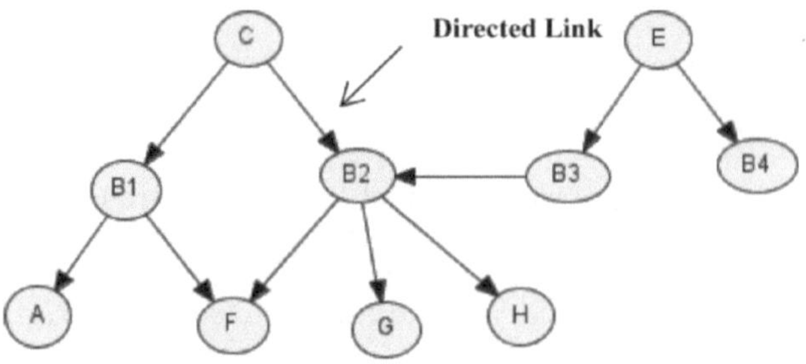

Figure 11. Bayesian network example.

4.2 *Definition of bayesian network*

A Bayesian Network is a concise graphical representation of the joint probability of the domain that is being represented by the random variables, consisting of (Russell & Norvig, 1995):

- A set of random variables that make up the nodes of the network.
- A set of directed links between nodes. (These links reflect cause-effect relations within the domain.)
- Each variable has a finite set of mutually exclusive states.
- The variables together with the directed links form a directed acyclic graph (DAG).
- Attached to each random variable A with parents $B_1, \ldots, B_n$ there is a conditional probability table $P(A = a \mid B_1 = b_1, \ldots, B_n = b_n)$, except for the variables in the root nodes. The root nodes have prior probabilities.

Figure 11 is an illustration of a simple Bayesian network. The arrows going from one variable to another reflect the relations between variables. In this example the arrow from C to B_2 means that C has a direct influence on B_2.

A Bayesian Network (BN) is a graphical and concise representation of a joint probability distribution of all the variables, taking into account that some variables are conditionally independent. The simplest conditional independence relationship encoded in BN is that a node is independent of any ancestor[1] nodes given its parents, i.e. that a node only depends on its direct parents. Thus, the joint probability of a Bayesian network over the variables $U = \{A_1, \ldots, A_n\}$, can be represented by the chain rule:

$$P(U) = \prod_{i}^{n} P(A_i = a_i \mid parents\,(A_i)) \tag{7}$$

where "parents (A_i)" is the parent set of A_i.

The difference between the general chain rule and Equation 7, chain rule applied to Bayesian networks is that in Bayesian Networks a variable is conditionally independent of their non-descendents, given the values of their parent variables, e.g. in the network of Figure 11 the variable A is conditionally independent of C given B1. It is this property that makes Bayesian Networks a very powerful tool for representing domains under uncertainty.

4.3 *Inference*

Since a Bayesian Network defines a model for variables in a domain and their relationships, it can be used to answer probabilistic queries about them. This is called inference.

---

1. Ancestor nodes of a node are all nodes that come prior to that node in topologic order, e.g. the ancestors of A are B1 and C.

The most common types of queries are the following:

- *A priori* probability distribution of a variable.

$$P(\mathbf{A} = \mathbf{a}) = \sum_{X_1} \ldots \sum_{X_k} P(x_1, \ldots, x_k, \mathbf{A} = \mathbf{a})$$

Where A is the query-variable and X_1 to X_k are the remaining variables of the network. This type of query can be used during the design phase of a tunnel for example to assess its probability of failure for the design conditions (geology, hydrology, etc).

- Posterior distribution of variables given evidence (observations). This query consists of updating the state of a variable (or subset of variables) given the observations (new information).

$$P(A = a \mid \mathbf{e}) = \frac{P(A = a, \mathbf{e})}{\sum_{X_1} \cdots \sum_{X_k} \sum_{A} P(x_1, \ldots, x_k, \mathbf{A} = \mathbf{a}, \mathbf{e})}$$

Where **e** is the vector of all the evidence, and A is the query variable and X_1 to X_k are the remaining variables of the network. This type of query is used to update the knowledge of the state of a variable (or variables) when other variables (the evidence variables) are observed. It could be used, for example, to update the probability of failure of a tunnel, after construction has started and new information regarding the geology crossed becomes known.

The most straightforward way to make inference in a Bayesian Network, if efficiency were not an issue, would be to use the equations above to compute the probability of every combination of values and then marginalize out the ones one needed to get a result. This is the simplest but the least efficient way to do inference. There are several algorithms for efficient inference in Bayesian Networks, and they can be grouped as follows: Exact inference methods and approximate inference methods. The most common exact inference method is the *Variable Elimination* algorithm that consists of eliminating (by integration or summation) the non-query, non-observed variables one by one by summing over their product. This approach takes into account and exploits the independence relationships between variables of the network.

Approximate inference algorithms are used when exact inference may be computationally expensive, such as in temporal models, where the structure of the network is very repetitive, or in highly connected networks.

4.4 *Learning algorithms*

Humans are normally better at providing structure than probabilities. Therefore, when possible, it is good to use data to obtain the conditional probability tables.

The structure is normally given by experts and the conditional probability tables can be estimated through available data. When there is a good amount of data available and not enough domain knowledge it is also possible to learn the network structure from data.

Learning is basically to search over a space of models to find the one that suits best. For this one has to define:

- The space of models
- Criteria or an objective function on models (i.e. What is the meaning of a model that suits one better).

One of the most common problems that one tries to solve in applying learning to BN is:

Density estimation. The idea is that the data were presumably generated according to some probability distribution $P_X(x)$. There is some process out in the world that is generating these data that are observed, and there is a joint probability distribution (of the data) $P_X(x)$. The goal is to estimate that probability distribution as well as one can, $P_X(x)$, i.e. as close to the reality as possible.

There are different versions of the problem of density estimation, which have to do with what is given. This can be:

1. **Parameter estimation**. One is given the variables and the structure of the model. The only thing left to do is parameter estimation, i.e. what are the probabilities that go into the probability tables.
2. **Structure learning**. One is given the variables only. In this case one will have to search over the space of possible structures as well as estimate the parameters.

The next sections will discuss these two types of learning. For more information on the other subjects see Pearl, J.,1988; Jordan, M., 1998; Jensen, F.V., 2001 and Cowell., R.G. et al., 2003.

A. Parameter estimation

Parameter estimation in a Bayesian Network is the task of estimating the values of the parameters of the conditional distributions for each node X, given *X*'s parents, from a data set (Jensen, 2001). The methods of parameter estimation can be grouped into main groups: maximum likelihood and Bayesian estimation.

A.1 Maximum likelihood

Maximum likelihood estimation (MLE) is statistical method used for fitting a statistical model to data, and provide estimates for the model's parameters. The principle of MLE is to find parameter values that make the observed data most likely (Kjaerulff, 2008; Jensen, 2001).

In order to illustrate the method of Maximum Likelihood in Bayesian Networks, consider the simple experiment of flipping a thumbtack. The outcomes of the experiment are heads or tails (see Figure 12). Let's say that one is ignorant about what one will get when flipping a thumbtack.

The simplest Bayesian network possible to illustrate this situation corresponds to a binomial variable, X where the values are either heads or tails, presented in Figure 13.

This BN has a probability table associated to it and it has only one parameter θ, the P (Heads). What would be a good way to estimate parameter θ, given some data D and some assumptions?

Imagine that the outcomes of the experiment are D = {x [1] = H, x [2] = T, x [3] = H, x [4] = H, x [5] = T}.

Assume that the elements of D are independent, i.e. that x [i] is independent of x [j] given θ and that θ does not change over time.

Unscientifically looking at the data one would say that a good estimator for θ could be as follows:

$$\hat{\theta} = 0.6 = \frac{number\ of\ heads}{number\ of\ tails + number\ of\ heads} \quad (8)$$

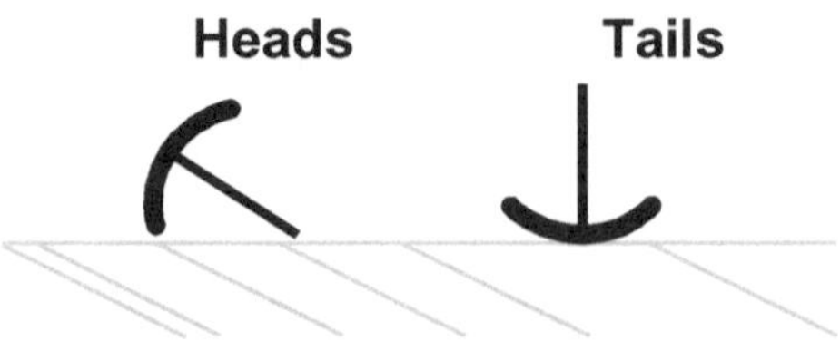

Figure 12. Thumbtack.

Figure 13. BN for thumbtack flipping.

This intuition of what θ should be is correct and it is possible to prove it mathematically, as will be shown next.

As mentioned before, the learning process is about defining a space of answers (models) and then deciding what makes an answer good, i.e. apply a criterion in order to determine which answer (model) is best. So the hypothesis space in this case is θ, a probability and therefore is going to be in the range [0, 1]. The criterion is to maximize the likelihood of the data given θ, i.e. find the model (θ) which makes the data as likely as possible.

Hypothesis space: θ ε [0, 1].

Criterion: Maximize likelihood of D. This is called the maximum likelihood criterion.

The likelihood of the data is the probability of getting the data assuming a value of θ. $L(D:\theta) = P(D:\theta)$[2].

Now applying this to the data D = {H T H H T}, what is the probability of getting that particular set of data (or the likelihood of the data)? It will be:

$$P(D:\theta) = \theta^3(1-\theta)^2$$

More generally if M_h and M_t are the number of heads and the number of tails observed:

$$P(D:\theta) = \theta^{M_h}(1-\theta)^{M_t}$$

Now what one needs to do is to find the θ that maximizes the Likelihood function. It is easier to take the log of the function before maximizing it:

$$Log(P(D:\theta)) = l(D:\theta) = M_h \log\theta + M_t \log(1-\theta)$$

The next step is to find the maximizing value of θ, by taking the derivative of $l(D:\theta)$ with respect to θ and setting it equal to zero.

$$\begin{aligned} &\frac{\partial l(D:\theta)}{\partial\theta} = 0 \\ &\Leftrightarrow \frac{M_h}{\theta} - \frac{M_t}{1-\theta} = 0 \\ &\Leftrightarrow M_h(1-\theta) - M_t\theta = 0 \\ &\Leftrightarrow M_h - M_h\theta - M_t\theta = 0 \\ &\Leftrightarrow \theta = \frac{M_h}{M_h + M_t} \end{aligned} \tag{9}$$

The intuitive result of Equation 8 has been mathematically proved to be correct by Equation 9.

The example considered is too simple with only one variable. In real problems, however, one is typically interested in looking for relationships among a large number of variables. In order to illustrate how this method can be applied to a case of more than one variable, consider the BN with known structure and 2 nodes presented in Figure 14, and assume the following:

- The existence of a data set D = $\{<v_1^1, v_2^1>, \ldots, <v_1^k, v_2^k>\}$
  Value of nodes in sample 1   Value of nodes in sample k
- The elements of D are independent given Model (M), i.e. x[i] is independent of x[j] given θ and that θ does not change over time.

---

2. The reason for having P (D : θ) instead of P (D | θ) is that θ in this model is not a random variable but a parameter.

P (X=0)	P (X=1)
θ_{X0}	θ_{X1}

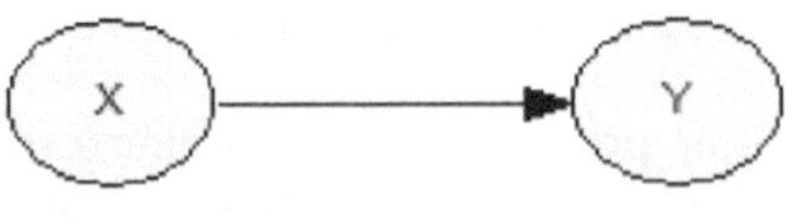

X	P (Y=0)	P (Y=1)
0	$\theta_{Y0\|X0}$	$\theta_{Y1\|X0}$
1	$\theta_{Y0\|X1}$	$\theta_{Y1\|X1}$

Figure 14. Bayesian network with two binary nodes.

The goal is to find the model M (in this case Conditional Probability tables) that maximizes the P (D|M), i.e. the probability of data occurring given the Model. This is known as the ***maximum likelihood*** model.

The parameters that one wants to determine are the probabilities in the probability tables associated with each node of the Bayesian network presented in Figure 14. The vector of parameters is the following:

$$\theta = <\theta_{x1}, \theta_{x0}, \theta_{y0|x0}, \theta_{y1|x0}, \theta_{y0|x1}, \theta_{y1|x1}>$$

Where,
$\theta_{x1} = P(X = 1)$
$\theta_{x0} = P(X = 0)$
$\theta_{y0|x0} = P(Y = 0 \mid X = 0)$
$\theta_{y1|x0} = P(Y = 1 \mid X = 0)$
$\theta_{y0|x1} = P(Y = 0 \mid X = 1)$
$\theta_{y1|x1} = P(Y = 1 \mid X = 1)$

The maximum likelihood function for these parameters given data set D is

$$L(D:\theta) = \prod_m P(X[m], Y[m]:\theta) = \prod_m P(X[m]:\theta)\, P(Y[m] \mid X[m]:\theta) \tag{10}$$

P (X [i]) only depends on θ_x and P(Y [i] | X [i]) only depends on $\theta_{y|x}$, Equation 10 can be simplified as follows:

$$\begin{aligned} L(D:\theta) &= \prod_m P(X[m]:\theta_X)\, P(Y[m] \mid X[m]:\theta_{Y|X}) \\ &= \prod_m P(X[m]:\theta_X) \prod_m P(Y[m] \mid X[m]:\theta_{Y|X}) \end{aligned}$$

This way one can choose θ_X to maximize $\Pi_m P(X[m]:\theta_X)$ and $\theta_{Y|X}$ to maximize $\Pi_m P(Y[m] \mid X[m]:\theta_{Y|X})$, independently. Note that the latter can be further decomposed as below:

$$\begin{aligned} \prod_m P(Y[m] \mid X[m]:\theta_{Y|X}) = &\prod_{m:X[m]=X0} P(Y[m] \mid X[m]:\theta_{Y|X0}) \\ &\times \prod_{m:X[m]=X1} P(Y[m] \mid X[m]:\theta_{Y|X1}) \end{aligned}$$

The final expression is:

$$L(D:\theta) = \prod_m P(X[m]:\theta_X) \prod_{m:X[m]=X0} P(Y[m] \mid X[m]:\theta_{Y|X0}) \prod_{m:X[m]=X1} P(Y[m] \mid X[m]:\theta_{Y|X1}) \tag{11}$$

Since it is a product of positive expressions, it can be maximized for each parameter separately and we do not need to make a joint optimization through all parameters. To make the maximization easier, normally one maximizes the log of the likelihood function.

$$\frac{\partial \log \prod_m P(X[m]:\theta_X) = \log(\theta_{X1}{}^{M[X1]} \times (1-\theta_{X1})^{M[X0]})}{\partial X} = 0$$

$$\Leftrightarrow \quad \frac{M[X1]\times \log(\theta_{X1}) + M[X0]\times \log(1-\theta_{X1})}{\partial X} = 0$$

$$\Rightarrow \quad \hat{\theta}_{X1} = \frac{M[X1]}{M[X1]+M[X0]}, \tag{12}$$

$$\Rightarrow \quad \hat{\theta}_{X0} = 1-\hat{\theta}_{X1} = \frac{M[X0]}{M[X1]+M[X0]}$$

Where M [Xi] are the counts of X = i (in this case i = 1 or 0)
The same way other parameters can be calculated:

$$\frac{\partial \log(\prod_{m:X[m]=X0} P(Y[m] \mid X[m]:\theta_{Y1|X0}))}{\partial X} = 0 \quad \Rightarrow \quad \hat{\theta}_{Y1|X0} = \frac{M[X0,Y1]}{M[X0]}$$

Note that this calculation is basically the same done to obtain Equation 12. So in a similar manner one will get the following results:

$$\hat{\theta}_{Y1|X1} = \frac{M[X1,Y1]}{M[X1]} \tag{13}$$

$$\hat{\theta}_{Y0|X0} = \frac{M[X0,Y0]}{M[X0]} \tag{14}$$

$$\hat{\theta}_{Y1|X1} = \frac{M[X1,Y1]}{M[X1]} \tag{15}$$

Based on Equation 10 to Equation 15 it is possible to conclude that the problem of learning in the case of several variables that are related, i.e. BN can be reduced mainly to the problem of learning one single variable. It is also possible to observe that the ML estimator is no more than the counts of the specific occurrence and dividing it by the number of all occurrences. Although very simple, this method has some shortcomings. A method that avoids some of the shortcomings of ML, and that enables one to include prior beliefs, is the Bayesian Estimation.

A.2 Bayesian estimation

In the Bayesian view, θ is the unknown value of a random variable θ, not a parameter like in ML. P (Θ = θ) is the prior probability distribution. If the parameter θ can be any value in the interval [0, 1], then P (Θ = θ) must be a continuous distribution between 0 and 1 and must integrate to 1. The beta distribution is a good candidate. This distribution is defined by two parameters α, β, such that:

$$P(\theta) = Beta(\theta \mid \alpha, \beta) = \frac{\Gamma(\alpha+\beta)}{\Gamma(\alpha)\Gamma(\beta)} \theta^{\alpha-1}(1-\theta)^{\beta-1} \tag{16}$$

where α, β > 0 are parameters of the beta distribution and Γ(·) is the Gamma function.

The beta distribution is convenient for several reasons. If Θ has a prior $Beta(\alpha,\beta)$ then after a data point is observed the posterior distribution of Θ is also a beta distribution. Imagine a random variable X that can take values 0 or 1. Suppose a data sample D is composed of only one observation, X = 0. Θ is a random variable that stands for the probability of X = 0 and its distribution varies between 0 and 1. Now assume that this prior distribution of Θ is a beta distribution with parameters α and β, $Beta(\alpha,\beta)$.

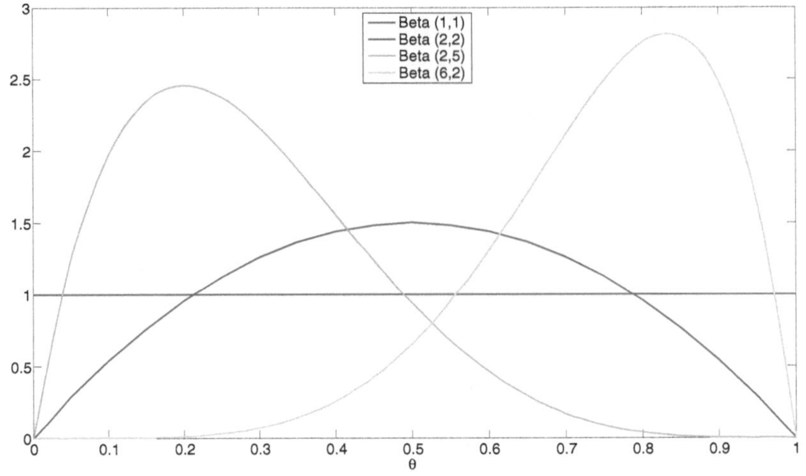

Figure 15. Shows how the beta distribution for different values of α and β.

The distribution a posteriori of Θ, after observing X = 0, will be equal to (applying Bayes' rule):

$$P(\theta | X=0) = \frac{P(X=0 | \theta) P(\theta)}{P(X=0)} \tag{17}$$

P(X = 0 | θ), which is equal to θ stands for the probability of X = 0 given the assumed model. P (θ) is $Beta(\alpha,\beta)$ distribution. Substituting P (θ) and P (X = 0 | θ), in Equation 17:

$$\begin{aligned} P(\theta | X=0) &= \alpha' \theta \, Beta(\alpha,\beta)(\theta) = \frac{\Gamma(\alpha+1+\beta)}{\Gamma(\alpha+1)\Gamma(\beta)} \theta \, \theta^{\alpha-1} (1-\theta)^{\beta-1} \\ P(\theta | X=0) &= \frac{\Gamma(\alpha+1+\beta)}{\Gamma(\alpha+1)\Gamma(\beta)} \theta^{\alpha} (1-\theta)^{\beta-1} = Beta(\alpha+1,\beta) \end{aligned} \tag{18}$$

The resulting probability distribution, i.e the posteriori distribution of P (θ) is also a Beta distribution with parameters (α + 1, β). So after observing X = 0 we have increased the parameter α by one. If X = 1 had been observed then the parameter β would have been increased by one (Remember that P (X = 1) = 1 – θ). Also the expectation of Θ with respect to the Beta distribution has a simple form:

$$\int \theta Beta(\theta | \alpha, \beta) d\theta = \frac{\alpha}{\alpha+\beta} \tag{19}$$

The problem one is interested in is to know what is the probability of X = 0 and/or X = 1 given the available data. To illustrate this problem imagine one is flipping some kind of biased coin and that one has a certain amount of observations. What one wants to know now is the probability of getting heads or tails the next time the coin is tossed. This is a problem of Bayesian updating that can be represented in a Bayesian Network (Figure 16).

One would like to estimate the probability that the next toss is heads, given what has been observed (given available data), i.e. $P(X[m+1] | D)$, where $D = X[1], X[2] \ldots\ldots X[m]$, are the available data.

To determine the probability that the next toss of the coin is heads, one averages over the possible values of θ (using the expansion rule of probability):

$$P(X[m+1] | D) = \int_0^1 P(X[m+1] | \theta, D) P(\theta | D). \; d\theta$$

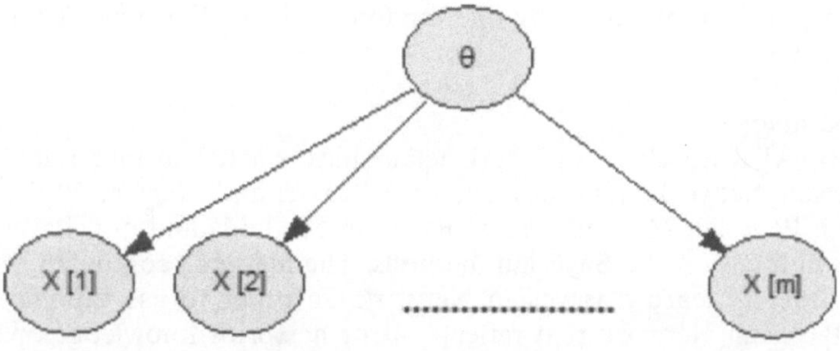

Figure 16. Bayesian Network model for estimating the parameter θ given the observed data.

since the elements of D are independent given θ:

$$P(X[m+1] \mid D) = \int_0^1 P(X[m+1] \mid \theta)\, P(\theta \mid D) \cdot d\theta$$

since the $P(X[m+1] \mid \theta) = \theta$, one will have:

$$P(X[m+1] \mid D) = \int_0^1 \theta\, P(\theta \mid D) \cdot d\theta \equiv E_{p(\theta|\mathbf{d})}(\theta)$$

Where $E_{p(\theta|\mathbf{d})}(\theta)$ is the expectation of θ with respect to the distribution $P(\theta \mid D)$. Applying Equation 19 one will get:

$$P(X[m+1] \mid D) = \alpha' \int_0^1 \theta^{M0} (1-\theta)^{M1}\, \theta\, d\theta = P(X[m+1] \mid D) = \frac{M_0 + \alpha}{M_0 + M_1 + (\alpha + \beta)} \tag{20}$$

where M_0 is the counts of X = 0 and M_1 is the counts of X = 1.

This is also called the Bayesian (or Laplace) correction. When using this correction, in the case M_0 and M_1 are equal to zero, i.e. if there are no observations the probability of the next toss given the available data (in this case none), and given a prior Beta (1,1) will be:

$$P(X \mid D) = \frac{M_0 + 1}{M_0 + M_1 + 2} = \frac{0+1}{0+0+2} = 0.5.$$

A prior of Beta (1, 1), i.e. uniform distribution, is called the "uninformed" prior, meaning that one believes that all values of θ between 0 and 1 have the same probability. As seen before the Beta distribution family provides a great range of priors.

The parameters α and β in the beta distribution can be seen has virtual counts. According to this idea, when the prior is equal to the uniform, βeta (1, 1), one is saying that our initial virtual count is one of each possible values. Basically one does not have strong beliefs and is saying the probabilities are P(X = 0) = P(X = 1) = 0.5. However it is possible that one has strong believes that the probabilities are P(X = 0) = P(X = 1) = 0.5, for example the probability that one get heads or tails when tossing a coin. In this case a probability distribution such as Beta (100, 100) is more adequate. In this case the "virtual counts" are 100 for each state, and what one is saying is that one has a strong belief that P(X = 0) = P(X = 1) = 0.5 (because one "virtually" observed 100 tails and 100 heads). For Networks like the one in Figure 20, the Bayesian prior must cover all parameters θ1, θ2, θ3, i.e. P(X), P (Y|X = 0), P (Y|X = 1), respectively. However P (θ1, θ2, θ3) = P (θ1) x P (θ2) x P (θ2), since we have assumed that the parameters are independent from each other. Based on this assumption, each parameter can be represented by one random variable. For more details please refer to Heckermen, 1997.

B. Structure learning
In recent years, AI researchers and statisticians have started to investigate methods for learning Bayesian Networks (Heckermen, 1997; Russel and Norvig, 2003). These methods range from Bayesian Methods, quasi-Bayesian Methods and non Bayesian methods. This section will focus on the Bayesian methods. The methods combine prior knowledge with data in order to learn a Bayesian Network. In order to use this method the user constructs a Bayesian Network that reflects his or her prior knowledge on the problem. This is called the prior network. The user will also need to assess her/his confidence on the prior network. Once the prior network has been determined, a *structure learning algorithm* will search for the "best" structure (including the respective conditional probability tables, which can be estimated using one of methods described in section A), i.e. the one that best fits the data.

Given a set of random variables the number of possible networks is well defined and finite. Unfortunately it grows exponentially with the number of variables. Although *Structure Learning* in Bayesian Networks is still a topic of research, there are several algorithms that have been developed and can be applied (Heckermen, 1997; Jensen, 2001, Russel and Norvig, 2003).

To specify a *structure learning algorithm* one must choose the following elements (the state space is known, i.e. the random variables are known):

- scoring function
- state transition operators
- search algorithm (for example A*, greedy hill-climbing, etc.).

Scoring functin (selection criterion)
The score of the network is used for model selection (i.e. some criterion is used to measure the degree to which a network structure fits the prior knowledge (if any) and the data. One of the most common criteria is the maximum likelihood (or log-likelihood). A penalty is normally introduced in order to account for overfitting (the most complex model is not always the most adequate). Figure 17 shows an example where using the most complex function is not the most adequate, i.e. one can use the higher polynomial curve to fit almost exactly the data points however this is clearly overfitting the data and will not fit correctly new data. In this case the 2nd order polynomial is the most adequate solution and not a more complex model.

There are different possible scoring criteria. A good scoring criterion is the so-called Bayesian score.

When trying to find the best graph, the probability of a certain graph structure given the data is what one wants to calculate. This is given by Bayes' Rule.

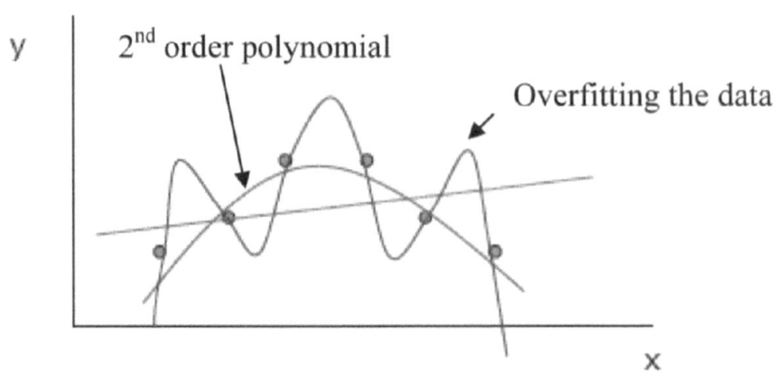

Figure 17. Example of overfitting.

$$P(G \mid D) = \frac{P(G,D)}{P(D)} = \frac{P(D \mid G)\,P(G)}{P(D)}$$

Since P (D) is just a normalizer and only depends on the data it can be ignored when comparing possible different graph structures. This way the numerator can be defined as the score for the graphs one wants to compare as follow:

$$S_G(G,D) = LogP(G,D) = LogP(D \mid G) + LogP(G)^3 \quad (21)$$

where,

$P(G)$ is the prior distribution on the graph structures, G, normally assumed to be a Dirichlet prior, and more specifically a uniform Dirichlet prior.

$P(D|G)$ is the marginal likelihood of the data given the structure which is equal to $P(D \mid G) = \int P(D \mid \theta_G, G) P(\theta_G \mid G)\, d\theta_G$.

The log marginal likelihood has the following interesting interpretation described by Dawid (1984). From the chain rule of probability,

$$P(D \mid G) = P(X[1])P(X[2] \mid X[1])...P(X[m] \mid X[1],...X[m-1])$$

Which looks like making successive predictions, i.e. approximately equal to the expected value of the P (X| G, D).

Imagine we have a Dirichlet prior, P (θ) ~ Dirichlet (α_h, α_t), $\alpha_h + \alpha_t = \alpha$ and the data are D = {H T T H T} $P(X[m] \mid X[1],...X[m-1]) = \frac{M_h^m + \alpha_h}{m+\alpha}$, where M_h^m is the number of heads up to element m in our data sequence.

$$P(D \mid G) = \int_{\theta_G} P(D \mid \theta_G, G) P(\theta_G \mid G)\, d\theta_G$$

$$P(D \mid G) = \frac{\alpha_h}{\alpha}\frac{\alpha_t}{\alpha+1}\frac{\alpha_t+1}{\alpha+2}\frac{\alpha_h+1}{\alpha+3}\frac{\alpha_t+2}{\alpha+4} = 0.017$$

For a P (θ) ~ Dirichlet (α_h, α_t) and D = {X [1]....X [M]}

$$P(D) = P(X[1],...X[m]) = \frac{(\alpha_h \cdot (\alpha_h+1)...(\alpha_h+M_h-1))(\alpha_t \cdot (\alpha_t+1)...(\alpha_t+M_t-1))}{\alpha \cdot (\alpha+1)....(\alpha+M-1)}$$

$$= \frac{\Gamma(\alpha)}{\Gamma(\alpha+M)}\frac{\Gamma(\alpha_h+M_h)}{\Gamma(\alpha_h)}\frac{\Gamma(\alpha_t+M_t)}{\Gamma(\alpha_t)}$$

Where $\Gamma(x) = (x-1)!$ for x integer and $\Gamma(x) = \int_0^\infty t^z e^{-t} dt = (z-1)\Gamma(z-1)$ for x real.

For a prior P (θ) ~ Dirichlet (α_1 … α_k) and D = {X [1]....X [M]}

$$P(D) = \frac{\Gamma(\sum_k \alpha_k)}{\Gamma(\sum_k(\alpha_k + M_k)}\frac{\Gamma(\alpha_h+M_h)}{\Gamma(\alpha_h)}\frac{\Gamma(\alpha_t+M_t)}{\Gamma(\alpha_t)}\prod_k \frac{\Gamma(\alpha_k+M_k)}{\Gamma(\alpha_k)} \quad (22)$$

One cannot assume that the data are independent because we do not know the parameters of the model that generated the data, because different network structures will entail different parameters. What one can say however is for a determined structure G, the data are independent given the parameters. This is the same case as the Bayesian parameter estimation described previously. Figure 17 illustrates this independence between data given the structure G and the parameters θ in the form of a Bayesian network.

3. The logs are just used to simplify the math.

As M goes to infinity (for Dirichlet priors) the log of margin likelihood can be approximated to the following equation:

$$\log P(D \mid G) \approx \log P(D \mid \hat{\theta}, G) - \frac{d}{2} \log M, \tag{23}$$

where d is the number of parameters in G, and $\hat{\theta}$ the estimator of θ.

This approximation is called Bayes Information Criterion (BIC) and was first derived by Schwarz (1978). In particular; Schwarz shows that Equation 4.42 for curved exponential models can be approximated using Laplace's method for integrals, yielding Equation 4.43.

The BIC approximation is interesting in several respects. First, it does not depend on the prior. Consequently, we can use the approximation without assessing a prior. Second, the approximation is very intuitive. It contains a term that measures how well the model predicts the data, $\log P(D \mid \hat{\theta}, G)$ and a term that penalizes the complexity of the model, $- d/2 \log M$.

Priors

To compute the relative posterior probability of a network structure, we must assess the structure prior P (G) and the parameter priors P (θ_G|G), unless we are using large-sample approximations such as BIC.

It has been assumed previously that the prior distributions on the parameters are Dirichlet distributions. A special case is assuming a uninformative prior with (α_i ....α_k) = (1,....,1). This is called the K2 metric. However this metric can lead sometimes to inconsistent results (Heckerman, 94).

In order to avoid inconsistencies one can use the so-called BDe prior. In this metric the user assigns a prior sample size M' and a prior distribution p' to the whole space. Then $\alpha_{X_i|Pa(X_i)} = M' p'(X_i \mid Pa(X_i))$. A very common choice for p' is the uniform distribution over the whole space. This particular case of the BDe metric is called the BDeu.

State transition operators

These are transition functions applied to the network structures to go between states (network structures) until reaching the final network structure, such as adding an arc, deleting an arc, or reversing the direction of an arc. They are basically used to transition from network to network during the search for the network with the best score.

For each state (each network structure), one takes the best guess regarding the parameters of that specific network structure given the data (for example using the maximum likelihood method) and determines the score of the network through a scoring function.

Figure 18 shows the typical transition operations (add, delete and reverse).

Search

Based on a defined scoring function that evaluates the "performance" of a certain Bayesian Network structure, the search is going to be reduced to a search for one or more structures that have high score. There are several search algorithms to perform the search. The most common are:

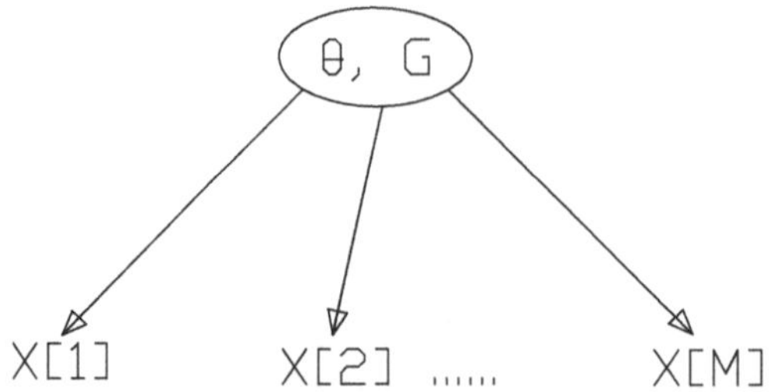

Figure 18. Bayesian Network describing the independence between data points given the structure and the parameters (G = BN structure; θ = conditional probability table parameters).

- Greedy Hill Climbing, which consists of considering every legal move (transition between network structures) and takes the one that yields the highest score.
- Random Hill Climbing, which consists of considering moves (transition between network structures) drawn at random and takes the one that yields the highest score.
- Thick-Thin Greedy Search, which consists of greedily[4] add single arcs until reaching a local maximum and then Prune back edges which don't contribute to the score.

These algorithms can be "stuck" in a local maximum and not return the optimal structure. There are several ways one can do to avoid this situation, such as random re-starts and simulating annealing, which consists of allowing the algorithm some "bad" moves but gradually decrease their size and frequency, in order to escape local maximum.

4.5 *Influence diagrams*

Bayesian networks can serve as a model of a part of the world, and the relations in the model reflect causal impact between events. However the reason we are building models is to use them when making decisions (i.e. the probabilities provided by the network are used to support some kind of decision making). Decision graphs or influence diagrams are an "extension" of Bayesian Networks. In addition to nodes for representing random variables, influence diagrams also provide node types for modeling alternatives and utilities. Besides chance nodes that denote random variables, and correspond to the only node type available in Bayesian networks, decision nodes are also modeled. A decision node indicates a decision facing the decision maker (similar to decision nodes in decision trees) and contains all alternatives available to the decision maker at that point. The third node type provided by these diagrams is the utility node. These nodes represent the utility function of the decision maker. In utility nodes, utilities are associated with each of the possible outcomes of the decision problem modeled by the influence diagram.

Directed links between nodes represent influences. Links between two chance nodes have the same semantics as in Bayesian networks. Other links in an influence diagram may also represent a temporal relation between the nodes involved. For example, a link from a decision node to a utility node not only indicates that the choice of action influences the utility, but also that the decision precedes the outcome in time.

Influence diagrams are useful in structuring a decision problem. While, for example, decision trees are more effective at presenting the details of a decision problem, influence diagrams more clearly show factors that influence a decision. Figure 19 illustrates a simplified scheme of an Influence Diagram. It is composed of two chance nodes ("Threat" and "Warning Device"), one decision node ("Decision") and a utility node ("Consequence"). In this specific example, the chance node "Threat" can represent the occurrence or not of a natural threat (for example a tsumani or a hurricane). The "warning device" chance node represents the fact that a warning alarm maybe issued or not. The decision node represents the decision between evacuation a population or do not evacuate. The utility node ("consequences") represents the consequences (expressed in utilities of the decision) in combination with the occurrence or not of the threat. The *warning device* issuing an alarm depends directly of the possibility of occurrence of the *threat*. The *decision* of evacuating or not evacuating the population will depend directly on the *warning device* issuing an alarm. Finally the consequences will depend on the decision taken and on whether or not the threat actually happens.

There are mainly four types of connections for structural influence in a decision graph. They are represented in Figure 20.

The first one (Figure 20a) is used when a Decision 1 affects the probabilities of event 1, i.e. Decision 1 is relevant for event 1. In Figure 20b the outcome of event 1 affects the probabilities of event 2, i.e. Event 1 is relevant for Event 2. This a typical Bayesian Network with no decision included. The type of connection in Figure 20c is used when Decision 1 occurs

4. Consider all the possible addition of arcs and choose the one that yields the best score.

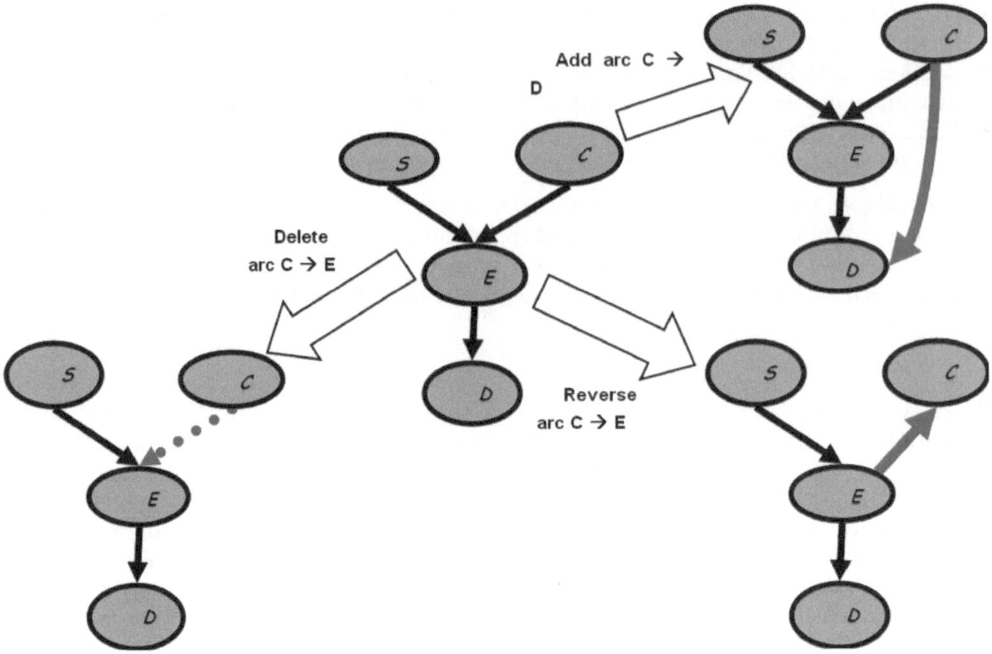

Figure 19. Typical transition operators (add, delete, reverse in white arrows) (Friedman and Goldszmidt, 1998)

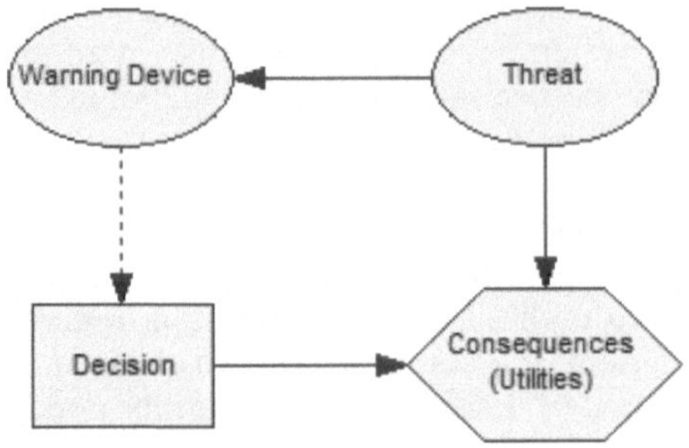

Figure 20. Influence diagram.

before Decision 2, i.e. Decisions 1 and 2 are sequential. Finally, Figure 20d represents a connection used when Decision 1 occurs after event 1. In this case the outcome of Event 1 is known when making Decision 1.

Besides the structural influences described in Figure 20, there are also value (utilities) influences such as the ones illustrated in Figure 21.

In Figure 21a) the value (or utility) depends on the (uncertain) event, for example a manufacturing cost depends on the (uncertain) availability of a certain input. In the second value influence (Figure 21b), a decision influences the value (or utility). For example a manager's decision influences the profit of a plant.

4.5.1 *Inference for influence diagrams*

The process of inference in an influence diagram consists of computing the expected utility associated with the different decisions or strategies. As in Bayesian networks there are two groups of algorithms that can be used to make inference in an influence diagram: exact and

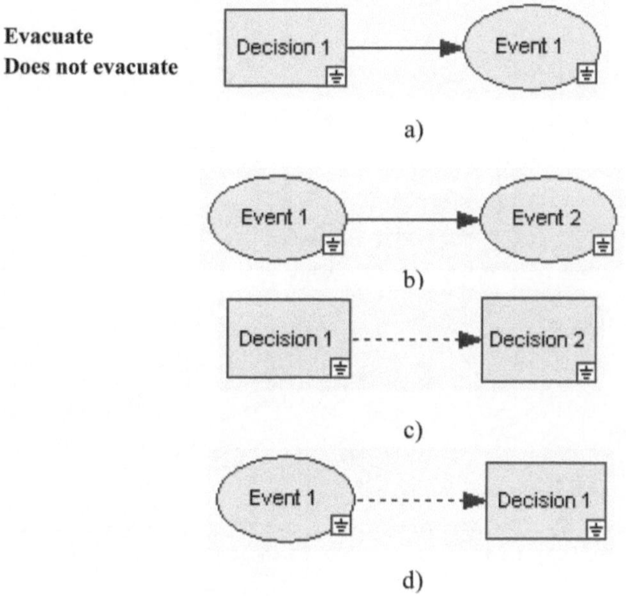

Figure 21. Influence diagram connections.

approximate. The most basic of way to solve an influence diagram is to unfold it into a decision tree and solve it. However if one wants to take advantage of the structure of an influence diagram and encoded conditional independences, one of the most common is the Variable Elimination algorithm for influence diagrams which has many similarities to the Variable Elimination technique described for Bayesian Networks. For more details reference is made to Jordan, M., 1998; Jensen, 2001.

4.5.2 *Examples*

4.5.2.1 *Bayesian network example (Faber, 2005)*

Figure 22 presents an example by Faber, 2005, regarding a problem related with the failure of a power supply system. The system is composed of an engine, a main fuel supply for the engine and electrical cables distributing the power to the consumers. In addition a backup fuel support a reserve fuel support with limited capacity was installed. The power supply system fails if the consumer is cut off from the power supply. This will happen if either the power supply cables fail or the engine stops, which is assumed to occur only if the fuel supply to the engine fails. All probabilities given in the network are for simplicity assumed to be annual probabilities.

Executing the Bayesian network will provide the probability structure for the different states of the system as illustrated in Figure 22.

One of the advantages of Bayesian Networks is to the ability to provide diagnostics support. By entering evidence into the Bayesian Network that the power supply has failed one, by conditioning the state of power supply on Fail we can back propagate (explaining away) this information in the BN and assess the most probable cause of the failure. The result is illustrated in Figure 24. It can be seen that the most probable cause of power supply failure is failure of the power distribution cables.

4.5.2.2 *Decision graph example (Einstein and Sousa, 2006)*

In this example a decision analysis problem with Bayesian network extended to decision graphs to include decision nodes and utilities will be illustrated in this example. Imagine a threat, such as natural hazard for example (landslide, hurricane, etc). For this example three different actions are considered: Passive countermeasure, Active countermeasure, No action.

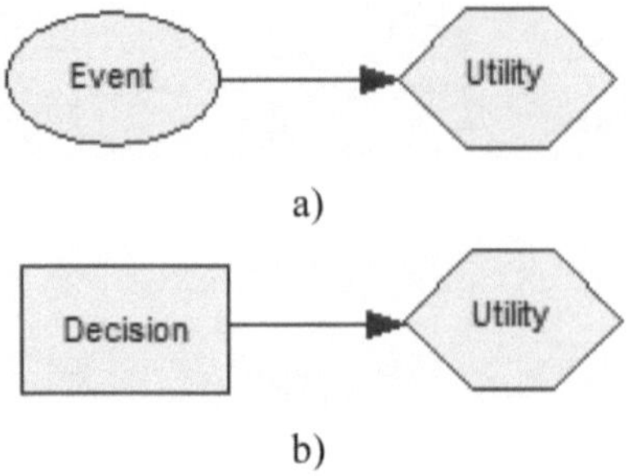

Figure 22. Value influence.

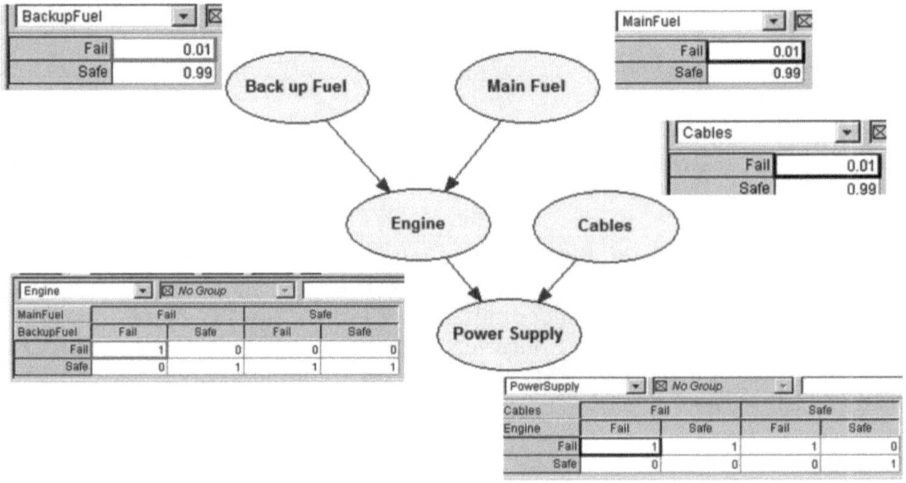

Figure 23. BN example from Faber, 2005.

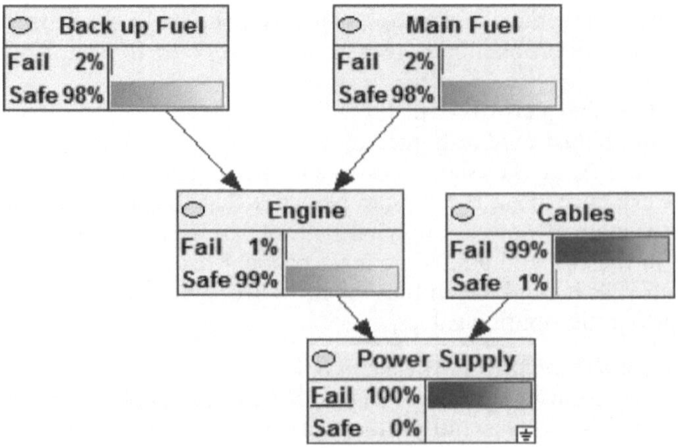

Figure 24. BN from Figure 23 executed.

If the threat materializes Damage may occur with different levels, which will depend on the weather passive, active countermeasure or no action are chosen. Associated to the damage levels are utilities and associated to the different measures are different costs. Figure 25 presents the decision graph that illustrates the problem.

The network has two chance nodes, one decision node and two utility nodes. The chance node *Threat* represents the probability of the threat happening. The chance node Damage represents the vulnerability, i.e. the fact that even if the threat materializes the consequences (Damage Levels) are uncertain. The Decision node represents the different options available: No Action, Passive countermeasure and Active countermeasure. The utility node *Utilities* represents the utilities associated with the different damage levels. The utility node *Cost* represents the costs associated with the different decisions.

A passive countermeasure will reduce the risk by reducing the vulnerability, meaning it the will reduce the probably that damage occur given the occurrence of the threat. A passive countermeasure is for example an evacuation during hurricane threat, as can be seen here in the figure. Countermeasure can be successful or not. An active countermeasure will reduce the risk by reducing probability of the threat. An example of active countermeasures are slope stability methods.

Solving the BN one will get that the active countermeasures are the preferred option in this case (Figure 26).

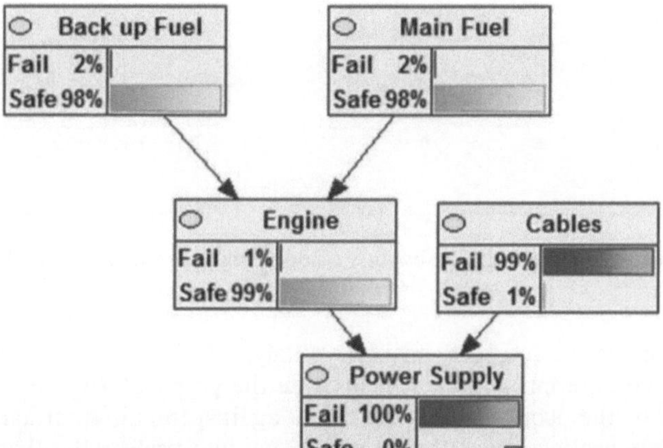

Figure 25. BN from Figure 23 executed after introducing evidence that power supply has failed.

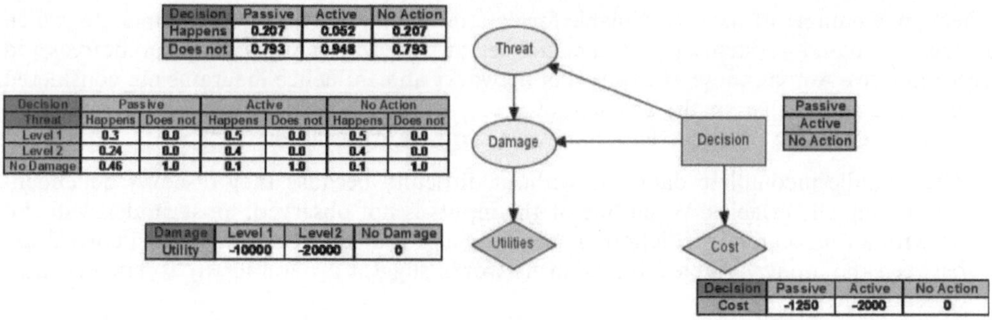

Figure 26. Decision graph (Influence diagram) for threat (Einstein and Sousa, 2006).

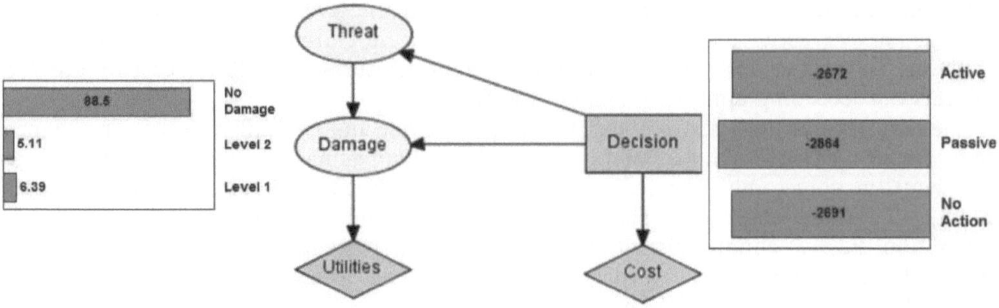

Figure 27. BN of Figure 26 executed (Einstein and Sousa, 2006).

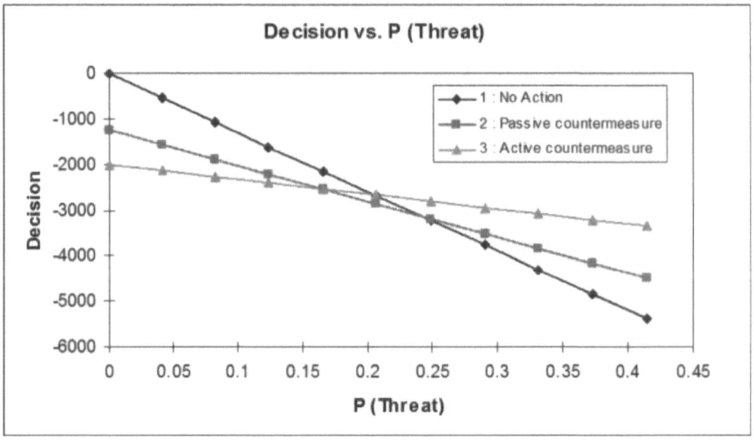

Figure 28. Sensitivity analysis on the probability of occurrence of the threat, P (Threat).

It is always beneficial to conduct sensitivity analysis to investigate the effect of different variables against the different actions. In this case the graph of Figure 27 shows the effect of the Probability of the occurrence of the threat against the different actions. As expected for low values failure probability of the threat No action is preferred, otherwise is the Active Countermeasure.

5 CONCLUSIONS

There are a number of models available for data representation and decision making, which include rule based—systems, artificial neural networks, Fuzzy-rules, fault and event trees and decision trees. Among these, the Bayesian networks and Influence diagrams are considered to be the most suitable for the problem of accidents during tunnel construction. The main reasons for choosing BN representation over the others are as follows:

- They handle incomplete data sets without difficulty because they discover dependencies among all variables. When one of the inputs is not observed, most models will end up with an inaccurate prediction. That is because they do not calculate the correlation between the input variables. Bayesian networks suggest a natural way to encode these dependencies.
- One can learn about causal relationships by using Bayesian networks. There are two important reasons to learn about causal relationships. This is worthwhile when one would like

to understand the problem domain, for instance, during exploratory data analysis. Additionally, in the presence of intervention, one can make predictions with the knowledge of causal relationships.

- Bayesian networks facilitate the combination of domain knowledge and data. Prior or domain knowledge is crucially important if one performs a real-world analysis; in particular, when data are inadequate or expensive. The encoding of causal prior knowledge is straightforward because Bayesian networks have causal semantics. Additionally, Bayesian networks encode the strength of causal relationships with probabilities. Therefore, prior knowledge and data can be put together with well-studied techniques from Bayesian statistics.
- Bayesian methods provide an efficient approach to avoid the over-fitting of data. Models can be "smoothed" in such a way that all available data can be used for training by using Bayesian approach.

In spite their potential to address inferential processes, there are however some limitations to Bayesian Networks:

- Depending on their size they may require initial knowledge of many probabilities. The results are very sensitive to the quality and extent of the prior knowledge, i.e. a Bayesian network is only as useful as this prior knowledge is reliable.
- Performing exact inference on a Bayesian network, as well as learning Bayesian networks from large amounts of data, can have a significant computational cost, since they are a NP hard tasks[5]. Approximate algorithms can be used in these situations.
- The restriction of the Bayesian Network to be acyclic can be an issue when modeling problems where feedback loops are common features.

Despite their limitations, from all methods, BN is the one with the ability to best represent problems in a complex domain of inherent probability and to provide project managers/designers/contractor with good understanding of the problem.

REFERENCES

Ang, A. H-S. & Tang, W.H. (1975). "Probability Concepts in Engineering Planning and Design", vol. 1, Basic Principles. New York, NY: John Wiley & Sons.

Baecher, G.B. (1972). "Site Exploration: A probabilistic approach". PhD Massachusetts Institute of Technology, Cambridge, MA.

Cowell., R.G., Dawid, A.P., Lauritzen, S.L. & Spiegelhalter, D.J. (2003). "Probabilistic Networks and Expert Systems", (Information Science and Statistics, 2003).

Darlington, K. (2000). "The essence of expert systems". Ed. Prentice Hall. 167 pages.

Dempster, A.P. (1968). "A generalization of Bayesian inference". Journal of the Royal Statistical Society, Series B 30 205–247.

Eskesen, S., Tengborg, P., Kampmann, J. & Veicherts, T. (2004). "Guidelines for tunnelling risk management": International Tunnelling Association, Working Group No. 2. Tunnelling and Underground Space Technology. Volume 19, Issue 3, May 2004, pages 217–237.

Faber, M.H. (2005). "Risk and Safety in Civil", Surveying and Environmental Engineering. Lecture Notes. Swiss Federal Institute of Technology, ETHZ, Switzerland. 394p.

Friedman, N. & Goldszmidt, M. (1998). Learning Bayesian Networks from Data. Presentation.

Jensen, F.V. (1996). "Introduction to Bayesian Networks". Taylor and Francis, London, UK, Springler-Verlag, New York, USA, 1996.

Jensen, F.V. (2001). "Bayesian Networks and Decision Graphs". Taylor and Francis, London, UK, Springler-Verlag, New York, USA, 2001.

Jordan, M. (1998). "Learning in Graphical Models", (MIT Press 1998).

Heckerman, D. (1997). "A Tutorial on Learning with Bayesian Networks". Data Mining and Knowledge Discovery, 1:79–119.

Einstein, H.H. & Sousa, R. (2006). Warning System for Natural Threats. ECI Conference on Geohazards, Lillehammer, Norway. Volume P7.

Henrion, M., Breese, J. & Horvitz, E. (1991). "Decision Analysis and Expert systems". AI Magazine Volume 12 Number 4.

Howard, R.A. & Matheson, J.E. (editors) (1984). "Readings on the Principles and Applications of Decision Analysis", 2 volumes, Menlo Park CA: Strategic Decisions Group.

Howard, R.A. (1966). "Decision Analysis: Applied Decision Theory," Proceedings of the 4th International Conference on Operational Research (1966) 55–77.

Kjaerulff, U. & Madsen, A. (2008). "Bayesian Networks and Influence Diagrams. A Guide to Constructiona and Analysis". Springer Ed.

Melle, W.V., Shortliffe, E.H. & Buchanan, B.G. (1981). "Rule-Based Expert Systems. The MYCIN Experiments of the Stanford Heuristic Programming Project". Technical Report Stanford School of Medicine. Published in 1981.

Mehrotra, K., Mohan, C.K. & Ranka, S. (1997). "Elements of Artificial Neural Networks". MIT Press, 344 pages.

Pearl, J. (1988). "Probabilistic Reasoning in Intelligent Systems: Networks of Plausible Inference", (San Mateo, CA: Morgan Kaufmann, 1988).

Russell, S. & Norvig, P. (2003) "Artificial Intelligence. A Modern Approach". Second Edition. Prentice Hall Series in Artificial Intelligence.

Shortliffe, E.H. (1976). "Computer based medical consultations: MYCIN", American Elsevier, 1976.

Silva, C. (2001). "Safety control in railway tunnels. Development of support methodologies and a knowledge based system" (in Portuguese). University of Porto, MSc Thesis, Porto, 267p.

Sonmez, H., Gokceoglu, C. & Ulusay, R. (2003). "An application of fuzzy sets to the Geological Strength Index (GSI) system used in rock engineering". Engineering Applications of Artificial Intelligence. Volume 16, Issue 3, April 2003, Pages 251–269.

Sousa, R.L., Sousa, L.R. & Silva, C. (2007). "Maintenance of tunnels and the use of AI techniques". 11th ISRM Congress, Specialized Session S5 on Maintenance and Repair of Underground Structures in Rock Masses, Lisbon, 6p.

Suwansawat & Einstein (2006). "Artificial neural networks for predicting the maximum surface settlement caused by EPB shield tunneling". Tunnelling and. Undergound. Space Technology. 21 (2) (2006), pp. 133–150.

Suwansawat, S. (2002). "Earth pressure balance (EPB) shield tunneling in Bangkok: ground response and prediction of surface settlements using artificial neural networks". PhD Thesis. Massachusetts Institute of Technology. Dept. of Civil and Environmental Engineering.

Zadeh, L.A. (1965). "Fuzzy sets", Information and Control 8 (3): 338–353.

Zadeh, Lotfi (1999). "Fuzzy Sets as the Basis for a Theory of Possibility". Fuzzy Sets and Systems 1:3–28, 1978. (Reprinted in Fuzzy Sets and Systems 100 (Supplement): 9–34, 1999.

Von Neumann, J. & Morgenstern, O. (1944). "Theory of Games and Economic Behavior". Princeton University Press. 625pp.

CO_2 Storage in Carboniferous Formations and Abandoned Coal Mines –
He, Ribeiro E Sousa, Elsworth & Vargas Jr. (eds)
 ISBN 978-0-415-62079-6

Risk associated to storage of CO_2 in carboniferous formations. Application of Bayesian networks

L. Ribeiro e Sousa
State Key Laboratory for GeoMechanics and Deep Underground Engineering, Beijing, China
University of Porto, Portugal

R. Leal e Sousa
Department of Civil and Environmental Engineering, Massachusetts Institute of Technology, Cambridge, USA

ABSTRACT: In this publication risk assessment involved in storage of CO_2 is presented, with a particular focus on cases where the injection is made into unminable coal seams and in abandoned coal mines. The risks associate with earlier stages of storage and during storage is analyzed in detail and different types of hazard scenarios are described. The importance of monitoring is emphasized in order to prevent risks. Some applications concerning the Risk Assessment of CO_2 injection processes and storage in carboniferous formations and contamination of aquifers by CO_2 are presented and analyzed using Bayesian Networks for different hazard scenarios. Finally, based on the applications of Bayesian Networks several conclusions were made.

1 GENERAL

Storage of CO_2 in deep, onshore and offshore geological formations, uses many of the technologies developed by oil and gas industry and has been proved to be economically feasible under specific conditions in oil and gas fields and saline formations (IPCC, 2005). CO_2 can also be stored in carboniferous formations, either in unminable coal seams or in abandoned coal mines. CO_2 should be safely injected and stored at well characterized and properly managed sites. Injecting carbon dioxide in deeply geological formations can be stored in underground for long periods of time. At depths below about 800–1000 m, CO_2 has a liquid-like density that permits the potential for an efficient use of the underground reservoirs in porous of sedimentary rocks.

Figure 1 illustrates the options for storing CO_2 in deep underground geological formations. Other geological options which may serve as storage sites include caverns in basalt, organic-rich shales and in salt.

Geological storage requires constructing facilities to capture large emission sources of CO_2 such as power plants for electricity production or cement, steel, ethanol, among others. It is also possible to act on fuel, that is, instead of proceeding to the capture of CO_2 after combustion of coal or natural gas can be drawn primarily engaged in industrial units, the carbon present in them. The captured CO_2 is then transported by pipelines or in ships, for underground storage sites. Most of the mechanisms related to this technology are not new, since they are already employed by the oil industry by contracting for management and distribution of natural gas, for some industries in the food sector, among others.

The environments impacts from geological storage of CO_2 can be integrated into two types of categories, i.e. local environmental effects and global effects from the release of stored CO_2 to the atmosphere. Global effects may be viewed as the uncertainty in the effectiveness of CO_2 storage. Local hazards arise from causes like direct effects of elevated gas-phase CO_2 concentrations in the shallow surface or near surface, effects of dissolved CO_2 on groundwater and effects caused by displacement of fluids by the injected CO_2.

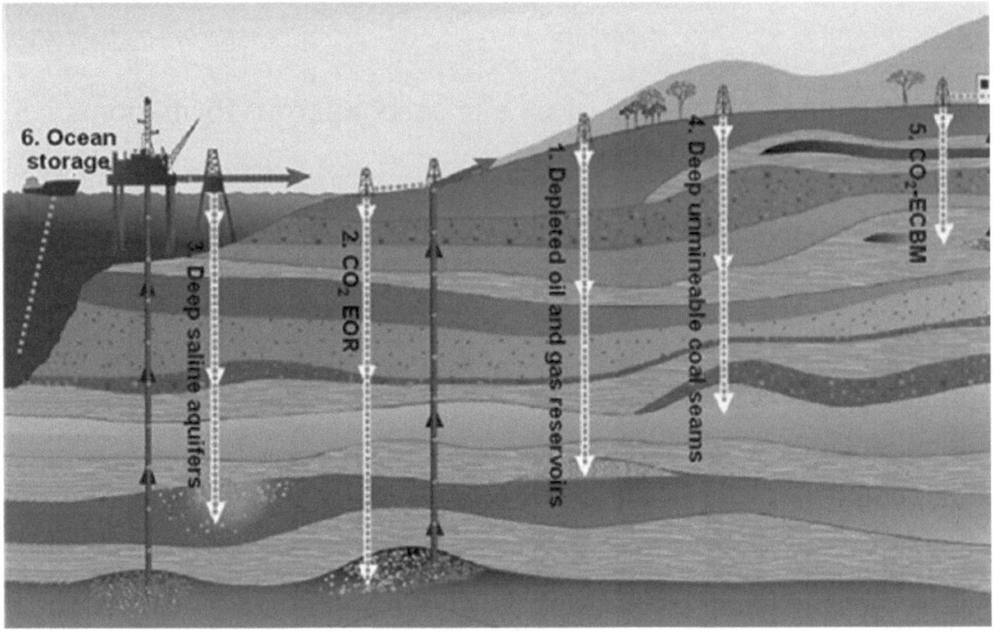

Figure 1. Options for storing CO_2 in deep underground geological formations (Adapt. from IPCC 2005).

There are different potential escape routes for CO_2 injected into geological formations exist. Risk assessment should be an integral element of risk management activities, like site selection, site characterization, storage system, design, monitoring and if necessary remediation. The methodology to be developed for Risk Assessment will be developed using Bayesian Networks (BN). BN are a graphical representation of knowledge for reasoning under uncertainty and they have become a popular representation for encoding uncertain expert knowledge in expert system. BN can be used at any stage of a Risk Analysis, and provide a good tool for decision analysis, including prior analysis, posterior analysis and pre-posterior analysis. Furthermore, they can be extended to influence diagrams, including decision and utility nodes in order to explicitly model a decision problem (Sousa, 2011a).

In 2010 the State Key Laboratory of GeoMechanics and Deep Underground Engineering from China University of Mining and Technology, Beijing, was awarded by a project on the field of Risk Assessment of CO_2 injection and sequestration in carboniferous reservoirs by the State Administration of Foreign Experts Affairs, from China. The importance of the project is based on the fact that China is the major producer of coal in the world. Therefore, there are several possibilities for selecting appropriate sites for the reservoirs, even in abandoned coal mines. Coal formations contain cleats that impart some permeability to the system. Between cleats, coal has a large number of micropores into which gas molecules can diffuse and be tightly absorbed. Gaseous CO_2 injected through wells will flow through the cleat system, diffuse in the coal matrix and be absorbed onto the coal micropore surfaces. If CO_2 is injected into coal seams it can displace gas methane enhancing coal bed methane recovery.

This chapter was developed in the framework of the referred project. It presents in section 2 technical aspects of injection and safety storage. In section 3 the Risk Assessment and Risk concepts are referred. After reviewing and discussing known hazards for carbon sequestration, section 4 analyze preventing risks by monitoring. Applications of BN are outlined for different hazard scenarios in section 5. Finally, some conclusions are drawn and presented in Section 6.

2 INJECTION AND SAFETY STORAGE

2.1 *Introduction*

CO_2 is a common constituent of the atmosphere, non-toxic. However, high concentrations can be dangerous (IEA, 2002). An uncontrolled release of CO_2 from an underground reservoir will not have long term effects, as can happen in cases of highly toxic or nuclear waste, once the CO_2 is diluted in air or water. Thus, slow migration of gas toward the surface, are not a direct threat to humans. However, high concentrations can be attained by a sudden release or by other process. Due to the higher density of CO_2 compared to air in case of leakage of large volumes can be created depressions or enclosures near the earth's surface, causing loss of consciousness or asphyxiation to humans who are in the vicinity (Piessens and Dusar, 2003).

The main risks of geological storage of CO_2 vary from place to place, given the heavy dependence on such factors as (IPCC, 2005; IEA, 2008; Vargas et al., 2011):

- the configuration of the storage facility, including the geological characteristics of the stratum selected;
- the heterogeneity of the sealing cap-rock;
- the heterogeneity of the mass taken as a whole (stratigraphic heterogeneity, existence of discontinuities, etc.);
- knowledge of the existence of injection/pumping wells abandoned nearby;
- the adequacy of the injection system;
- changing biogeochemistry;
- geomechanics weathering (generation of cracks and fractures);
- methods of abandonment of the wells when the reservoir reached the limit.

Duguid et al. (2007) suggest as one of the first requirements to be met by a site candidate for the reservoir, is to have several layers of sealing, so that the system is redundant and it is possible to can make early detection of potential problems. The latter argument is easily understood by observing Figure 2, which shows a measuring instrument in a layer above the first level of protection. If CO_2 escape, the system gives an indication to the authorities, and while the problem is not resolved the secondary layers of protection are in charge of retaining leakage.

In accordance to the publication of IPCC (2005), commercial projects of CO_2 storage in large scale should be adopted if it is assumed that the location is well chosen, designed, operated and monitored properly. The data available from existing projects suggest that is very likely that the fraction of stored CO_2 that is trapped in the first 100 years, is over 99%; and it is likely that the fraction of stored CO_2 that is trapped in the first 1000 years is over 99%.

2.2 *Risks associated with earlier stages to storage*

In this section the factors that should be taken into account in assessing the risks in CO_2 storage projects in coal formations are presented and evaluated.

The various stages leading up to the store itself, cause changes in the state of stress and strain, which, in turn, may generate flow paths through which the CO_2 can escape due to the discontinuities (pre-existing or not), such as faults or other fractures. Associated with the existence of faults may occur seismic episodes, which can also bring more risk to the CCS project.

To be able to understand the influence that the entire storage system has in the rock mass, it is necessary to study each phase separately. If there is sequestration of CO_2 in a given formation and with methane gas production, distinct phases may be considered (Myer, 2003):

- drilling and completion of wells;
- formation dewatering and methane production;
- CO_2 injection with or without secondary production of methane.

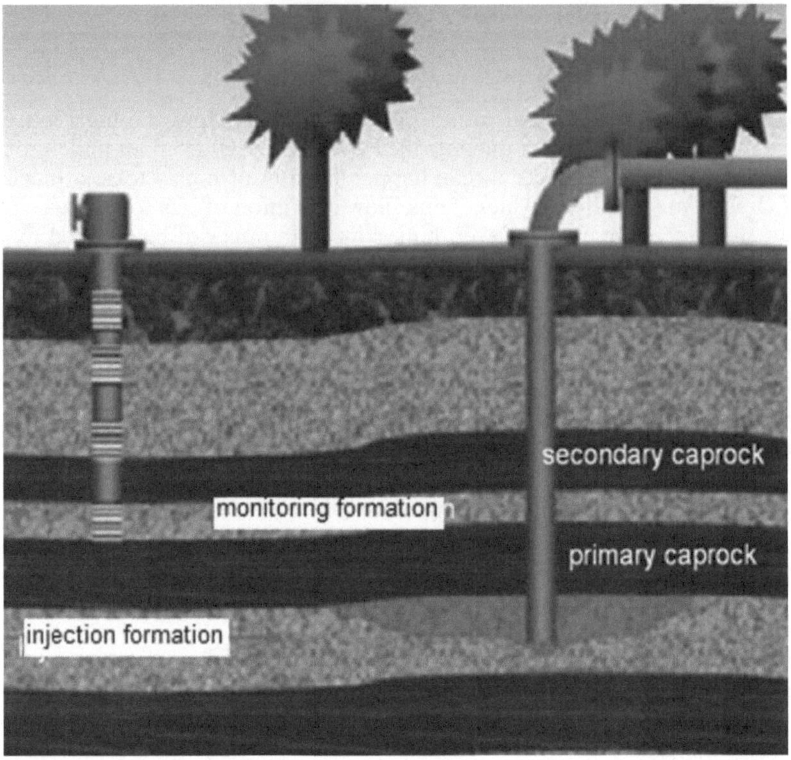

Figure 2. Reservoir with multiple layers of sealing (Adapted from Duguid et al., 2007).

If the storage of CO_2 is expected to be held in unminable coal seams, never considered by the mining industry (due to great depths, the lack of profitability of the project, or to poor safety conditions for workers), it is necessary to carry out wells with withdrawal of water, and possibly advantageous to the extraction of methane adsorbed on coal if intended to store CO_2 during drilling. In abandoned coal mines, the production wells may no longer be a crucial stage of analysis. The water withdrawal in principle also no longer decisive in the analysis, once the mass has been disturbed and it will be a new balance with all the galleries inside.

Wellbore stability is a geomechanical problem which can be encountered during drilling. Rock failure and the displacements associated with wellbore instability generate potential leakage paths. These drilling issues and the main causes of instabilities are analyzed in detail in the publication of Myer (2003). The risk of leakage will be minimized by cementing the case. Two constructive methods are conventionally used in the execution of wells: Cased hole wells and openhole cavity wells. Figure 3 illustrates schematic diagrams for the coalbed methane wells. The risks associated to the two methods are analyzed in detail in the publications of Myer (2003) and Gomes (2010).

Injection of CO_2 for enhanced methane production and sequestration will increase pore pressures in the coal seam. If pore pressures exceeded pre-development levels there is a risk that slips will occur. This is conceptually illustrated in Figure 4.

The causes for geomechanical problems, their consequences, the risks and their factors are summarized in the Figures 5 to 8, respectively for wells totally cemented, wells partially cemented with cavities, formation dewatering and methane production and for CO_2 injection with or without methane production.

2.3 *Risks associated to the storage*

The geological storage of CO_2 means that the CO_2 will be retained for hundreds or thousands of years. Therefore, it is necessary to carefully evaluate all potential escape mechanisms.

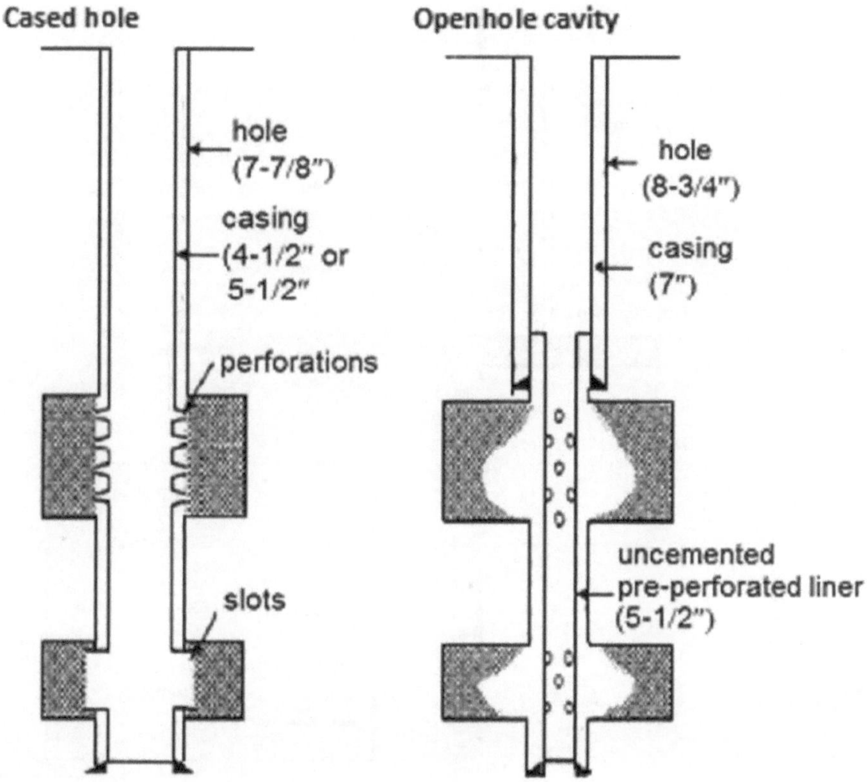

Figure 3. Schematic diagrams of core hole and cavity completions for coalbed methane wells (Myer, 2003).

The mechanisms that may occur in unminable coal seams and takes place in abandoned mines are presented. In terms of risk, abandoned mines require major rehabilitation work, checking the conditions for sealing of wells and shafts, and the removal of all materials that might react with carbon dioxide. The existence of wells abandoned or not, in the vicinity of the reservoir is an important issue to be analyzed in terms of safety. Figures 9 and 10 make a summary of some possible leakage of CO_2.

In general, the CO_2 retained by adsorption on the surface of coal remain in the deposit, even without caprocks, unless the pressure in the coal mine is reduced through mining. If the pressure drops suddenly, any excess of CO_2 from coal can flow freely through one of the mechanisms described previously in the Figures. Therefore it is necessary to ensure that after storing CO_2 the coal is never mined (Gomes, 2010).

In Figure 11 a scheme of storing CO_2 in carboniferous formations with Enhanced Coal-Bed Methane Recovery (ECBMR) is presented.

An abandoned coal mine when used as a reservoir can be seen as a very long gallery, with curves, as shown in Figure 12 (Piessens and Dusar, 2003). The storage capacity is much greater than in the case of unminable coal seams. This is due partly to the large area of contact between coal and CO_2 which enhances the adsorption phenomenon, but mainly also because of large void volume constituted by the massive volume mined. The use of abandoned coal mines for CO_2 storage is consequently a good option, particularly in China (He, 2011), where the number of abandoned mines is relatively large.

Moreover, the complex geometry of a coal mine can also be translated by a sealed container vertical upwards, according to an idealization of Piessens and Dusar (2003), (Figure 13). In a coal mine CO_2 can be stored in the voids, in dissolution in water at the mine, or adsorbed on the coal matrix. However, coal mines suitable for CO_2 sequestration should not be flooded.

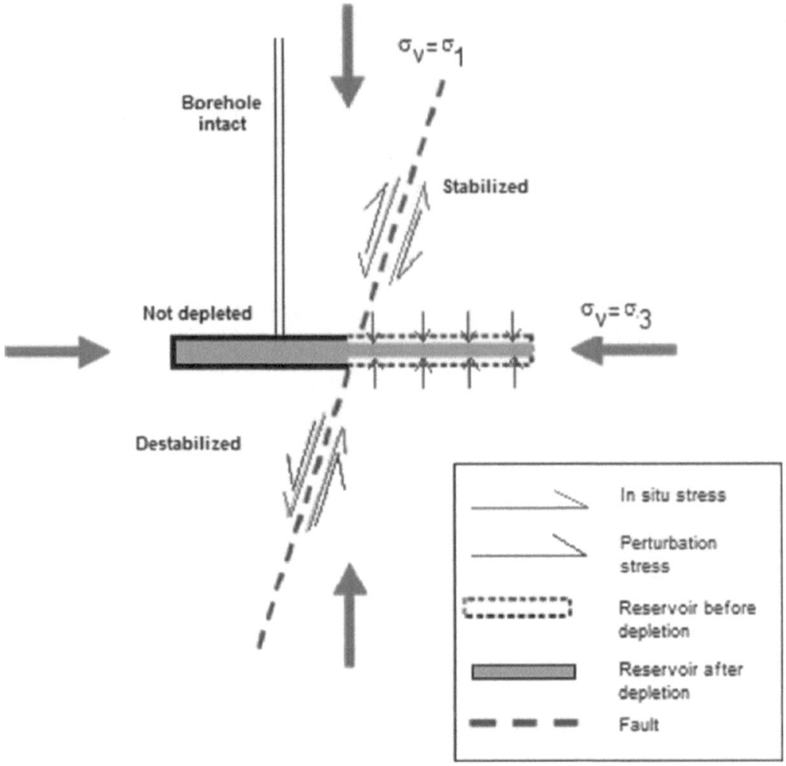

Figure 4. Displacements in the fault of a reservoir (Adapted from Desroches et al., 2006).

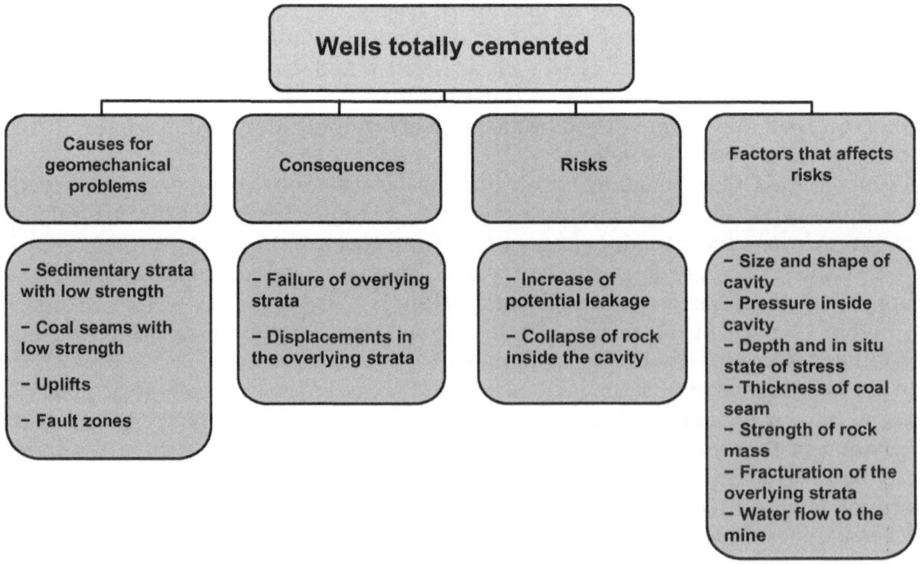

Figure 5. Wells totally cemented—diagrams with geomechanical problems, consequences, risks and factors.

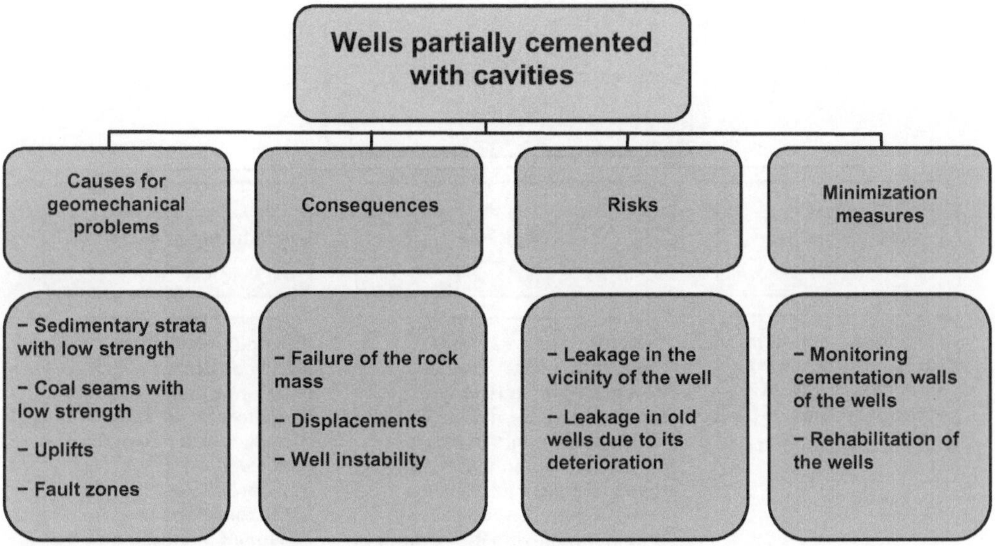

Figure 6. Wells partially cemented with cavities—diagrams with geomechanical problems, consequences, risks and minimization measures.

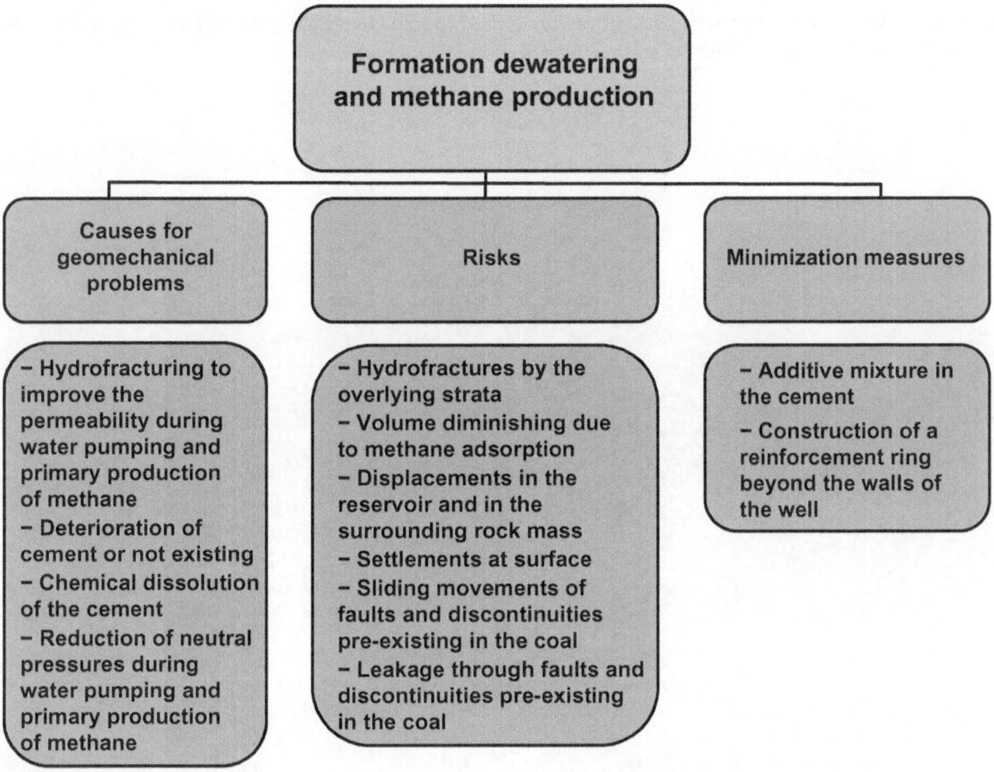

Figure 7. Wells partially cemented with cavities—diagrams with geomechanical problems, consequences, risks and minimization measures.

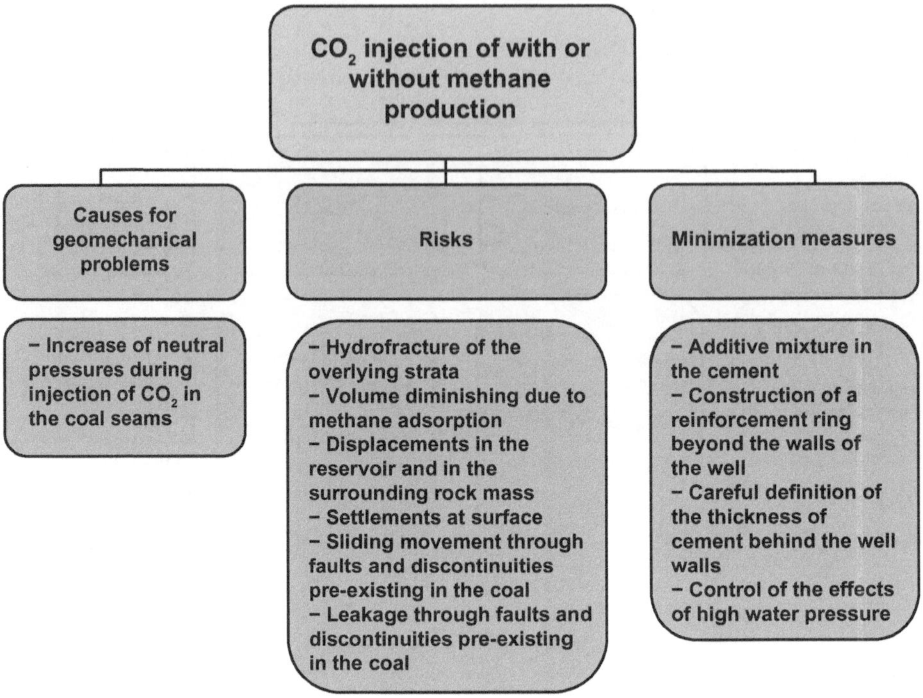

Figure 8. CO_2 injection with or without methane production—diagrams with causes for geomechanical problems, risks and minimization measures.

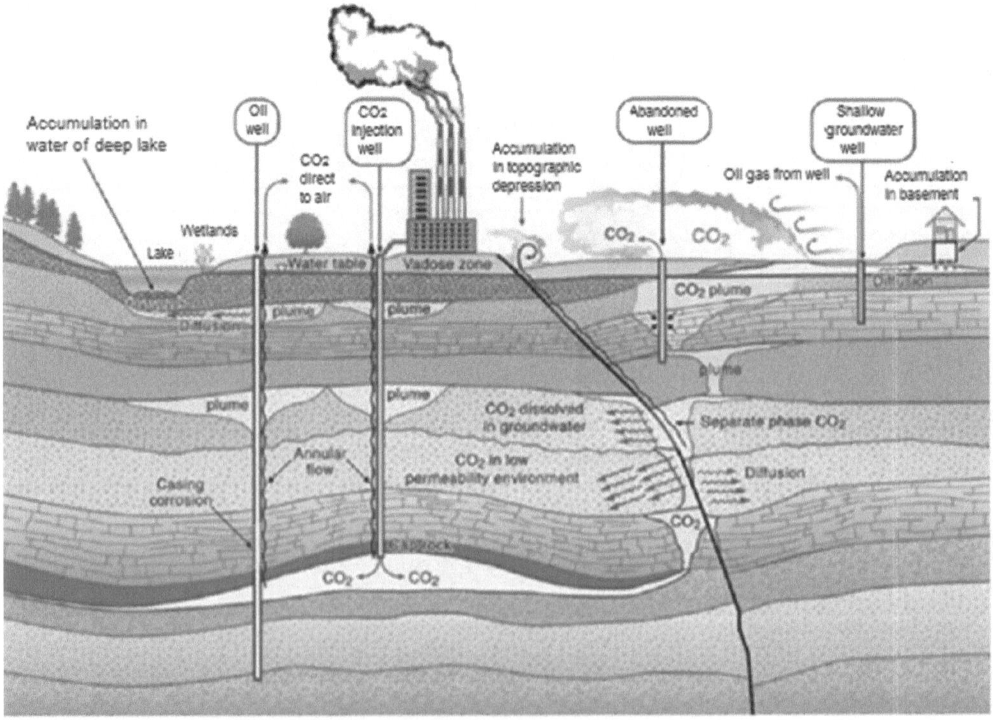

Figure 9. Potencial escape mechanisms (Benson, 2005).

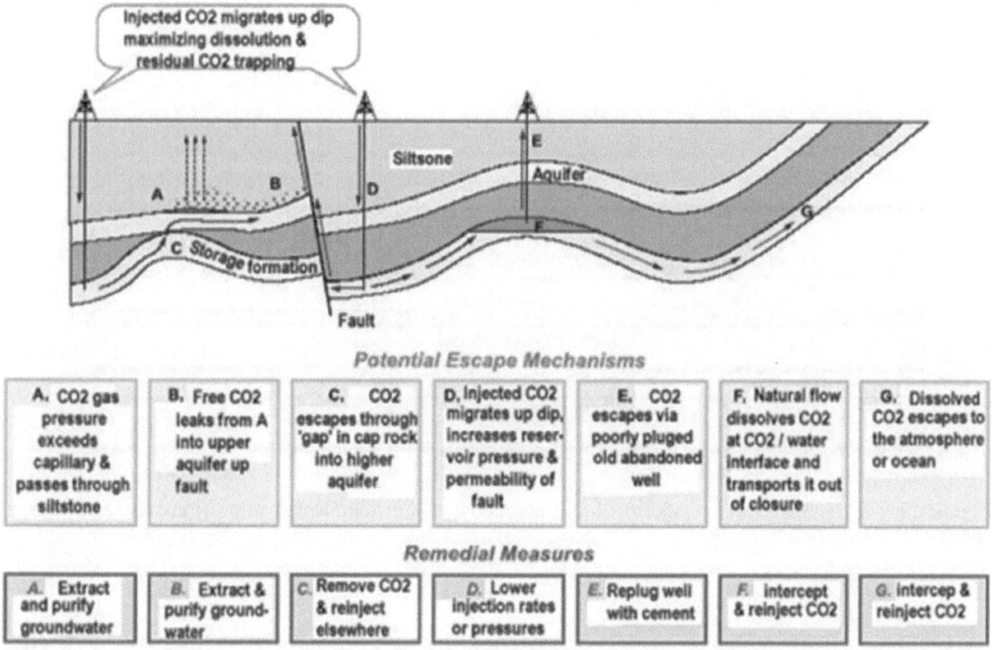

Figure 10. Potential escape mechanisms for CO_2 into saline formations (IPCC, 2005).

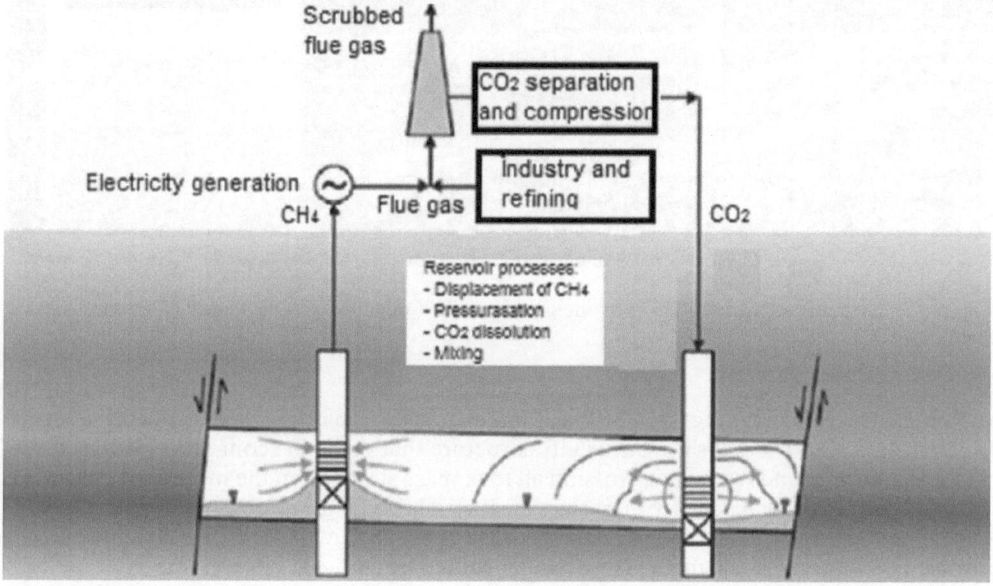

Figure 11. Storing in carboniferous formations with ECBMR (IEA, 2008).

So either it is a mine in which there is not entrance of water (good sealing strata) or, what is the most likely case, you have to put the CO_2 under high pressure. Note that in the first situation, the initial pressure of such reservoirs will be low (near atmospheric pressure), which means that the initial state of pressure is in great unbalance with the hydrostatic gradient. In the second situation, it is necessary to ensure that the sealing caprocks, despite being deformed due to the pressurization of the cavity should be able to resist this action, without open cracks or there are sliding along existing faults (Piessens, 2011).

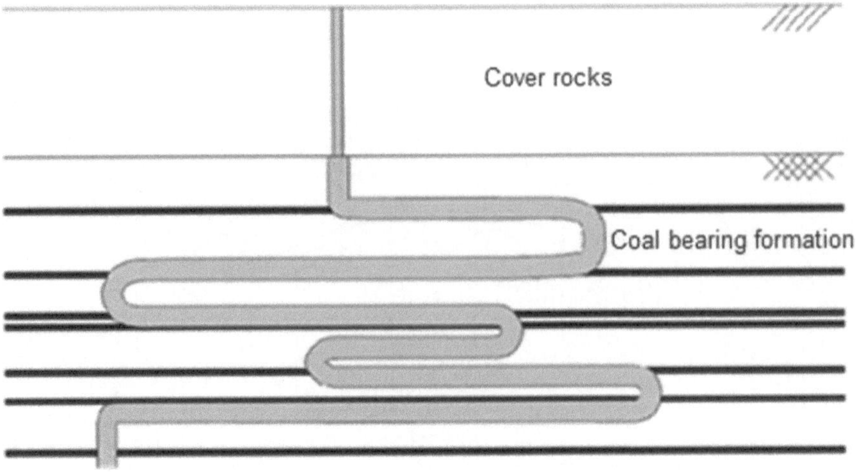

Figure 12. Simplified representation of the geometry of an abandoned mine (Piessens and Dusar, 2003).

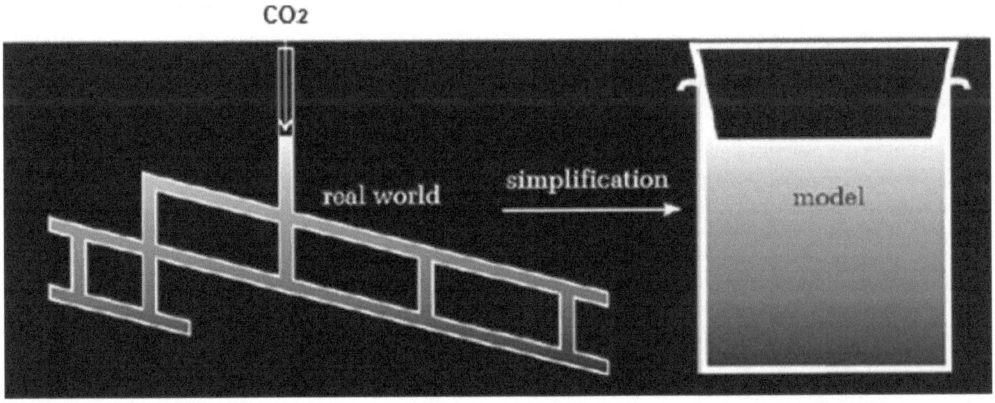

Figure 13. Schematic representation of the geometry of a mine: Model vs. reality (Piessens e Dusar, 2003).

The existence of pumping wells or injection of fluids is a major source of potential escape problems of CO_2. The wells are linear infrastructure that make the connection between surface and underground reservoirs, crossing all rock mass strata, even the most impervious. An eventual path to the leakage of CO_2 is then created. The sealing caprock of the well, the walls of well, and the annular area of interface with the walls, the first layer of cement case and the involved rock mass are the main elements that should be studied.

In the presence of water, CO_2 becomes carbonic acid, which can affect the integrity of the casing cement, or even the first cement layer that lies between the walls and the rock mass. Figure 13 shows the appearance of an intact specimen of cement and another in which the acid reacted with the cement. Then the resistance of the cement can be affected. Should be considered an extra thick wall, and the introduction of additives to the cement should be considered (IEA, 2008).

Figure 14 shows potential escape paths of CO_2 along injecting or pumping wells. In abandoned wells the type of escape mechanisms along the walls is similar to the wells still in operation. In (a) focuses on the flow through the interface of the well casing and cement layer on

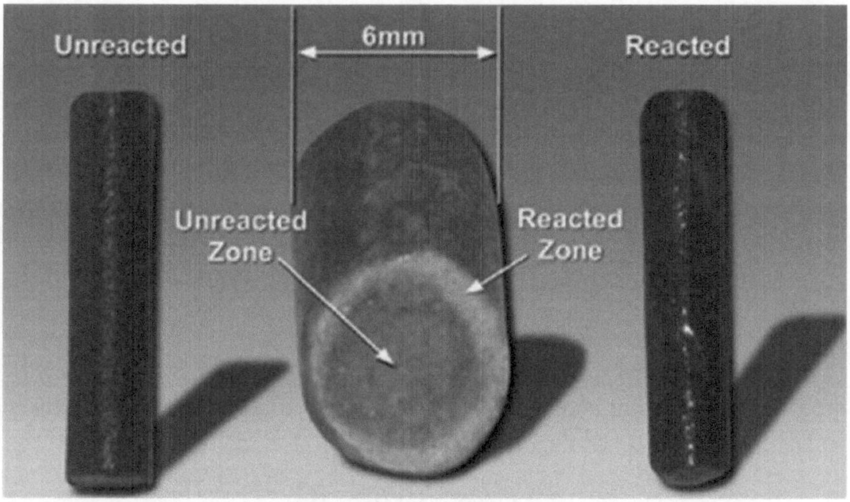

Figure 14. Effect of CO_2 in the cement (Celia et al., 2006).

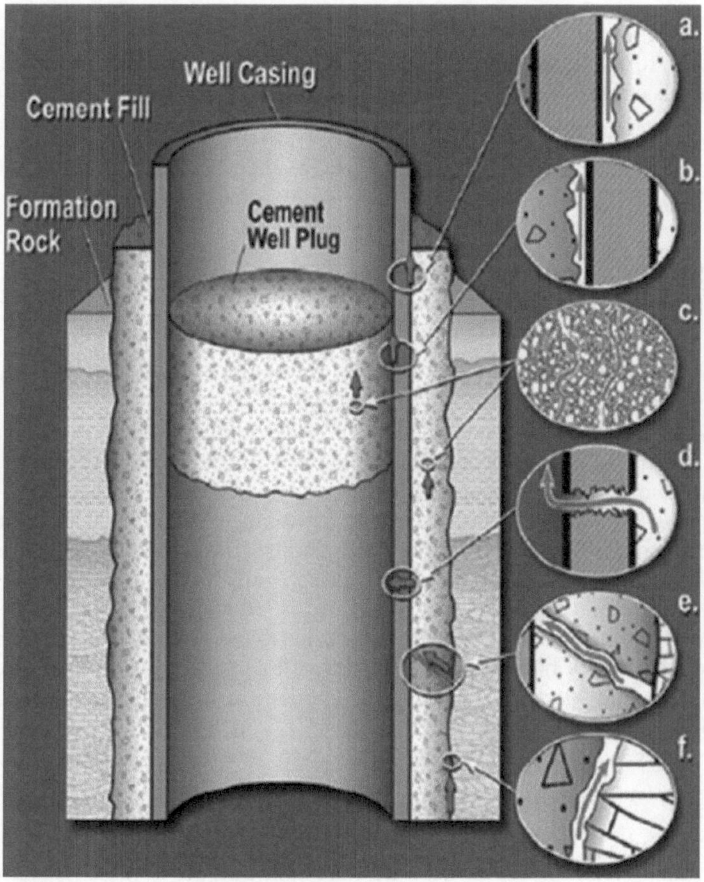

Figure 15. Potential escape pathways along wells (Celia et al., 2006).

the inside face of the coating. Since both materials are very permeable, runoff is very focused in the vertical direction. In (b), there is an escape mechanism similar to (a), but between the casing and cement element that makes the closing hole. In detail (c) you can view the mechanism of percolation of CO_2 through the cement seal. In (d) and (e) flow crossing the finish layer of concrete and masonry is represented. Finally, in (f), is represented another way of leakage, this time between the cement and the strata surrounding the well.

3 RISKS ASSOCIATED

3.1 *General*

Risk Assessment and mitigation strategies are developed with the goal of avoiding major problems described before. There are many definitions for risk. More generally for an undesirable event E with different consequences, vulnerability levels are associated and the risk can be defined as (Einstein, 1997):

$$R = P[E] \times P[C \mid E] \times u[C] \tag{1}$$

where
R is the Risk
$P[E]$ is the hazard, i.e. the probability of the event
$P[C \mid E]$ is vulnerability of event E
$u[C]$ is the utility of consequences C.

More generally, for different failure modes E_j, with which different consequences and hence vulnerability levels are associated, expected risk can be defined as (Sousa, 2010):

$$E[R] = \sum_j \sum_i P[E_j] \times P[u(C_i) \vdots E_j] \times u(C_i) \tag{2}$$

where
$P[u(C_i) \mid E_j]$ is the vulnerability to the failure mode j,
$P[E_j]$ is the probability of failure mode j and
$u(C_i)$ is the utility of consequence i.

For risk evaluation it is necessary to identify the tools or models to be used to represent this existing knowledge and perform a decision analysis.

Risk Assessment and Risk Management for CCS systems requires an evaluation of what hazard is possible and of the assessment of the likelihood of the harmful effects. Risk Assessment starts with the hazard identification which refers to identifying the major possible hazards with focusing on the likelihood of extent of damage. After the hazard identification, Risk Characterization is followed that involves a detailed assessment of each hazard in order to evaluate the risk associated to each hazard (Price et al., 2008).

Based on studies presented in several publications (Myer, 2003; IPCC, 2005; Price et al., 2008; Sousa, 2010), nine hazard identification scenarios were characterized (Table 1). Once the risks associated with a hazard have been identified the decision maker develops a basis for their evaluation and when necessary to develop and to carry actions to reduce the risks (Price et al., 2008).

3.2 *Leakage of CO_2 from pipelines or pumping stations and from shipping*

CO_2 from power plants or other industrial facilities can be transported to storage sites by pipelines. For any transportation option there are calculable and perceivable risks (Sousa, 2011b).

CO_2 pipelines provide direct route to harmful human exposure or to harmful impacts on animals and plants by producing a local high concentration of CO_2 and generating exposures

Table 1. Hazard identification scenarios.

Hazard	Description
H_1	Leakage of CO_2 from pipelines or pumping stations
H_2	Leakage of CO_2 from shipping
H_3	Slow and steady leakage of CO_2 from geological storage
H_4	Fast and large discharge of CO_2 from geological storage
H_5	Leakage from geological storage to groundwater
H_6	Leakage of CO_2 from geological storage to fossil fuel assets
H_7	Leakage of CO_2 that eliminates the benefits of geological storage
H_8	Induced fracturing or seismicity
H_9	Leakage from abandoned coal mines

sufficient to harm or kill people, plants and animals (IPCC, 2005). While an important risk precautions can be taken to minimize the likelihood of a major pipeline rupture.

In the USA exist several long-distance CO_2 pipelines, with special emphasis to the Cortez pipeline with 808 km, to the Sheep Mountain pipeline (660 km) and to the Weyburn pipeline with 330 km. Measures to be taken in order to minimize the risks from CO_2 pipelines include (IPCC, 2005: Price et al., 2008):

- To localize pipelines away from populous areas
- To avoid pipelines near populated valleys where leaking CO_2 could accumulate to dangerous levels
- To monitor pipelines against corrosion and to monitor regularly for leaks
- To install safety valves to shut off the pipeline in the case of a large leak
- To consider adding odorant to CO_2 to allow people to notice small leaks.

Leakage of CO_2 from shipping they can occur from different ways, namely through collision, foundering, stranding and fire (IPCC, 2005). The accidents can occur to badly maintenance of the ships and to crew by inadequately trained people, as well as by system failures and human errors.

Carbon dioxide tankers and terminals are clearly much less at risk from fire, but there is an asphyxiation risk if collision should rupture a tank. The risk can be minimized by making certain that high standards of construction are applied. An accident to a liquid CO_2 tanker might release liquefied gas to the surface of the sea. CO_2 would behave differently from LNG because liquid CO_2 in a tanker is not so cold as LNG. Its interaction with sea could be complex. Some of the gas would dissolve in the sea, but some would be released to the atmosphere. With little wind and temperature inversion CO_2 gas might lead to asphyxiation and might stop the engines.

The risk can be minimized by careful planning of routes and by high standards of training and management (IPCC, 2005).

3.3 *Slow and fast leakage of CO_2 from geological storage*

Leakage of CO_2 from the geological reservoir can produce two types of hazard depending how low or fast is the leakage (Price et al., 2008).

For slow and steady leakage of CO_2 from geological storage, the releases are too small to cause significant deaths or injuries. However the leakage can cause local problems including human fatalities.

For fast and large discharge of CO_2 from geological storage, as occurred in 1986 at Lake Nyos in Cameroon, where about 1700 persons and 3500 cattle were killed when the lake release a large amount of CO_2, can cause a large scale of fatalities, although it is rare the occurrence.

Possible actions and measures for these hazards are referred in detail in Price et al. (2008).

3.4 *Leakage from geological storage to groundwater*

CO_2 migration from a storage reservoir to the surface potentially affects shallow groundwater used for potable water and industrial and agricultural needs. Dissolved CO_2 forms carbonic acid, altering the pH of the solution and potentially causing indirect effects, including mobilization of (toxic) metals, sulphate or chloride; and possibly giving the water an odd odor, color or taste. In the worst case, contamination might reach dangerous levels, excluding the use of groundwater for drinking or irrigation (IPCC, 2005).

Among other measures referred in Price et al. (2008), it is relevant to develop appropriate inspection methodologies coupled with the use of dynamic BN for risk analysis (Sousa, 2010).

3.5 *Leakage of CO_2 from geological storage to fossil fuel assets*

Underground injection of CO_2 at high pressures can lead to seepage of fossil deposits through faults and other discontinuities or through not well sealed wells. The contamination of the fossil reservoir induces a severe economic risk since the contamination decrease the value of the fossil fuel. The probability of occurrence of this hazard is similar to the probability of leakage to the groundwater.

Actions to reduce risks from leakage to fossil fuels can be:

- To select reservoir sites that are likely to retain their CO_2 for at least thousands of years and
- To select sites that are not near fossil fuel assets.

3.6 *Leakage of CO_2 eliminating the benefits of geological storage and induced seismicity*

Leakage from a reservoir returns CO_2 into the atmosphere. The sequestration of CO_2 is for a long period of time, then when a leakage with a faster rate eliminates the benefits of the geological storage, inducing additional costs.

Some actions can be performed in the case of inadequately sealed wells, namely finding and plugging such wells. Wells can be monitored in order to ensure that they are adequately sealed and additional activities can be performed to better seal the wells.

Geological carbon sequestration into porous rock masses at high pressure can also induce fracturing and movements along faults. The resulting stresses can fracture the surrounding rock. They may pose two types of risks, brittle failure and associated microseismicity thus providing pathways for CO_2 migration; and fault activation that can induce earthquakes large enough to cause damage (IPPC, 2005). So far, only moderate earthquakes have occurred due to injection.

Eventual actions to reduce risks from induced fracturing or seismicity are referred in Price et al. (2008).

3.7 *Leakage from abandoned coal mines*

In coal mines slow migration towards the surface are not a direct threat to humans and nature. However high concentrations can be reached by a sudden and temporary release of CO_2. Because is more denser than air it could be up high concentrations in depressions and confined areas near the surface and cause problems to humans, which is a known risk that happened in volcanic lakes. Leakage may also occur along infrastructure, case of wells, and faults. The effect of active faults on sealing properties of the overburden is a important issue to be considered in terms of safety.

A technical obstacle for injection of CO_2 in abandoned coal mines is the low initial reservoir pressure.

The feasibility of CO_2 sequestration in coal mines and eventual actions to be considered in order to reduce the risk are presented in the publication of Piessens (2011).

4 PREVENTING RISKS BY MONITORING

In order to prevent risks monitoring is needed. Measurements of certain parameters should be made in order to assess the behavior of the CO_2 system. The results of the monitoring must compared with ones predicted by modeling and risk analysis. The models can be updated after careful interpretation of a set of observed results.

Monitoring is performed for various purposes, such as (IPCC, 2005):

- to ensure and to document the volume injected into wells, specifically to monitor the conditions of the injection well and measuring the rates of injection, as well as the pressures on the top of the well and in the formation;
- to verify the amount of injected CO_2 that was stored by different mechanisms;
- to optimize the efficiency of the storage project, through the knowledge of the volume storage, the most appropriate injection pressures and the need for drilling new wells;
- to demonstrate, with appropriate monitoring techniques, that CO_2 is still contained in the intended storage formations;
- to detect leaks and to provide an early warning of any occurrence, so that the situation can be remedied by appropriate mitigation measures;
- to know the integrity of wells that are being used or are abandoned;
- to calibrate and verify models for determining the performance;
- to detect microsismicity associated to storage projects.

Before storage is necessary to make measurements of the most relevant parameters to be controlled and to characterize the site, in order to know the initial situation (baseline) which will be used in future comparisons. It is convenient to perform several in situ testing campaigns over different seasons, since some properties have a natural variability. This need is particularly felt when remote sensors are used, for example seismic sensors. This is particularly true of seismic and other remote-sensing technologies, where the identification of saturation of fluids with CO_2 is based on comparative analysis. Monitoring the initial situation is also a prerequisite for geochemical analysis, where anomalies are identified relative to background concentrations (Solomon, 2006; Gomes, 2010).

Measurements of CO_2 injection is a common practice in oil and gas fields, and the instruments for this purpose are available in the market. Measurements are made by gauges at the wellhead injection or in the vicinity of the injection tube. The accuracy of measurements depends on a number of factors described by IPCC (2005). For CO_2, accurate estimation of the density is very important for improving measurement accuracy. Small changes in temperature, pressure and composition can have large effects on the density.

Measurements of injection pressure at the surface and in the rock formations are also usually performed. Gauges are installed in most injection wells through holes on the surface piping near the wellhead. Measurements of pressure in the well are routine. A wide variety of pressure sensors are available and are adequate to monitor pressures at the wellhead or in the rock formations. The data are continuously available. The surface pressure gauges are often linked to shut-off valves that will stop or reduce the injection pressure to a certain limit if the pressure exceeds a predeterminated safe value or if there is a drop in pressure as a result of a leak (IPCC, 2005). The relatively recent fibre-optic pressure sensors and temperature sensors are available. These new systems should provide more reliable results, as well as better control of the well.

The current state of technology is more than sufficient to meet the needs of monitoring rates of injection, and the pressures of training and the top of the hole. Combined with temperature measurements, the data provide information on the state of CO_2 (supercritical, liquid or gaseous) and precise values of the quantity of CO_2 injected. This information may be used for verification and possible updating of the model adopted.

Figure 16 presents a methodology that can be used by monitoring for the long term integrity analysis of a well in terms of risk evaluation.

The way the CO_2 is distributed and moves underground can be monitored in several ways. Table 2 summarizes various techniques and how each can be applied to CO_2 storage projects. The applicability and sensitivity are different from place to place, from reservoir to reservoir.

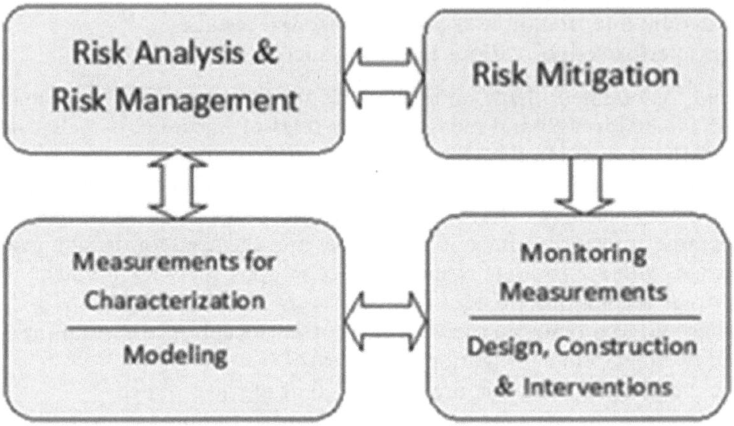

Figure 16. General methodology for integrity analysis of a well.

Table 2. Summary of direct and indirect techniques that can be used to monitor CO_2 storage projects (Adapted from IPCC, 2005).

Measurement technique	Measurement parameters	Example applications
Introduced and natural tracers	– Travel time – Partitioning of CO_2 into brine or oil – Identification of sources of CO_2	– Tracing movement of CO_2 in the storage formation – Quantifying solubility trapping – Tracing leakage
Water composition	– CO_2, HCO_3^-, CO_3^{2-} – Major ions – Trace elements – Salinity	– Quantifying solubility and mineral trapping – Quantifying CO_2-water-rock interactions – Detecting leakage into shallow groundwater aquifers
Subsurface pressure	– Formation pressure – Annulus pressure – Groundwater aquifer pressure	– Control of formation pressure below fracture gradient – Wellbore and injection tube condition – Leakage out of the storage formation
Well logs	– Brine salinity – Sonic velocity – CO_2 saturation	– Tracing CO_2 movement in and above storage formation – Tracking migration of brine into shallow aquifers – Calibrating seismic velocities for 3D seismic surveys

(Continued)

Table 2. (Continued).

Time-lapse 3D seismic imaging	– P and S wave velocity – Reflection horizons – Seismic amplitude attenuation	– Tracing CO_2 movement in and above storage formation
Vertical seismic profiling and crosswell seismic imaging	– P and S wave velocity – Reflection horizons – Seismic amplitude attenuation	– Detecting detailed distribution of CO_2 in the storage formation – Detecting leakage through faults and fractures
Passive seismic monitoring	– Location, magnitude and source characteristics of seismic events	– Development of microfractures in formation or caprock – CO_2 migration paths
Electrical and electromagnetic techniques	– Formation conductivity – Electromagnetic induction	– Tracking movement of CO_2 in and above the storage formation – Detecting migration of brine into shallow aquifers
Time-lapse gravity measurements	– Density changes caused by fluid displacements	– Detect CO_2 movement in or above storage formation – CO_2 mass balance in the subsurface
Land surface deformation	– Tilt – Vertical and horizontal displacements using interferometry and GPS	– Detect geomechanical effects on storage formation and caprock – Locate CO_2 migration pathways
Visible and infrared imaging from satellite or planes	– Hyperspectral imaging of land surface	– Detect vegetative stress
CO_2 land surface flux monitoring using flux chambers or eddycovariance	– CO_2 fluxes between the land surface and atmosphere	– Detect, locate and quantify CO_2 releases
Soil gas sampling	– Soil gas composition – Isotopic analysis of CO_2	– Detect elevated levels of CO_2 – Identify source of elevated soil gas CO_2 – Evaluate ecosystems impacts

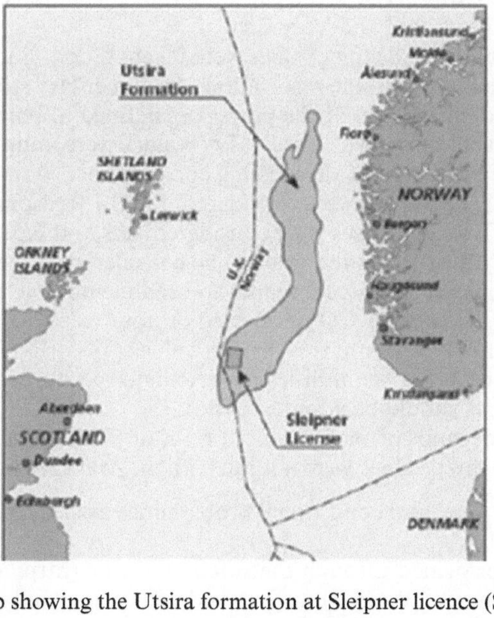

Figure 17. Location map showing the Utsira formation at Sleipner licence (Solomon, 2006).

A monitoring case study is the Sleipner Gas field, in the middle of the North Sea. It is referred as a reservoir where has been injected 1 Mt CO_2 per year since September 1996 (IPCC, 2005, Solomon, 2006), (Figure 17). The CO_2 is injected into salt water containing sand layer, called Utsira formation, 1000 m below sea bottom.

In 1999 the project started monitoring CO_2 behavior and established a baseline for a first seismic survey. The project is being carried out in three phases (Phases 0, 1 and 2). Last phase involved data interpretation including monitoring and model verification. The fate and transport of the CO_2 plume in the storage formation has been monitored successfully by seismic time-lapse surveys. Work at Sleipner demonstrated that conventional, time-lapse, p-wave seismic data can be a successful monitoring tool for CO_2 injected into a saline aquifer with CO_2 accumulation (Eiken et al., 2000; Sousa, 2011b).

5 APPLICATION OF BAYESIAN NETWORKS (AND DECISION GRAPHS)

5.1 *General*

In this section examples of BN and DBN (Dynamic Bayesian Networks) are presented to illustrate their potential use for risk analysis in CO_2 Injection Processes. The first example was developed for a situation where one wants to determine whether or not is beneficial to Inject CO_2 carboniferous formations at a certain location (section 5.2). The example presented on Section 5.2, CO_2 is based on hazards H_4 and H_5 defined previously. In this example the decision maker is looking at different mitigation measures (for reducing the leakage of CO_2), assessing the risk of each option and choosing the one that minimizes it. Finally in section 5.3 an example of a DBN is presented to illustrate the use of DBN couple with results of a monitoring system. This section follows an article published in the Journal of Rock Mechanics and Geotechnical Engineering (He et al., 2011).

5.2 *Risk analysis for storage of CO_2*

For the Risk Analysis due to CO_2 injection in carboniferous formations a BN was develop as presented in Figure 18. The involved variables are associated to:

- Sedimentary strata conditions over the carboniferous formations. 3 values were adopted for the formations: Good, Bad and Very Bad.
- Coal seams characteristics. 3 distinct values were taken: Good, Bad and Very Bad.
- Combined characteristics due to the association of sedimentary strata and coal seams. The values were attributed in function of the properties defined to both formations.
- Geomechanical characteristics of the wells. Two values were adopted for the shaft: Good State and Bad State in function of the existing corrosion.
- Corrosion of the well. Two levels were considered: Level 1 (reasonable) and Level 2 (bad).
- Existence of faults. Two hypotheses were considered: Yes and No.
- Escape of CO_2. For this situation the value to be considered of the combined characteristics of both formations involved (coal seams and sedimentary strata), of the existence of wells and faults, and of course if CO_2 is injected or not.
- Injection of CO_2. For this situation two distinct values (Yes or No) were considered.
- Utilities (Consequences). For the utilities the calculated result permits to conclude if the rehabilitation measures should be adopted or not.
- The calculated Risk depends of the existence or not of CO_2 escape and of the existence of faults. The following three values were adopted: High, Average and Low.

In Tables 3 to 10 the local and conditional probabilities associated to each variable of the BN are represented.

Applications were performed through the software Genie (http://genie.sis.pitt.edu/downloads.html). Two hypotheses (A and B) were considered as assigned in Table 11.

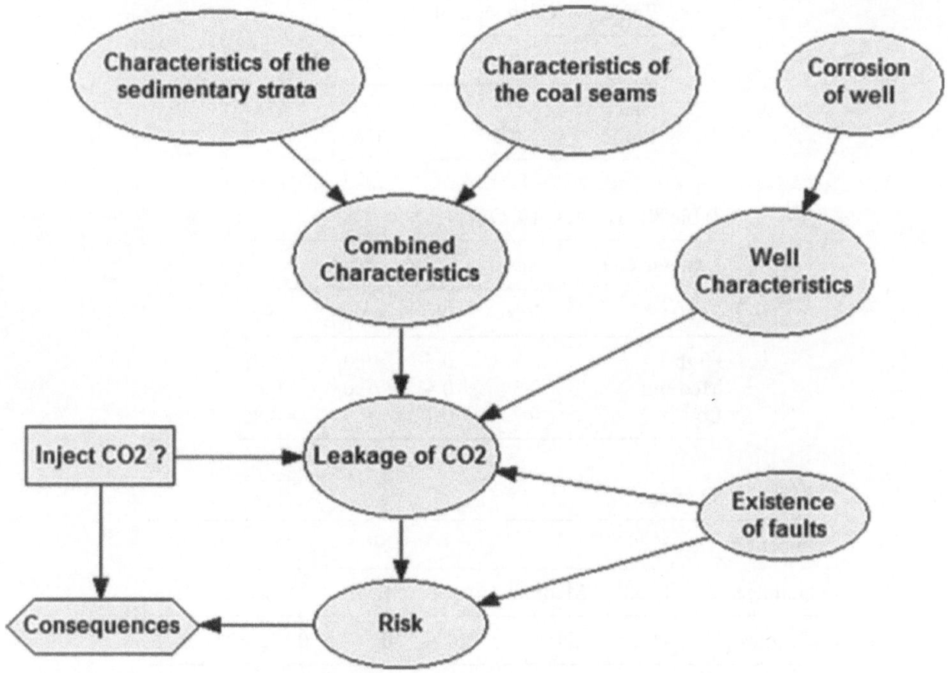

Figure 18. BN for risk analysis of storage of CO_2.

Table 3. Sedimentary strata characteristics.

Good	0.333
Bad	0.333
Very Bad	0.333

Table 4. Coal seam characteristics.

Good	0.333
Bad	0.333
Very Bad	0.333

Table 5. Combined characteristics of the sedimentary strata and coal seam.

Sedimentary cap char.	Good			Bad			Very bad		
Coal seams char.	Good	Bad	Very bad	Good	Bad	Very bad	Good	Bad	Very bad
Good	1	0	0	0	0	0	0	0	0
Bad	0	1	0.7	1	1	0.3	0.3	0.3	0
Very Bad	0	0	0.3	0	0	0.7	0.7	0.7	1

Table 6. Characteristics of the wells.

Corrosion	Level 1	Level 2
Bad	0.7	0.4
Good	0.3	0.6

Table 7. Corrosion of wells.

Corrosion	Level 1	Level 2
Bad	0.7	0.4
Good	0.3	0.6

Table 8. Escape of CO_2.

Leakage CO_2	Yes		No	
Faults	Yes	No	Yes	No
High	0.7	0.5	0.05	0.01
Medium	0.3	0.4	0.05	0.03
Low	0	0.1	0.9	0.96

Table 9. Damage.

Inject CO_2?	Yes			No		
Damage	High	Medium	Low	High	Medium	Low
Utilities	-40	-20	20	0	0	0

Table 10. Utilities.

Inject CO_2?	Yes											
Comb. char	Average				Bad				Very bad			
Well char.	Good		Bad		Good		Bad		Good		Bad	
Faults	Yes	No	Yes	No	Yes	No	Yes	No	Yes	Yes	Yes	No
Yes	0.1	0.01	0.6	0.2	0.3	0.2	0.7	0.6	0.6	0.5	0.8	0.7
No	0.9	0.99	0.4	0.8	0.7	0.8	0.3	0.4	0.4	0.5	0.2	0.3

Table 11. Different hypothesis considered in the BN.

Hypothesis	Sedimentary strata	Coal seams	Wells	Corrosion	Existence of faults
A	Good	Good	Good	–	–
B	Good	Bad	–	Level 2	–

For hypothesis A, Figure 19 shows the resulting diagram with probabilities calculations. The results demonstrate clearly that is beneficial to inject CO_2 in the coal seams. For the hypothesis B, Figure 20 shows that resulting BN diagram recommended not inject CO_2 in the coal seams.

Another BN is presented in Figure 21 when active faults are considered. The consequences in this situation can imply the existence of induced earthquakes.

5.3 *Contamination of aquifers by CO_2*

Contamination of aquifers corresponds to the hazard H_5—Leakage from geological storage to groundwater, accordingly to the different hazards defined in Table 1. CO_2 injected into the ground will be dissolved into water including pore water between grains or minerals in the

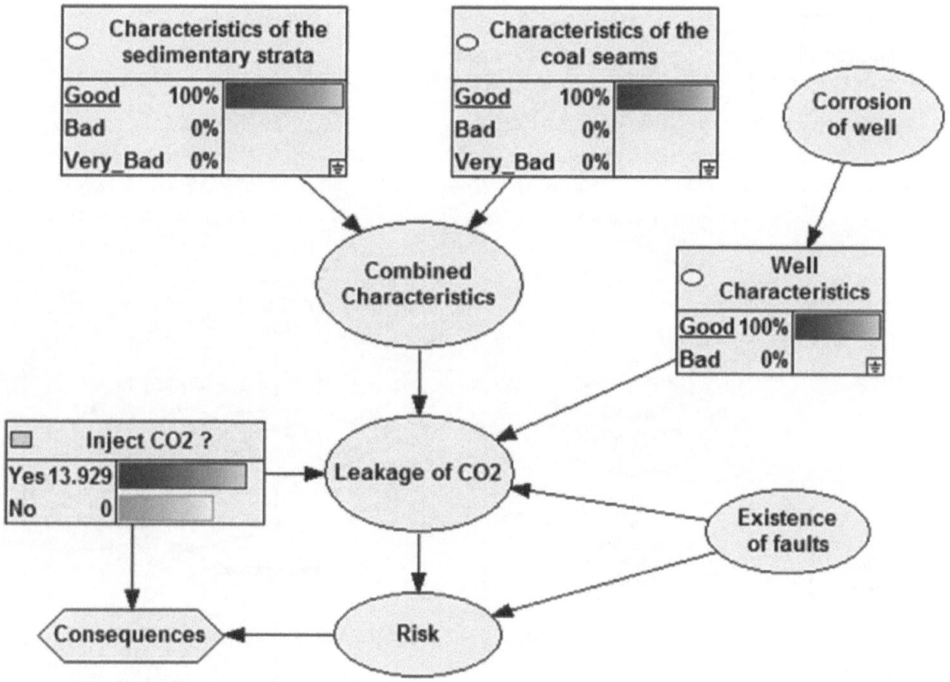

Figure 19. Diagram for hypothesis A.

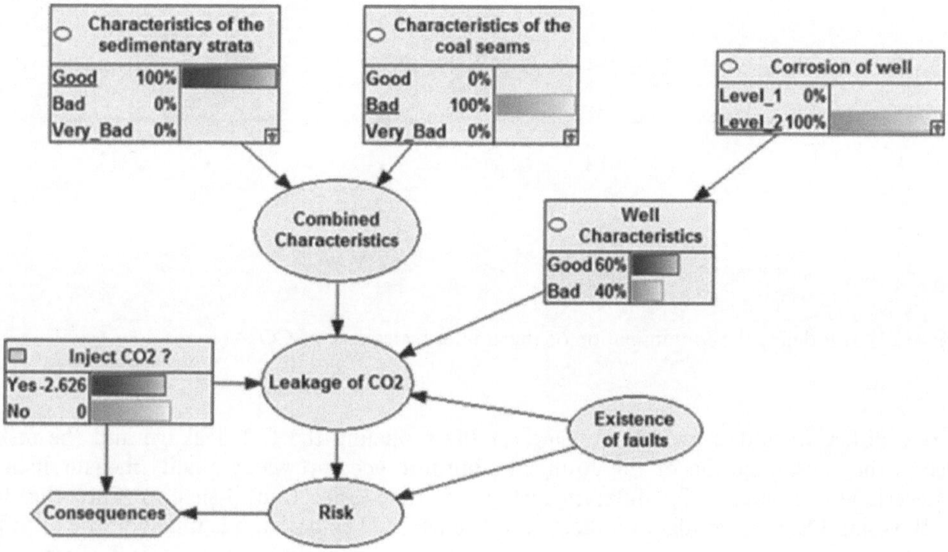

Figure 20. Diagram for hypothesis B.

geologic formations. Dissolution into water can be problematic. The water will be acidified which allow it to degrade geological formations and the water saturated with CO_2 is not useful for drinking water.

In order to deal with this situation BN were developed in order to analyzing the situation. DBN were adopted in order to evaluate the situation of contamination of aquifers due to the leakage of CO_2.

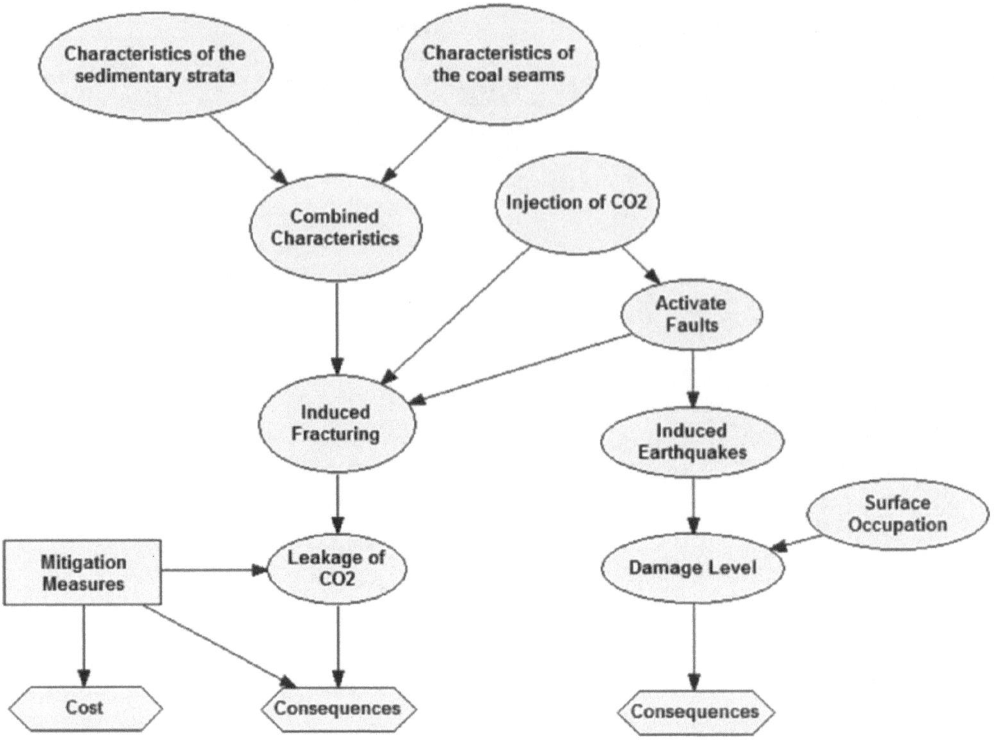

Figure 21. BN for risk analysis of storage of CO_2 with the existence of active faults.

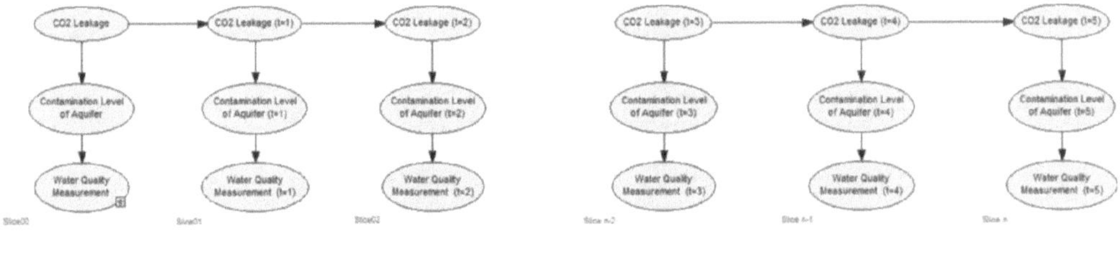

Figure 22. Modeling the contamination of the aquifer by leakage of CO_2.

Two different models were built. One for the modeling the CO_2 leakage and the influence in the contamination of the aquifer taking into account water quality measurements, as described in Figure 22 for different instants of time (Slice00 until slice n). The other for the Bayesian Decision model as indicated in Figure 23. The decision is made on the optimal remedial measures solution for the problem that can pass through the decision of no more injecting CO_2.

The way the model works is as follows:

- Step 1: Observation (Water quality measurement) is made at time t_0 and entered into the network.
- Step 2: The evidence is propagated through the network at time t_0, and the Probability of Leakage is determined.
- Step 3: The evidence is propagated through into the future, and the Probability of Leakage in the next slices of time is determine.

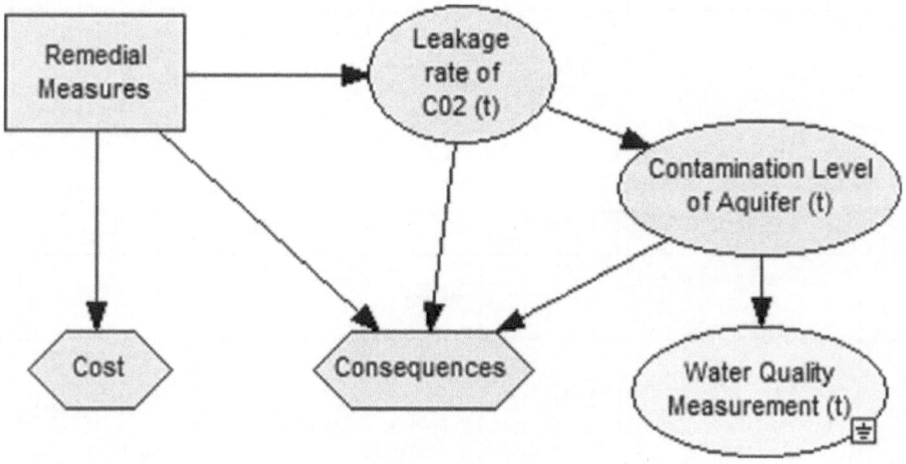

Figure 23. Bayesian Decision model based on the water quality measurements.

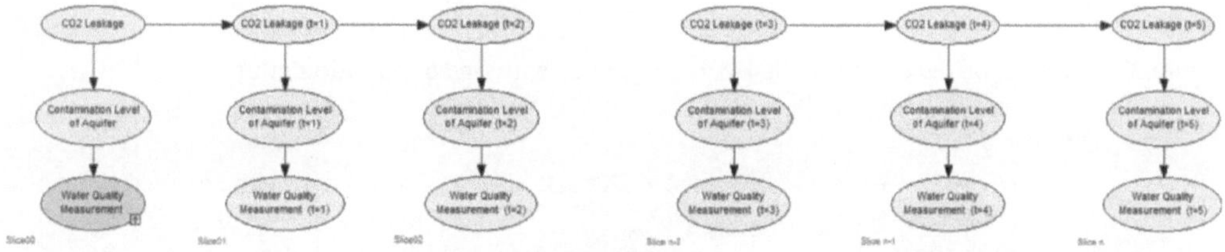

Figure 24. Step 1. Evidence: Water measurement = Good.

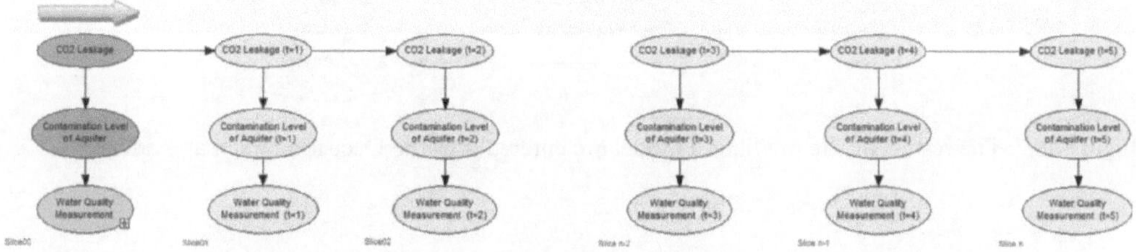

Figure 25. Step 2. DBN results: P (Cont. Level = high) = 0.09; P ($C0_2$ Leakage rate = high) = 0.10.

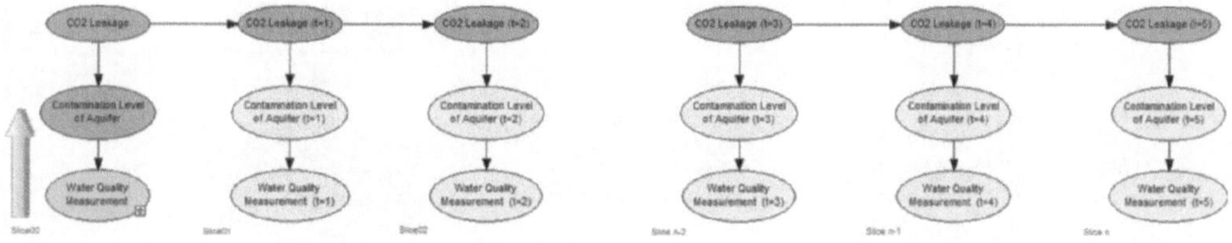

Figure 26. Step 3. Results shown in Figure 28.

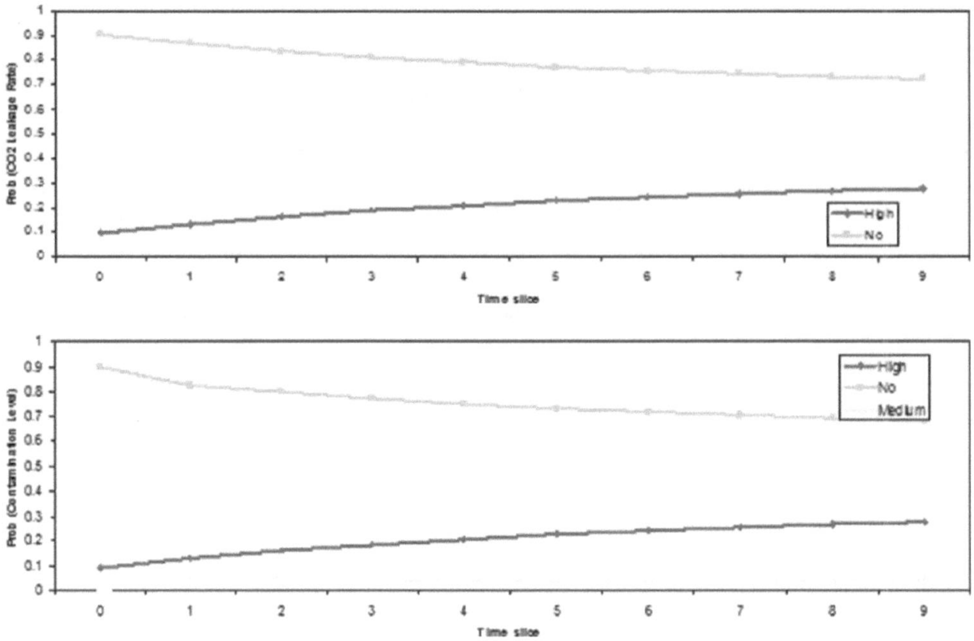

Figure 27. Results of the execution of BN of Figure 25 with one observation at time t_0.

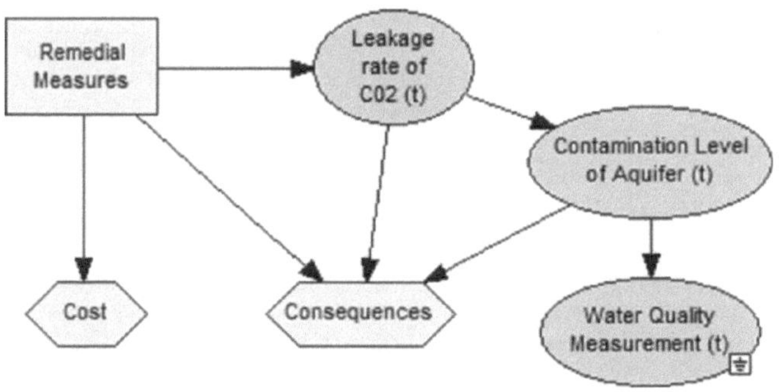

Figure 28. The results of the prediction model are entered into the Decision model as evidence.

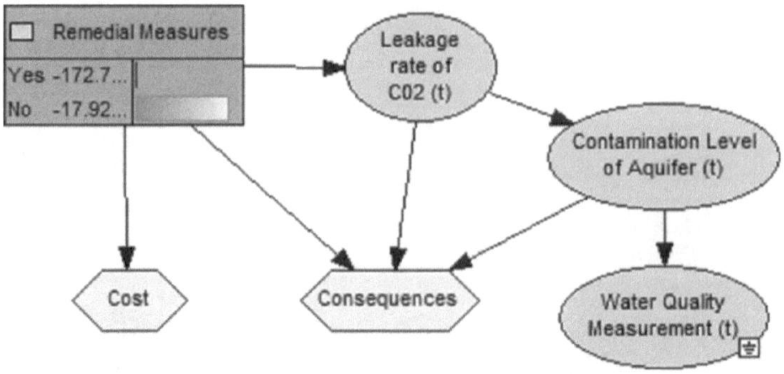

Figure 29. Execution of the Decision model of Figure 24.

Once the prediction model has been executed, one uses its results to determine the optimal remedial measure (which can be to do nothing, i.e. no remedial measure), by minimizing the risk. Figure 28 shows the decision model with evidence (coming from the prediction model) entered into the network. The results of the execution of this model are presented in Figure 29. The results show that the best decision given the water measurement at time t0 is not to apply a remedial measure.

These steps are then repeated for each slice of time.

6 CONCLUSION

This article describes briefly the most effective technologies for CCS projects. Geologic carbon sequestration presents the possibility to reduce emissions of CO_2 into the atmosphere at low cost compared to many other options. China became the largest emitter of CO_2 and has exceptional conditions to storage CO_2 in carboniferous formations, particularly in abandoned coal mines.

Geologic carbon sequestration entails risk that may be large and significant. However risks can be limited or reduced. Development of methodologies for risk evaluation based in BN were made and some relevant applications were performed with particular emphasis to the development of DBN for the hazard related to the leakage from geological storage to groundwater.

Based on the applications of BN several conclusions can be made:

- In the Risk Management, BN are a powerful tool in the Decision Analysis including priori and posteriori analyses.
- BN presented the extension of influence diagrams including the use of decision nodes and also utilities nodes.
- BN allow to combine the knowledge of experts and available data through statistical methods.
- The beneficial use of DBN in decision processes involving time is very relevant as the application made demonstrated.

REFERENCES

Benson, S. (2005). Carbon Dioxide Capture and Storage: Overview with an Emphasis on Geological Storage. Tutorial Presented at the AGU Annual Meeting, 5/Dezembro/2005, San Francisco.

Celia, M. et al. (2006). Implications of Abandoned Wells for Site Selection. Princeton University.

Desroches, J., Jammes, L. & Berard, T. (2006). Building a Mechanical Earth Model for Storage Integrity. CO_2SC Symposium, Lawrence Berkeley National Laboratory, Berkeley, 20–22/Março/2006.

Duguid, A. et al. (2007). MMV Technologies for Effective and Efficient Monitoring of Geologic Carbon Capture and Storage Projects. GWPC/US EPA CO_2 MMV Workshop, 16/Fevereiro/2008, New Orleans.

Eiken, O., Brevik, I., Arts, R., Lindeberg, E. & Fagerbik, K. (2000). Seismic monitoring of CO_2 injected ito a marine aquifer. SEG Calgary 2000, Int. Conf., paper RC–8.2.

Einstein, H. (1997). Landslide risk: Systematic approaches to assessment and management. In Cruden, A. and Fell, R. (editors). Landslide Risk Assessment, Balkema, Rotterdam, pp. 25–50.

Elsworth, D., Wang, S., Izadi, G., Kumar, H., Liu, J., Lee, D.-S., Mathews, J. & Pone, D. (2011). Complex process couplings in systems pushed far-far-from equilibrium: Applications to deep geologic sequestration and energy recovery. *Int. Workshop on CO_2 Storage in Carboniferous Formations and Abandoned Coal Mines*, Ed. He, Sousa, Elsworth and Vargas. Beijing, pp. 55–68.

Gomes, A. (2010). CO_2 injection processes in carboniferous formations (in Portuguese). MSc Thesis, University of Porto, Porto, 116p.

He, M. (2011). Considerations on CO_2 storage in abandoned coal mines in China. Int. Workshop on CO_2 Storage in Carboniferous Formations and Abandoned Coal Mines, Taylor & Francis, Ed. He, Sousa, Elsworth & Vargas, Beijing, pp. 25–36.

He, M., Sousa, R., Sousa, L., Gomes, A., Vargas A. Jr., & Na, Z. (2011). Risk assessment of CO_2 injection processes and storage in carboniferous formations: a review. Journal of Rock Mechanics and Geotechnical Engineering. Vol 3(1) pp. 39–56.

IEA—International Energy Agency. (2002). Carbon Dioxide Chemistry and Properties. http://www.ieagreen.org.uk.

IEA—International Energy Agency. (2008). CO_2 Capture and Storage—A Key Carbon Abatement Option. IEA, Head of Communication and Information Office, Paris.

IPCC (2005). Carbone dioxide and storage. IPCC Special Report, Cambridge University Press, 431p.

Myer, L. (2003). Geomechanical risks in coal bed carbon dioxide sequestration. Lawrence Berkeley National Laboratory, Earth Sciences Division, Berkeley, 24p.

Piessens, K. (2011). CO_2 storage in abandoned coal mines. Pressure constraints. Int. Workshop on CO_2 Storage in Carboniferous Formations and Abandoned Coal Mines, Taylor & Francis, Ed. He, Sousa, Elsworth & Vargas, Beijing, pp. 179–200.

Piessens, K. & Dusar, M. (2003). CO_2-Sequestration in Abandoned Coal Mines. Royal Belgian Institute for Natural Sciences, Geological Survey of Belgium.

Price, P., McKone, T. & Sohn, M. (2008). Carbon sequestration risks and risk management. Lawrence Berkeley National Laboratory, Environment Energy Technologies Division, Berkeley, 19p.

Qu Hongyn, Liu Jishan, Zhong Wei, Pan Zhejun & Connell Luke. (2011). Multiphysics of coal-gas interactions. Something old, something new and something very new. *Int. Workshop on CO_2 Storage in Carboniferous Formations and Abandoned Coal Mines*, Ed. He, Sousa, Elsworth and Vargas. Beijing, pp. 69–93.

Solomon, S. (2006). Carbon dioxide storage: Geological security and environmental issues – case study on the Sleipner gas field in Norway. Bellone Foundation, Oslo, 20p.

Sousa, L.R. (2011b). Present Day Conditions in the World of Carbon Capture and Storage (CCS) Projects. Int. Workshop on CO_2 Storage in Carboniferous Formations and Abandoned Coal Mines, Taylor & Francis, Ed. He, Sousa, Elsworth & Vargas, Beijing, pp. 1–23.

Sousa, R.L. (2010). Risk analysis for tunneling projects. MIT, Cambridge, 589p.

Sousa, R.L. (2011b). Methodologies for Risk Analysis and Decision Making. Int. Workshop on CO_2 Storage in Carboniferous Formations and Abandoned Coal Mines, Taylor & Francis, Ed. He, Sousa, Elsworth & Vargas, Beijing, pp. 125–152.

Vargas E.A. Jr., Velloso R.Q., Ribeiro W.N., Muller A.L. & Vaz L.E. 2011. Considerations on the numerical modeling of injection processes of CO_2 in geological formations with emphasis on carboniferous formations and abandoned coal mines. *Int. Workshop on CO_2 Storage in Carboniferous Formations and Abandoned Coal Mines*, Ed. He, Sousa, Elsworth and Vargas. Beijing, pp. 105–123.

CO_2 Storage in Carboniferous Formations and Abandoned Coal Mines – He, Ribeiro E Sousa, Elsworth & Vargas Jr. (eds)
© 2012 Taylor & Francis Group, London, ISBN 978-0-415-62079-6

The conceptual model for an abandoned coal mine reservoir

K. Piessens
Royal Belgian Institute of Natural Sciences, Geological Survey of Belgium, Brussels, Belgium

ABSTRACT: The concept of storing CO_2 in abandoned coal mines formed the subject of studies already about 10 years ago, but the work mainly remained at the conceptual level and was detailed for only a few actual case studies. This is partly because optimal storage constraints for coal mines require deep and well sealed mines. The jar pot model provides a clear outline of these constraints, is an important aid to qualitatively projecting the pressure evolution in a coal mine after injection has stopped, and provides the backbone for general, but reliable calculations of the storage potential of generic or case-specific coal mines. Particular attention is given in this paper to evaluating the sensitivity of capacity estimates in function of the depth of a coal mine reservoir, and variations in reservoir pressure and temperature. Shallow to intermediate coal mines will usually have a storage capacity that is too low to be of economic interest. Deep mines, sealed at depths of 500 m or more, may be of interest if their top seal can withstand a sufficient amount of overpressure (up to 30% above hydrostatic pressures). Small differences in the pressure and temperature conditions in such mines may result in shifts between gaseous and liquid like CO_2, and as such have large effects on the estimated storage capacity. Capacity estimates for ultra-deep mines, sealed at depths below 800 m, are more stable, although errors due to uncertainties on pressure and temperature gradients are significant.

1 INTRODUCTION

Coal mines are only rarely discussed as potential reservoirs in Carbon Capture and Storage (CCS) schemes. The IPCC Special Report on Carbon Capture and Storage (Metz et al., 2005), for example, a document of 431 pages, only spends 16 lines on this type of reservoir. It is also not a reservoir that is appearing in any of the current pilot or demonstration projects on CO_2 Geological Storage (CGS). An important element for the lack of attention is that the geological and mining context needs to be correct to make storage in coal mines feasible.

A second reason is that the concept of coal mine storage was never sufficiently developed. When given superficial attention, several arguments can easily be given for or against coal mine storage. Advocates may point out the geographic coincidence of CO_2 intensive industry with coal mine basins, a situation that may persist decades after closure of the coal mines, and consequently the very short transport distances when coal mines would be used for CO_2 storage. Putting an abandoned mine to use is also valorization of underground infrastructure. Opponents are likely to point out that coal mines are relatively small compared to other geological storage options, such as aquifers, and that they are further complex, difficult, relatively shallow and pose additional safety issues due to e.g. subsidence induced fracture patterns.

A proper judgment of coal mines as potential reservoirs should be based on a proper understanding of these particular reservoirs. The foundation for such an advanced understanding is given in this publication by presenting the conceptual model for abandoned coal mine reservoirs. The conceptual model allows making early quantifications of the storage capacity, identifying risks, knowledge gaps and other points of concern and defining the optimal geological and mining generic context. As such it should also provide a proper basis for reservoir simulations of detailed case studies.

2 PROPOSED CONCEPTS

The amount of peer review literature that deals with CO_2 storage in coal mines is very limited and has been covered by a few authors only (Piessens and Michiel Dusar 2004a; Piessens and Michiel Dusar 2004b). Nevertheless, the option of using abandoned coal mines for CO_2 storage is not new, and was discussed in first assessment studies such as the GESTCO project (Christenson and Sam Holloway 2003; May et al., 2003), that evaluated the overall storage potential in Europe. These first assessments have in common that the coal mine option was discussed based on analogies with aquifer storage concepts. There are however fundamental differences that have lead to incorrect assumptions, as will be shown in the subsequent paragraph. A second comparison could be based on the practical experiences that exist with storing natural gas (CH_4) in coal mines, and we will consequently discuss this approach and again highlight the major differences.

2.1 *Aquifer storage concept applied to coal mines*

First assessments of geological reservoirs are often based on a rules-of-thumb approach, some of which have become so well embedded that they tend to be applied without careful evaluation. One such rule with quite some reputation is the 800 m cut-off depth for CO_2 storage. The rule is based on the critical point of CO_2 which is situated at 7.6 MPa and 31.1°C. Below the critical pressures, CO_2 will be gaseous or gas like and more difficult to store in terms of the required space (low density) and higher mobility (e.g. larger density contrast with formation water). 7.6 MPa corresponds approximately to the hydrostatic pressure commonly found at depths of 800 m and hence the rule to consider only reservoirs at depths below this limit. This rule is commonly taken one step further by assuming that the density of CO_2 can be approximated by some average value below the 800 m line. An additional, implicit element in this reasoning is accepting that the pressure in the reservoir is approximately in equilibrium with the hydrostatic pressure of the formation water at that depth.

The 800 m rule determines to a large part the estimated capacity of a geological reservoir, since the latter depends, like the capacity of any volume, on two basic parameters: the volume that is available, and the density of the fluid to store. The volume is estimated from parameters such as thickness and extent of the aquifer, effective porosity, compressibility factors, and accessibility factors. Aspects such as solubility are usually out of scope of the first assessments.

The 800 m rule is known to be inaccurate due to its sensitivity to the geothermal gradient, and is claimed to have resulted in overestimates in e.g. the Australian storage atlas (Spencer et al., 2010). This can also be seen in figure 1, where the isochors of pure CO_2 are shown as a function of pressure and temperature. It can easily be seen that the density of CO_2 can vary significantly at depths of 800 m, but also 1000 m and deeper due to a different geothermal gradient.

The first studies that have evaluated coal mines (e.g. May et al., 2003), realized that the coal mine reservoirs under consideration were only partly deeper than 800 m. In their conceptual approach, they divided the coal mine into two parts: one deeper and one shallower than 800 m. For the deeper ones a constant, liquid like density was assumed, and for the shallower one a more gaseous density. The important erroneous assumption is that the boundary between high and low density CO_2 is placed at a fixed, a-priori defined depth. It is important to realize here that the pressure inside a coal mine filled with CO_2 is not identical to that of hydrostatic pressure, because hydrostatic pressure requires a continuous column of water (such as is present for example in the pores of geological formations). If a coal mine is filled with CO_2, then the pressure is determined by CO_2 and should hence be called a CO_2 static gradient (cf. Lu 2010). The density of CO_2 is in geologic pressure and temperature conditions lower than that of water, and the pressure increase will therefore be lower. The transition between low and high density CO_2 will therefore occur at a different, usually deeper level. The exact position is however more difficult to calculate due to the sensitivity of the density of CO_2 for temperature.

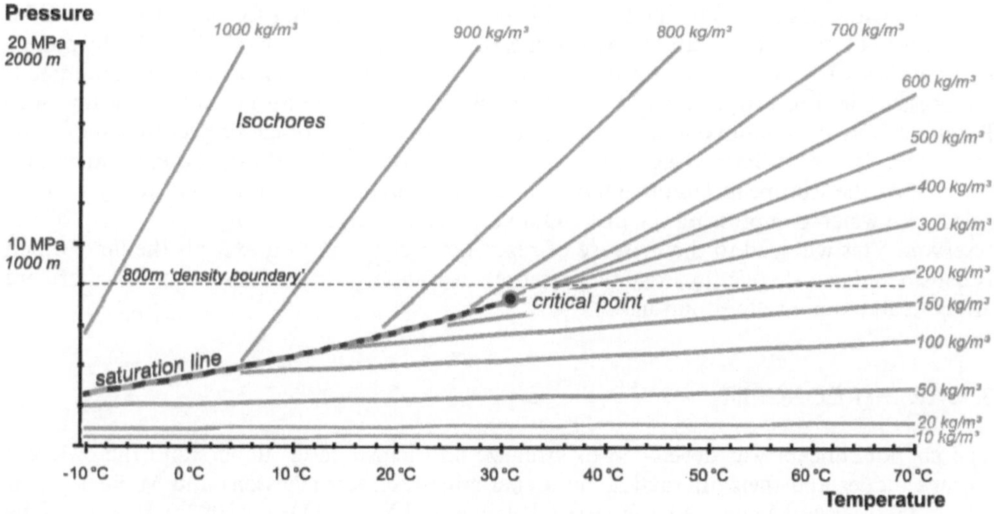

Figure 1. Density of pure CO_2 in function of pressure and temperature. Approximate depths corresponding to the pressures are indicated for reference, as well as the 800 m 'rule of thumb' boundary between high and low density of CO_2. Although widely spread, this rule should be used with extreme care. Temperatures at 800 m are generally between 30 and 50°C, allowing for density variations at this depth between 250 and 700 kg/m³, meaning that depth alone is not sufficient to even approximately determine the density of CO_2. Thermodynamic data based on Span and Wagner (1996).

2.2 *Natural gas storage concept applied to CO_2 storage in coal mines*

The storage of natural gas is practiced worldwide in order to compensate for seasonal or shorter incompatibilities in supply and demand. The large number of projects, over 400 in the USA alone, results in a significant amount of expertise. Different types of reservoirs are and have been used. This includes coal mines, although the experience is limited to only three projects (Peila and Pelissa 1995; Shi and Durucan 2005; D.J. Evans 2008; Lord 2009; Lu 2010) which are currently all abandoned. The longest one in operation was the Leyden coal mine in Colorado (USA) which has been in operation from 1961 to 2001 (D.J. Evans 2008). Two other projects are located in Belgium, the Anderlues and Péronnes coal mines, which were respectively operated between 1980 and 2000, and 1982 and 1996. The Leyden project was abandoned after leakage of methane into the surrounding aquifer was confirmed. Contrary to what is stated in literature (D.J. Evans 2008; D.J. Evans 2009a; D.J. Evans 2009b; D.J. Evans and Chadwick 2009) based on inaccurate referencing to (Piessens and Michiel Dusar 2004a), the Anderlues and Péronnes storage sites were closed due to purely economic reasons.

The storage of natural gas in coal mines shares several aspects with that of storing CO_2 in mines. The effect of adsorption of the stored gas onto the coal still in place plays for example in an analogue way. We will further use this to demonstrate that such effects can be expected to attribute to the storage capacity of a coal mine in real time. It should however be presumptuous to conclude from successful natural gas storage projects that CO_2 can be safely stored. The essential difference between natural gas storage projects and CO_2 storage projects is the dimension of time. Natural gas is normally not stored longer than a season, or in the extreme case of strategic reserves, not for longer than a few decades. It is therefore possible to control the reservoir conditions by human intervention. For all three coal mine storage projects, this meant that a safety rule could be adopted to keep the reservoir pressure at all times below hydrostatic pressure. The goal of such rule is to provide additional certainty because the escaping gas would have to migrate from a low pressure reservoir to the higher pressured host formations, which is called hydrodynamic trapping (Lu 2010). The operators of the Leyden coal mine have been accused of violating this rule, thus breaching the hydrodynamic trap.

The pressure gradient has however a side effect, in the way that formation water will be encouraged to flow towards and into the mine, which therefore needs to be kept dry by pumping activities. CO_2 should, from a human point of view, be stored permanently. Assuming that the pressure in a reservoir is kept below hydrostatic is therefore an unrealistic assumption. Pumping activity cannot be guaranteed for the duration of a storage project that should be 'permanently', as it is termed e.g. in the CCS Directive of the EU (European Parliament and Council of the European Union 2009). Without pumping, the mine will gradually become filled with water, compressing the free CO_2 in an increasingly small volume at the top of the reservoir. This will lead to the increase of reservoir pressure until it exceeds the limit which is considered as safe (e.g. hydrostatic pressure). An underpressured reservoir will therefore always lead to an unsafe situation.

3 JAR POT CONCEPT

The jar pot concept was developed to visualize and hence better understand the different storage modes and their interactions in a coal mine reservoir (Piessens and Michiel Dusar 2003a; Piessens and Michiel Dusar 2003b; Piessens and Michiel Dusar 2003c). It is especially useful at visualizing how the reservoir will evolve through time, and hence what the most critical points will be. Based on this qualitative understanding, the jar pot model is a useful guide for making quantitative calculations, which can be used as generic estimates or first estimates of actual coal mines (Piessens and Michiel Dusar 2002; Piessens and Michiel Dusar 2004a) and integration in geothermal schemes (Piessens and Michiel Dusar 2004b) which will not be further discussed in this paper. These quantification is embedded in the simulator CO_2-VR.

3.1 *The residual volume*

The jar pot model is a static model that simplifies a coal mine reservoir into a 6 containers or sub-reservoirs. It is conceptual in the sense that these containers bear similarities with a geotechnical reality, also in their interaction with each other, but are sufficiently simplified to avoid detailed calculations. It is therefore most useful for assessing the viability of using coal mines for storage purposes in general, and for making first evaluations of actual case studies.

The different containers and other elements are visualized in figure 2. The central element in the model is the residual space of the abandoned coal mine, which is the sum of the voids produced by mining, minus the volume lost by backfilling techniques, collapse, subsidence and other anthropogenic and natural processes. In the model, this volume is visualized as the 'jar' which will usually be or may become partly flooded. The complex design of a coal mine, basically consisting of mined panels connected by sub horizontal galleries and vertical shafts, is not an element of the model, because it is assumed that due to the high conductivity of the main galleries, the different lateral parts of the mine are sufficiently connected to be in pressure equilibrium. The vertical extension, between the deepest mined levels and upper levels or shaft filling (whichever is shallowest) will play an important role. Both elements, horizontal equilibrium and a large vertical extent, are visualized by the jar pot.

Part of the residual space may be dry, in which case it will dominantly be filled by methane, especially for well sealed mines that have been abandoned for some time. In most cases the mine will at least be partly flooded by formation water originating from the surrounding host-rock. This is integrated in the jar pot model, where the lower part can be filled with water, while the upper part remains dry.

The jar in figure 2 is sealed. This sealing corresponds to both the geological seal overlying the coal sequence, potentially disturbed by faults related to subsidence over the mined area, and the man-made plugs of the shaft. In the model, the potentially complex behavior of the seal is strongly simplified into two basic parameters: the minimum depth of the seal (where the reservoir pressure will be highest, as will be shown further) and a maximum pressure at which CO_2 can be stored safely. In the real world, this pressure constraint could determined

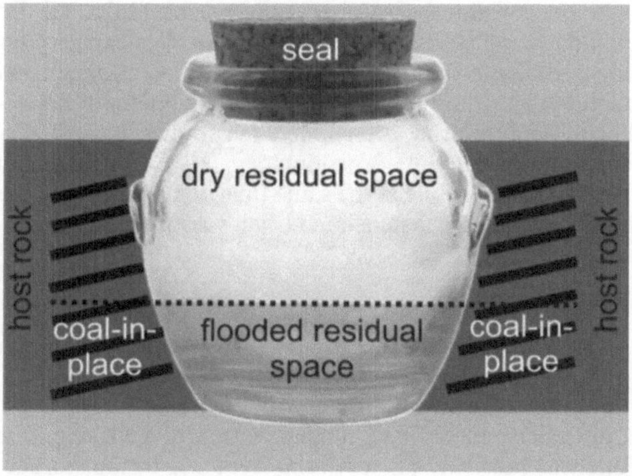

Figure 2. The different elements of the jar pot model, with the partially flooded residual space corresponding to the jar. Coal-in-place is adjacent to the jar. The natural and man-made sealing of the reservoir is represented by the seal of the jar, which can move in respect to the reservoir pressure to simulate the effects of absidence.

by e.g. the critical pressure for fault-valve behavior, the break-though pressure of the sealing formation for CO_2, etc.

When the pressure in a reservoir is restored, the area which was earlier subject to subsidence will partly recover (absidence). This is e.g. commonly observed in (overexploited) aquifers in which the pressure head is allowed to recover, but it may also occur over areas which have suffered from subsidence due to deep mining activities (Pöttgens and van Herk 2000; Devleeschouwer et al., 2008). The observation that the area over which absidence occurs corresponds neatly to the former area of subsidence demonstrates that the origin of absidence is primordially one of decompaction by restoration of the hydrostatic pressure of the collapsed mine panels and of the permeable rocks in connection to the mine. The pressure increases following storage of CO_2 will lead to pressures above the initial hydrostatic pressure, and hence enhance absidence. The related volumetric changes of the residual volume are taken into account in the model by allowing for a vertical movement of the jar seal in function of the reservoir pressure. The volume increase visualized by pushing the jar seal upward corresponds in terms of capacity to an increase in the volume of both the flooded and non-flooded parts of the residual volume. The movement shown in animations is not to scale, and will normally be in the centimeter to lower decimeter range.

3.2 *Coal in place*

When a coal mine is closed, coal reserves remain in place. These may be actual reserves that were not accessed for economic or due to the not optimal planning of the mining activities. Reserves may also occur at depth levels that are out of reach for commercial mine operations or in coal seams that are too thin to allow viable exploitation. Also coal close to shafts or galleries, or in pillars will not be mined. Finally also the separation of coal from the host rock is not a perfect process, leading to a certain percentage of coal in e.g. back-filled panels.

The presence of coal will add to the storage capacity of a coal mine by the adsorption of CO_2. The model simplifies the real-world situation by distinguishes only two groups of coal in place. The first group is coal in or close to the mined area, which can with some certainty be expected to take active part in the storage process. The capacity of this group will further be referred to as 'adsorption capacity'. The second group is coal that is found at a larger lateral distance or deeper level, or in compartments that do not readily communicate hydrodynamically with the coal mine, e.g. due to impermeable faults. The storage potential of this group is termed 'additional storage capacity'. This storage capacity will only be considered to attribute only partially to the storage potential and on a longer time frame than that of the injection activities. These are included because the line between accessible and non-accessible coal will be difficult to draw, even in detailed studies. Introduction of the concepts of adsorption capacity and additional adsorption capacity encourages making conservative calculations based on the former, while at the same time demonstrating that the capacity may be larger if CO_2 migrates further or faster than initially assumed.

3.3 *Storage modes*

The jar pot model identifies three storage modes for CO_2: free-space (CO_2 dominated gas mixture in the residual space), solution (CO_2 into solution in formation water in the residual space) and adsorption (CO_2 adsorption onto coal in place).

The model does not take into account mineralization of CO_2 (e.g. due to carbonation) because it will normally at best only on very long term add to the storage potential in sedimentary settings.

Fine grained rocks such as silt or shale constitute large volumes in many coal sequences, and these rocks are known to possess a potentially significant adsorption potential as well. This adsorption potential is not explicitly taken into account in the jar pot model because reliable data on their adsorption potential is usually lacking. It is also not clear, in terms of volumes or time, to what degree they will add to the overall storage capacity of a coal mine. If such data becomes available, then it can easily be accommodated for in the model.

3.4 *Hosting parameters*

The capacity calculations of the different storage containers will rely on a correct estimation of the volume of these sub-reservoirs and a proper equation-of-state (EOS) for a CO_2 dominated gas mixture, solution of CO_2 in formation water, or adsorption of CO_2 onto coal. Any of these EOS requires an accurate estimation of the pressure and temperature.

The temperature in the reservoir is considered to be in equilibrium with the regional geothermal gradient. This is usually a fair assumption even when mines have been cooled for decades passively by ventilation or with more active techniques. The geothermal temperature is calculated in CO_2-VR from a temperature at a known depth (usually the average surface temperature at depth 0) and a fixed geothermal gradient. CO_2-VR is very flexible, and standard equations can readily be modified where required in function of the available data.

The jar pot model assumes that the hydrostatic pressure governs outside of the residual volume and the coal in place (for calculation of the adsorption isotherms) and in the overburden and seal of the mine. Regarding coal, it makes abstraction of the reserves that are present in close contact with the residual volume, e.g. in gob in the mined panels, and they are drawn outside of the jar. However, the pressure conditions and the water level for the coal in place are assumed to be identical as those calculated for the residual space.

Assuming a hydrostatic gradient in the formations overlying the mined area is a logical consequence of assuming the presence of a sealing formation. Keeping the mine dry during mining activities should not have affected the overlying water tables.

3.5 *Jar pot pressure*

The pressure gradient in the residual space is a more complex issue, and should, as has been mentioned earlier, not be considered to be in equilibrium with the hydrostatic pressure in the surrounding hosting formations. Instead it is dictated in the non-flooding part by the density of the CO_2-mixture, which vertically integrates to a pressure gradient. The density of CO_2 is at geologic temperature conditions always higher than that of water. The resulting pressure gradient will therefore be less steep (the pressure will increase less with depth) than that of the hydrostatic pressure that governs in the hosting formations (Piessens and Michiel Dusar

2002). Similar differences between pressure gradients in and outside of the reservoir are seen in reservoirs for natural gas (Lu 2010).

CO_2-VR calculates the CO_2-pressure gradient in a discrete, recursive approach using pressure at a certain depth to calculate density, and density to estimate the pressure at the levels above or below this depth (see Piessens and Michiel Dusar 2003a for details). It thus avoids the mathematically complex operation of integrating the equations of state for gas mixtures, and at the same time gives the flexibility to interchange between different equations of state for calculating the density of CO_2.

The CO_2 pressure gradient can in principle be calculated if the volume of the container and the amount of CO_2 present is known, but given the uncertainties on these values more reliable gradients are obtained by assuming a known pressure at a certain depth. In principle any combination can be given, but there are a few boundary conditions that are particularly useful because they correspond to real-world equilibrium conditions. The most obvious one is assuming the maximum pressure allowed by the seal at the top of the coal mine, which corresponds to the calculation of the maximum capacity of the coal mine. Another is defining the level of flooding of the coal mine to determine the corresponding total capacity and maximum reservoir pressure. This second approach not only makes use of the CO_2 pressure gradient in the non-flooded part, but also of the hydrostatic pressure gradient in the flooded part of the mine. Indeed, where the pressure gradient in the non-flooded part is low compared to the hydrostatic pressure increase, the pressure gradient in the flooded part will (nearly) be parallel to that in the host rock, but not necessarily at the same level. The effect of this will be discussed in chapter 6.

3.6 *Container boundary conditions*

The jar pot model implicitly assumes some basic rules by which the different reservoir elements interact, more specifically how CO_2 and formation water can flow between the different containers.

CO_2 is assumed to be injected in the residual space. This may be a gallery or a collapsed panel. From here, it is assumed to migrate immediately through the whole residual space (the jar), where it will occupy the free space or move into solution. It will also without time delay occupy the coal in place. Such an assumption is corroborated by the pressure evolution in coal mines used for methane storage, which will be discussed in the next chapter. The assumption of immediate migration and equilibration is important because of the static nature of the model.

The walls of the jar therefore form no boundary for CO_2, but they are assumed to slow down the infiltration of formation water into the mine. Again there is physical evidence for such an assumption in the sense that especially deep coal mines may take several decades to become completely flooded again, which is relatively long compared to the anticipated duration of CO_2 injection in coal mine reservoirs.

Migration of CO_2 and formation water is lateral, or potentially allow for migration between the residual space and deeper, unmined reserves. The seal is assumed to be perfect, therefore excluding vertical migration towards the overlying strata and the surface.

4 CAPACITY ESTIMATIONS

The jar pot model identifies two main containers: the residual space (jar) and the surrounding coal in place. The volume calculation of the residual space is split in that of the actual and the additional residual space, because they are based on completely different estimation techniques.

4.1 *Residual space volume*

Residual space volume is relatively easily defined by the net amount of rock volume extracted, corrected by the amount of volume that is lost due to subsidence.

$$V_{residual} = (V_{extracted} - V_{backfilled}) \cdot n_{residual}$$

With:
$V_{residual}$: *residual volume*
$V_{extracted}$: *rock volume extracted*
$V_{backfilled}$: *volume of rock used in back-filling techniques*
$n_{residual}$: *residual volume fraction.*

The accuracy of the estimation of the residual space of an abandoned coal mine will depend on the data that was collected during its operation. The record of the amount of coal excavated is usually quite accurate, including the quantities of other rocks brought to the surface. The amount of material used for backfilling, and details on the backfilling techniques, can also be taken into account when they are available. For a few existing mines in-situ measurements of residual space volumes fractions have been derived. It is clear that these parameters vary for different parts of the mine, and the calculation can detailed for well documented mines.

In most cases the extracted volume will be known in sufficient detail for the whole of a mine. It is the estimation of the correct residual volume fraction that is more difficult, as these may vary significantly in function of the coal basin and the mining techniques. Most of these values, of which an overview is given in table 1, originate from grey literature and the underlying methodology, data and accuracy is not always well defined. Errors or

Table 1. Overview of different residual volume fractions found in or derived from literature (Berding 1952; Kunz 2000; Labasse 1965; Peele 1952; Malolepszy and Ostaficzuk 1999; Schultz 1998; van Tongeren and Laenen 2001).

Residual volume fraction (%)		Remarks	Author
Limburg-Campine-Ruhr coal mines (7 ± 3)%	5–10%	Ruhr basin. Collapsed panels and infrastructure. Depth > 800 m.	Kunz, 2000
	5%	(Campine) Belgian coal mines. Back-filled areas.	Labasse, 1965
	4.5%	The Netherlands, Limburg, abandoned panel from Maurits mine. 40% back-filled, 60% collapsed. Depth ~500 m. Value recalculated from original data.	Berding, 1952
	8%	Campine coal mines. Collapsed panels.	van Tongeren & Laenen, 2001
	5.5%	Campine coal mines. Back-filled panels.	van Tongeren & Laenen, 2001
Other European coal mines (20 ± 5)%	20%	Southern Belgian coal mines of Anderlues and Peronnes.	van Tongeren & Laenen, 2001
	18%	Nord-Pas-de-Calais basin.	van Tongeren & Laenen, 2001
	20%	Silesian coal mines (Poland). 50% back-filling, 50% collapsed. Water saturated. Possible contact with porous sandstones.	Malolepszy & Ostaficzuk, 1999
Non-European coal mines	12%	American coal mines in Carboniferous. Depth ~400 m.	Peele, 1952
	75–90%	Leyden mine, Colorado, Denver. Room-and-pillar, completely collapsed, 15 m of roof rocks affected, no evidence of subsidence. Depth 250 m. Value calculated from original data. Overestimation because no data on rock extraction.	Schultz, 1998

uncertainties in the residual volume fraction will have large effects on the total estimated capacities.

An alternative approach is to calculate the residual volume from the mined volume and the total subsidence. In principle, the collapse of mined voids will eventually result in subsidence at the surface. As such, the residual volume can be calculated as:

$$V_{residual} = V_{extracted} - V_{subsided}$$

With:

$V_{subsided}$: *volume of subsidence (area times average subsidence)*

This approach of course requires an accurate topographic survey before and after mining activities. These can be accurately and relatively easily be performed to date with satellite measurements (e.g. PS-InSAR techniques). Such data sets however do not date back sufficiently, and also show that subsidence may persist in some mining regions for decades after closure, further hampering its use to obtain accurate estimates.

In a few cases, the effective porosity of back-filled or collapsed panels may have been determined. This is very useful data, but one should not make the mistake to equivilate effective porosity with the residual volume fraction. It is in principle possible to calculate the residual volume from effective porosity, but it requires that the corresponding volume of rock affected by subsidence of a certain panel can be determined and that the effective porosity of the original rocks is low. This is visualized in figure 3, together with the corresponding formulas.

4.2 *Additional space volume*

Although the largest part of the subsidence over a mined area can be explained from collapse of the underground space that was mined out, part of it is also due to the pressure decrease resulting the dewatering of the mine and the surrounding formations during the mining activities. This secondary subsidence is an elastic phenomenon and will result in absidence or ground-level recovery in the area that was formerly subjected to subsidence. This behavior is rarely monitored because it rarely or never causes damage. The ground level movement is around an order of magnitude smaller than subsidence related phenomena, not differential and more gradual.

The few studies that have looked into it have demonstrated a clear coincidence with the former area of subsidence and the recovering groundwater level (figure 4; Pöttgens and van Herk 2000), and the observations are well explained by the theory of poro-elasticity (Pöttgens and van Herk 2000; e.g. Sroka, Tajdus, and Preusse 2011). Central in this theory is that the observed absidence is a direct effect of the increased volume of the (collapsed) mine

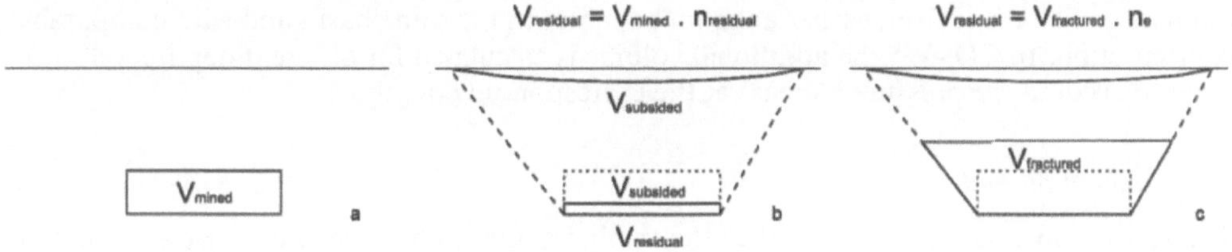

Figure 3. Illustration of the difference between residual volume fraction and effective porosity (n_e). (a) shows the initial void created by mining, which due to subsidence is decreased to the residual volume in (b) by a factor that is equal to the residual volume fraction. Effective porosity is not a parameter of the mined volume, but describes the effective porosity of the tock volume that was affected in a brittle way (fractured) by subsidence, such as is shown in (c). If this volume can be determined, and if the effective porosity of the initial rock is low, then the residual volume can also be calculated from this parameter. This has e.g. been done on the basis of the data in Labasse (1965).

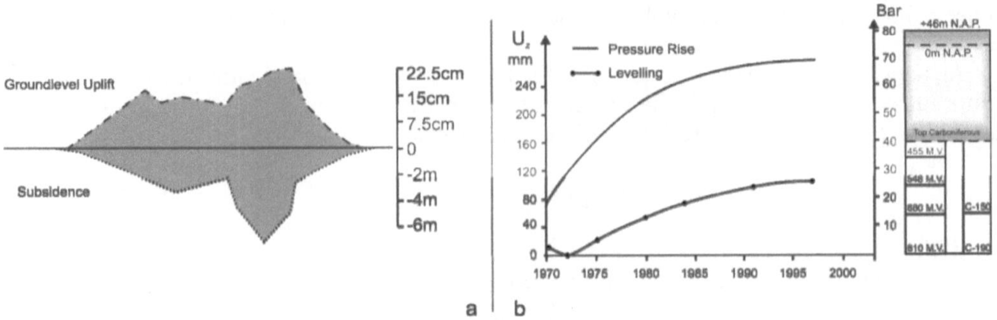

Figure 4. Data from topographic leveling at the Oranje-Nassau mine. (a) shows that subsidence due to mining activities was over 6 m. After closure of this mine, ground level uplift up to 22 cm was recorded. Note that the areas of subsidence and groundlevel uplift form a perfect match. (b) shows the relation between the ground level recovery (dotted line) measured at one location, with the registered ground level rise in the coal mine. (after Pöttgens and van Herk 2000).

workings at depth due to the increasing pore water pressure. The concept of additional space volume uses these properties to estimate the volume increase in function of the pressure in the residual space of the mine reservoir. This may potentially exceed the original hydrostatic pressure, as will be shown further.

The poro-elastic theorem can be applied directly and in detail, but a more direct calculation is possible for areas for which subsidence and (expected) absidence are known. In those instances the linear relation between pressure and volume increase, which is a central element of the theory of poro-elasticity, can be used directly to estimate the effect of pressure in terms of volume increase using the following formula:

$$V_{additional} = V_{residual} \cdot \frac{P_{reservoir}}{P_{hydrostatic}} - \frac{h_{absidence}}{h_{subsidence}}$$

With:
$V_{additional}$: *additional volume created by increase of reservoir pressure*
$P_{reservoir}$: *pressure in the mine when filled with* CO_2
$P_{hydrostatic}$: *normal hydrostatic pressure in the mine reservoir*
$h_{absidence}$: *ground level recovery after restoration of the hydrostatic pressure*
$h_{subsidence}$: *observed subsidence over the mined area*

The ground level recovery $h_{absidence}$ is not known for the mine under evaluation, but a reliable estimate can be made from earlier abandoned mines in the same basin and with comparable configuration. In CO_2-VR the additional volume is calculated for different depth levels, taking into account the pressure increase at the corresponding depth.

4.3 *Amount of coal-in-place*

The amount in coal that remains after the mining activities have ceased is also a value that can be calculated in detail if sufficient information is available. There will at least be estimates on the reserves of coal that were not mined. Other data, such as on thin coal seams or coal not mined for technical reasons, will add to this number. More difficult is to judge how much of this coal will be accessible for the CO_2 injected in the residual volume of the coal mine. It is for this reason that a difference is made between readily accessible and additional adsorption capacity, as has been mentioned earlier.

5 EQUATIONS OF STATE

Three storage modes are discriminated in the jar pot model, each to be described by a proper equation of state: CO_2 in free space, in solution and adsorption onto coal.

5.1 *Free space*

CO_2 in free space will occur as a gas, a liquid or in supercritical state. For correctly estimating the capacity, it is important to reliably calculate the density of CO_2 which is a function of temperature, pressure and the composition of the gas mixture. The conditions in a coal mine will often approach critical conditions and the EOS therefore needs to be accurate for densities close to that of the critical point and the saturation line. This requirement rules out the use of the ideal gas law, but also of many of the enhanced forms of this law, using for example empirical cubic corrections.

The most accurate EOS are those formulated for pure substances. CO_2 in coal mines is therefore currently considered to be pure, although it can be expected that even when pure CO_2 is injected it will become contaminated in the reservoir with non-negligible amounts of CH_4, N_2, Ar, O_2, CO, light hydrocarbons and other gasses. CO_2-VR uses by default the EOS of Span and Wagner (1996), although Duan and Sun (2003) can alternatively be selected.

5.2 *Solution*

CO_2 is relatively well soluble in water due to the hydration of CO_2 to form bicarbonate, although the storage potential is usually about an order of magnitude smaller than in free space. The quantity of CO_2 in solution is calculated as a function of pressure, temperature and salinity. The equation of state used is Henry's law, which is corrected for non-ideal behavior and non-standard conditions with activity coefficients to account for the effects of temperature and salinity. Also the vapor pressure of water is taken into account to correct total pressure into partial CO_2 pressure. Piessens and Dusar (2002) have compared data and equations from several authors (Takenouchi and Kennedy 1965; Barta and Bradley 1985; Enick and Klara 1990; Rumpf and Maurer 1993; S. Holloway 1996) to optimize the equation and the activity coefficients. Reliable predictions are obtained between 25 and 200°C below 10 MPa for salt-free systems, and between 50 and 200°C below 10 MPa at all salinities. Error analysis indicates an accuracy better than 10% or 2 g/kg. The reliability of this interval is about 95% for salt-containing systems, and better for salt-free systems. Details on the complete set of equations as it is implemented in CO_2-VR can be found in (p. 18 in Piessens and Michiel Dusar 2002).

5.3 *Adsorption*

The adsorption behavior of CO_2 onto coal can in general be reliably described using Langmuir based equations. CO_2-VR uses dual-site Langmuir equations that are more accurate over larger pressure ranges. The adsorption behavior however also depends on temperature, the presence of water and the presence of other gases, potentially already adsorbed onto coal. Also the coal properties strongly affect the adsorption behavior. This range of properties makes it difficult to predict (or even describe) the adsorption behavior in well controlled experiments on finely grinded coal samples. Adsorption on in-situ coal will additionally be influenced by the swelling behavior of coal which may dramatically reduce the permeability in confined conditions and as such effect the migration of CO_2 into the coal.

The magnitude of these effects is difficult to estimate, and the model assumes that due to the complex structure of a mine the contact with the coal-in-place is sufficiently large that CO_2 will readily migrate into coal in spite of swelling and other effects. The adsorption behavior of coal can then be expected to be similar to that of small-scaled experiments on grinded samples. This hypothesis is supported by the observed pressure behavior of mines that have been used for CH_4 storage. Those showed a pressure pattern that can only be explained by the adsorption and desorption of CH_4 (figure 5). These pressure cycles also show that equilibrium conditions can be approached within months after seasonal production of injection of CH_4.

Predicting the adsorption behavior of CO_2 onto coal accurately is currently not possible without empirical work on relevant coal samples taken from different seams from the coal basin under consideration. This data is however not readily available for all case studies or in generic studies. Therefore a different technique is used to estimate the adsorption behavior of coal based on experiments on active coal, with correction factors derived from basic properties of natural coal. Active coal is used as a reference because different experiments have demonstrated the reproducibility of the isotherms, even when performed on different samples with different experimental set-ups (Coppens 1967; Dreisbach F., Staudt R., and Keller J.U.; Van Der Vaart et al., 2000), and because the isotherms can be accurately described using Langmuir coefficients. Also the effect of temperature on adsorption is sufficiently studied, and the Langmuir coefficients can be reliably corrected to predict the effect of temperature on the overall adsorption behavior in a pressure range relevant for CO_2 storage in coal mines. A best fit of the isotherms in function of pressure and temperature, together with the experimental data, is shown in figure 5.

These isotherms form the basis for estimating the adsorption potential with CO_2-VR, but need to be corrected (reduced) for natural coal. This is done using the semi-empirical relation shown in figure 6, which shows a decrease of the adsorptive properties of natural coal in function of an increase in volatile matter. The quantity of volatile matter is corrected for ash-content and water. The equation assumes that even the theoretical situation of natural coal free of volatile matter will absorb only approximately one third of the CH_4 that would be absorbed on active coal. This fraction corresponds to the ratio of the internal surface of an average natural coal versus active coal (Van Krevelen 1993; Van Krevelen; Haan 1998 and

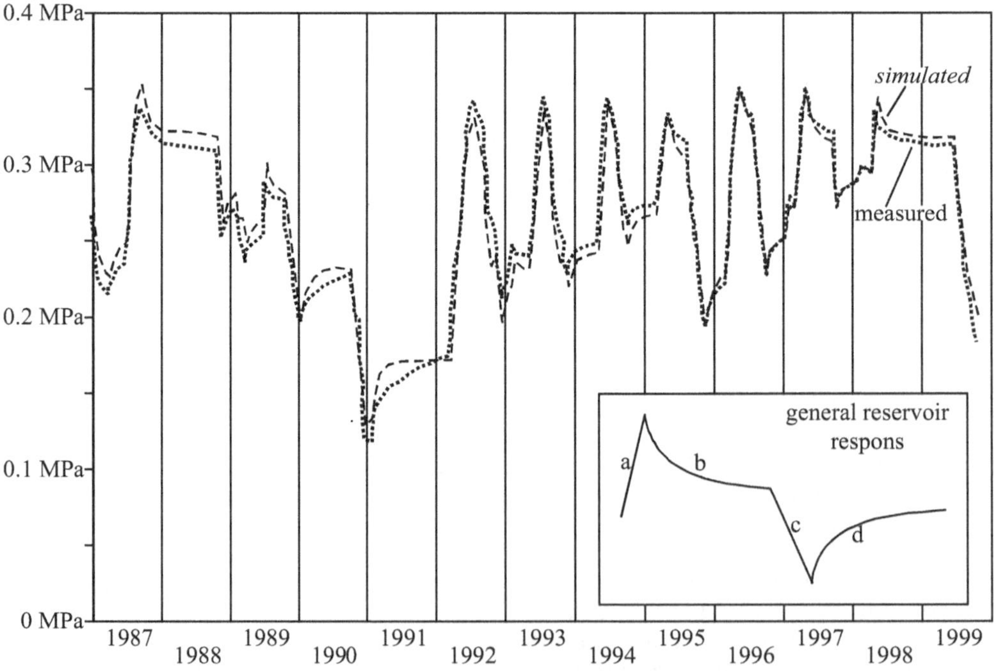

Figure 5. Reservoir pressure of the Anderlues storage facilities (Belgium) from 1987 to 1999. The inset shows the relation of the reservoir pressure to injection-production. (a) Injection resulting in a steep increase of pressure. (b) After injection has stopped, the reservoir pressure decreases gradually as CH_4 becomes adsorbed to coal. (c) Extraction results in a sharp decrease of reservoir pressure. (d) After extraction, the reservoir pressure increases again as CH_4 becomes desorbed from coal.

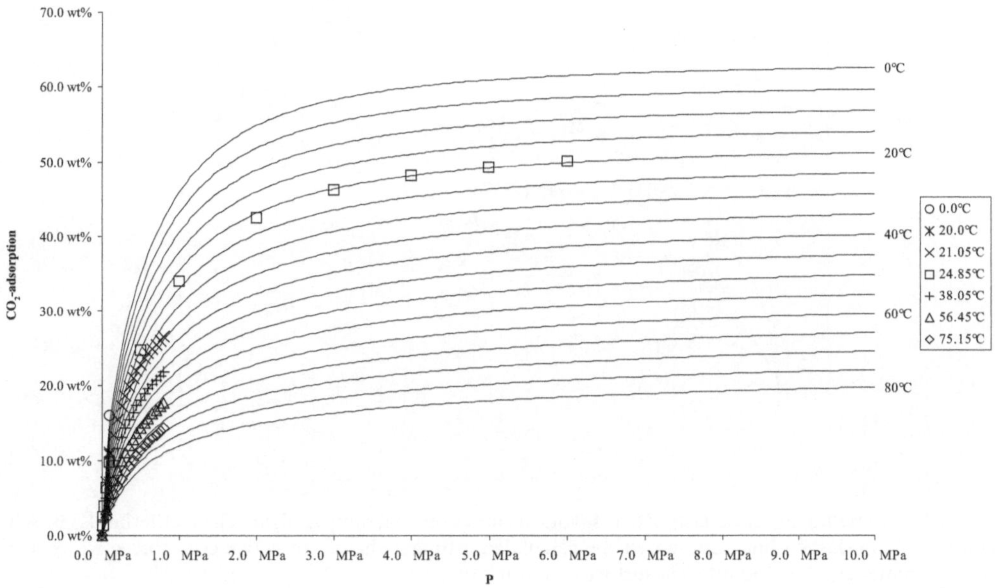

Figure 6. Data on adsorption of CO_2 on active coal from different authors (Coppens 1967; Dreisbach F., Staudt R., and Keller J.U.; Van Der Vaart et al., 2000) at different pressures and temperatures, plotted on the best fit dual-site Langmuir isotherms as used in CO_2-VR.

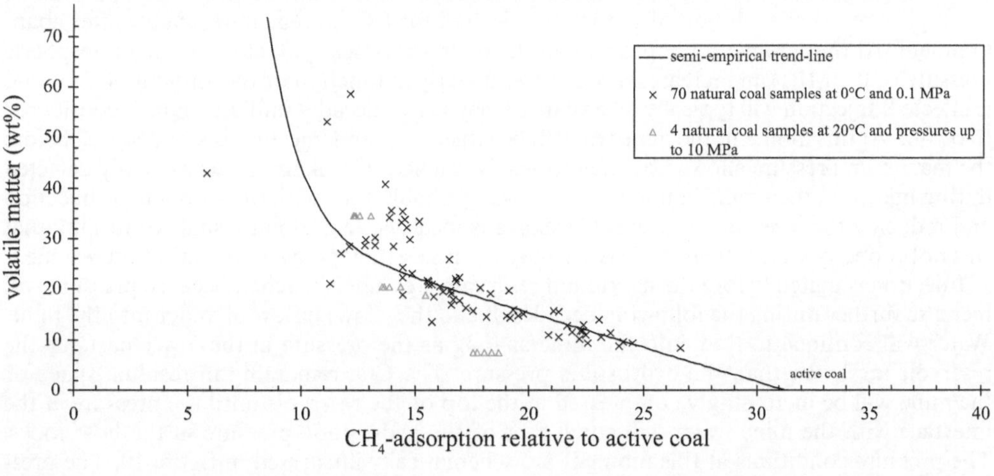

Figure 7. Data on CH_4 adsorption on natural coals corrected to standard conditions from Coppens (1967) plotted on a semi-empirical curve that is used to correct the adsorption behavior on active coal to natural coal. This relation is used in case no accurate experimental data is available.

product information on www.norit.com). The relation is not linear because overlap of volatile matter is assumed in the equation, and adsorption of CO_2 on volatile matter is assumed, leading to an upward bend in the curve towards the theoretical end member of a coal consisting of 100% of volatile matter.

Figure 8 shows how the estimated adsorption behavior compares to that of three experimentally determined isotherms (Busch et al., 2003). The predictions are in range of the observed curves, and can be used for first order evaluations. It is however clear that experimental data is required for more detailed evaluations.

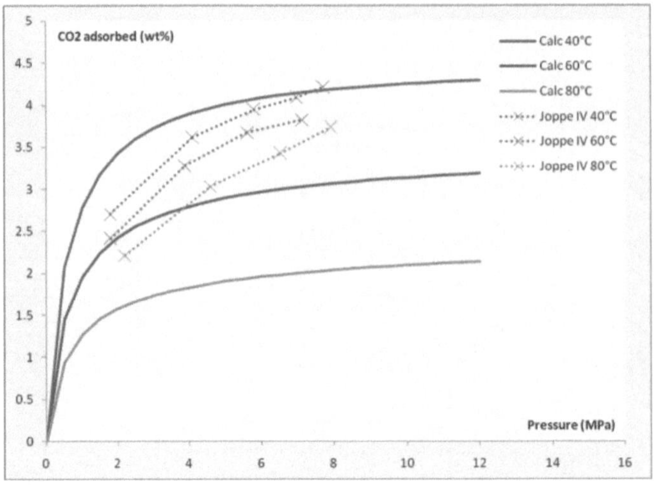

Figure 8. Experimentally determined adsorption curves on coal samples from The Netherlands (Busch et al., 2003) in dashed lines, compared the calculated estimates based on active coal corrected for the volatile matter content and internal surface of natural coal.

6 PRESSURE EVOLUTION AND LEAKAGE SCENARIOS

The pressure evolution in a coal mine reservoir is different from e.g. an aquifer that is used for CO_2 storage, but can be easily appreciated from the jar pot model. Figure 9 shows a generic, but typical pressure evolution of a mine that is used for CO_2 storage immediately after abandonment. At that moment, the pressure in the reservoir is nearly 0 (close to the atmospheric pressure of 0.1 MPa) an increases more or less linearly in function of the amount of CO_2 that is injected. Injection will typically take years to maybe two decades until the mine is considered to be full. At this moment, the mine will still be largely dry, and injection is stopped well before the maximum pressure allowed by the top seal is reached. If leakage is unexpectedly detected during injection, then remediating measures will probably start with the stopping of injection and reducing the reservoir pressure. If leakage is localized, e.g. along a shaft or drilling that was not properly sealed, then the leakage may be successfully plugged and injection resume.

Injection is halted before the maximum reservoir pressure is reached because pressure will increase further during the following decades due to the (slow) inflow of water into the mine. Water will continue to flow into the mine as long as the pressure in the lower parts of the reservoir are lower than the hydrostatic pressure. The CO_2 remaining in residual space of the mine will be increasingly compressed at the top of the reservoir until the pressure at the interface with the mine water will equal that of the hydrostatic pressure in the host rocks. The pressure conditions at this moment are schematically illustrated in figure 10. The pressure in the flooded part of the mine (level B to C) is equal to that of the hydrostatic pressure in the host rocks. The pressure gradient in the non-flooded part is different because of the lower density of CO_2. The reservoir pressure in the dry part is equal to hydrostatic pressure at point B, and increase to pressures above hydrostatic in the shallower levels. The inflow of formation water, which was driven by lower than hydrostatic pressures in the reservoir, will therefore come to an end. At this point, the maximum pressure condition is reached because the free CO_2 will no longer be compressed by the inflow of water. Instead, CO_2 will now start to migrate laterally out of the reservoir thereby reducing the vertical column of CO_2 and hence the pressure at the top seal. It is an essential element of coal mine storage that the pressure will typically reach a maximum only decades after the injection of CO_2 has stopped. If the pressure evolution is misjudged and increases during this time above the maximum allowed pressure, then leakage of CO_2 can be expected. Monitoring of the mine reservoir therefore needs to be continued until the reservoir pressure reaches its maximum.

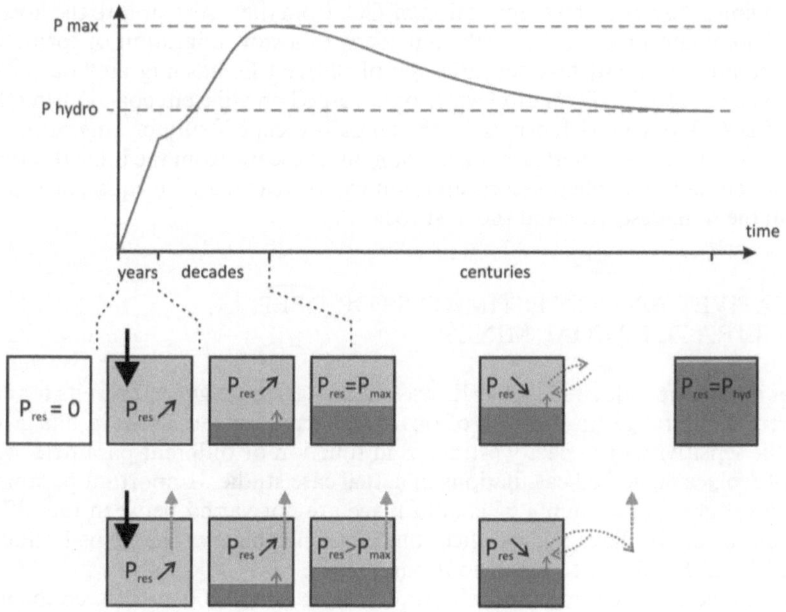

Figure 9. The pressure evolution of a typical coal mine, with an approximately linear increase of pressure during injection (black arrow), a further increase of the reservoir pressure due on the inflow of formation water until the maximum reservoir pressure is reached ($P_{res}=P_{max}$), after which the reservoir pressure will slowly decrease due to the migration of CO_2 from the reservoir into the host rock. The up grey arrows indicate that leakage will most probably occur from the reservoir, except when CO_2 starts moving out and away from the reservoir.

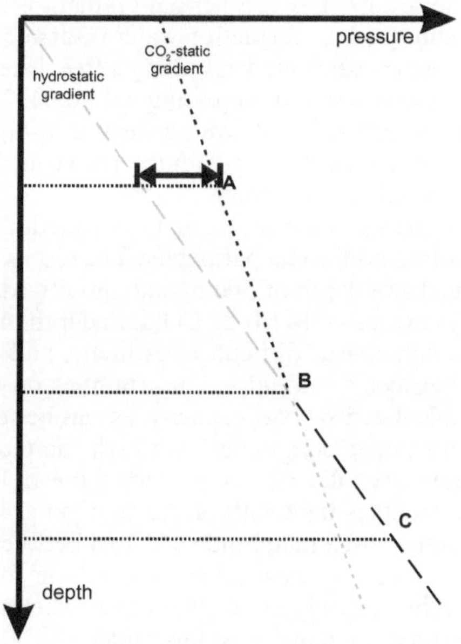

Figure 10. Schematic presentation of the effect of the different slopes of the hydrostatic and the CO_2-static pressure gradient at the maximum pressure situation. At this moment, the pressure in the reservoir will be equal to hydrostatic in the flooded part of the residual space, and above hydrostatic in the dry part. The pressure difference between the reservoir and the hydrostatic pressure will be highest in the shallowest part of the reservoir, this is at the top seal, which makes this the most vulnerable part in terms of potential leakage.

In a following phase, the slower migration of CO_2 from the reservoir into the host rock will become the dominant process. This will coincide with a slow migration of formation water into the reservoir which may take centuries. Until all free CO_2 has migrated out of the reservoir to become dissolved in formation water or adsorbed on adjacent coal. During this lateral migration, free CO_2 or CO_2 rich brines may potentially escape if suitable migration routes are encountered, but this is in general less likely than direct escape from the mine reservoir at earlier stages. Eventually a stable pressure situation will be reached with equal, near hydrostatic pressures in the mine reservoir and the host rock.

7 INDICATIVE CAPACITY ESTIMATES FOR DEEP AND ULTRA DEEP COAL MINES

The jar pot model, embedded in CO_2-VR, was used to generate generic results for coal mines with different depth ranges in function of one variable parameter. These results give an indication of the sensitivity of capacity estimates in function of different parameter values, but they do not replace dedicated calculations in actual case studies. Important parameters such as reservoir volume and amounts of coal in place are not varied between the different calculations, but of course have a large effect on the total capacity. The focus in the following comparison is on the effect of less obvious parameters.

Calculations were made for a mine with a vertical extent of 200 m between the lower level and the shaft seal, although the depth of the mined levels is a variable with the depth in the graphs referring to the shallowest point of the reservoir (e.g. a depth of 800 m corresponds to a mine between 800 and 1000 m deep). Further default values are a mined volume of 10 million m^3, a residual volume fraction of 10%, the area affected by subsidence/absidence of 50 million m^2 and an expected absidence of 3 cm after restoration of hydrostatic pressure. The maximum allowed overpressure above hydrostatic on the top seal is 30%. Adsorption on coal is not taken into account in these comparative calculations to facilitate interpretation of the differences in the results. The geothermal gradient is set at 0.03 K/m starting at 10°C at surface level. The salinity of the formation water is set at 5 wt% total dissolved solids (TDS). The hydrostatic gradient is calculated assuming a free water table at a depth of 10 m, with density starting from pure water and increasing with 0.0437 kg/m^3 every meter due to increasing salinity. The calculations are made with a vertical resolution of 10 m. The default parameters are not necessarily typical for a coal mine reservoir. It is in fact a rather small reservoir with a moderate ground level recovery.

Most of the results show the variation of capacity in function of the depth of the mine reservoir and the variation of an additional parameter. The discussion will focus on the comparison of a deep mine sealed at a depth of 500 m, and an ultra deep mine sealed at a depth of 800 m. Capacity is always expressed in Mt of CO_2 and depth in meters.

The residual volume fraction can be difficult to estimate, and can vary according to the situation and the authors between 5 and 90%. Uncertainties on this parameter can have a considerable effect on the calculated storage capacity, as can be seen in figure 11. The effect is especially large for the ultra deep coal mines, although the factor by which the capacity increases is identical (e.g. capacity increases by a factor 5 for an increase of 5 to 25%). The effect that dominates these results is the depth of the coal mine. This results in a non-linear increase of the storage capacity with a magnitude of 3 to 4 between a deep and an ultra deep mine. This immediately shows that shallow mines will only in rare cases be interesting for CO_2 storage because of their intrinsically low capacity even when their volume would be very large. Deeper mines on the other hand are most favorable.

An even more explicit effect of pressure on capacity is shown in figure 12 where the tolerable amount of overpressure is varied between 0 (only solution storage) and 50% (a probably unrealistic high value). The lines of equal capacities are closely spaced in a certain area of the graph. This corresponds to the pressures conditions at which the CO_2 in the reservoir changes from a gaseous (or low-density supercritical) state to liquid (or high-density supercritical). A deep mine at 30% of overpressure will only be in the transition between gaseous

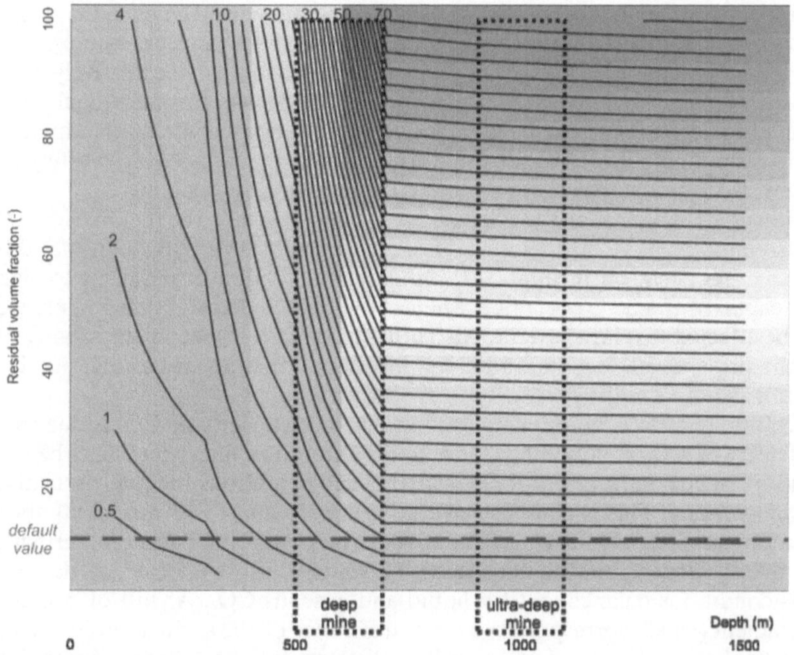

Figure 11. The capacity of a coal mine in function of its depth and the residual volume fraction. As can be seen from table 1, the estimates on residual volume fractions vary between 5 and 90%. Calculations for this generic mine were made assuming a default value of 10%, which may be rather conservative. This graph particularly shows the benefits of using deep and ultra deep mines for the storage of CO_2.

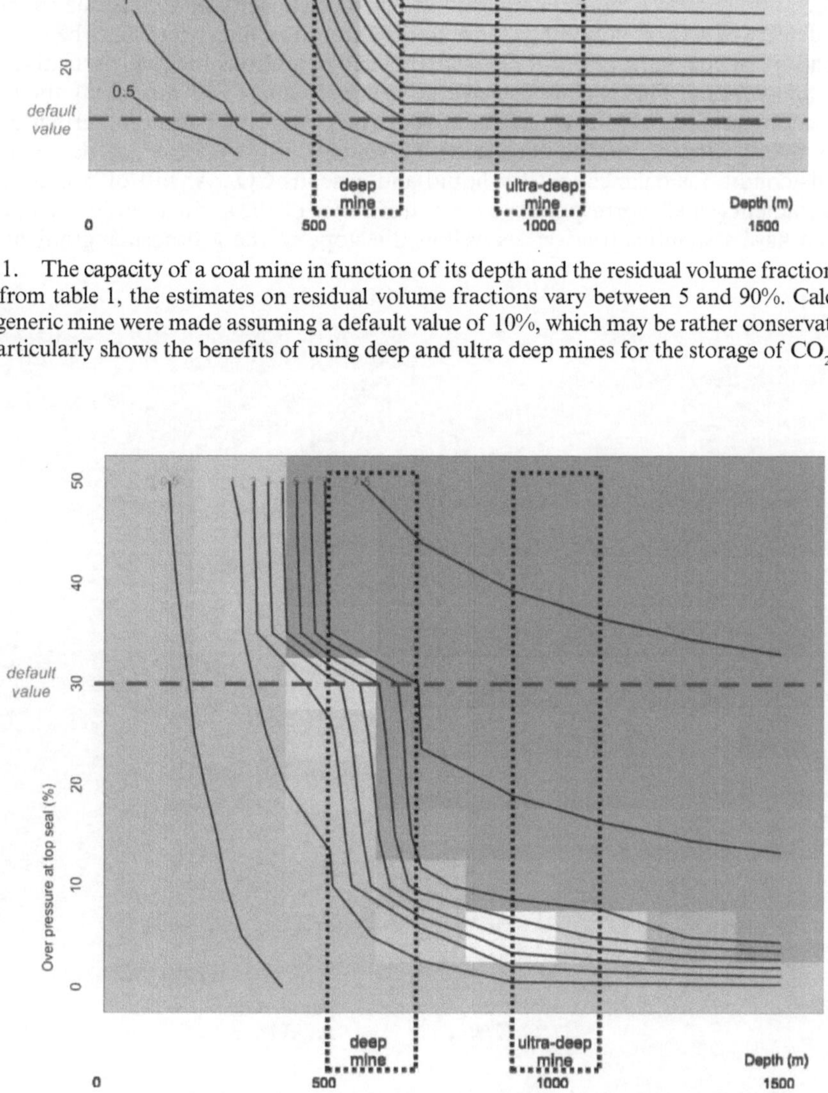

Figure 12. The capacity of a coal mine in function of its depth and the maximum overpressure at the top seal. Especially the capacity of deep mines is sensitive to changes in the maximum reservoir pressure because small changes may induce phase transitions between liquid and gaseous CO_2.

and liquid-like CO_2, and in fact contain both (see e.g. Piessens and Michiel Dusar 2004a for a discussion on the depth distribution of the different phases). At lower pressures, the reservoir will only contain gaseous CO_2 and its capacity will correspondingly be relatively low. Pressure conditions at which CO_2 will be present in a liquid, high density form are much more readily achieved in ultra deep mines, where overpressures of 10% above hydrostatic can be sufficient. This element is not only favorable for the storage capacity, but also for the safety considerations in view of the pressure evolution outlined in the previous chapter.

Since pressure is expressed in function of hydrostatic pressure, the regional level of ground water is also important. In arid regions, the ground water table may be deeper than 100 m below surface. As is shown in figure 13, a difference of 50 m is sufficient to significantly reduce the capacity of a deep coal mine, whereas the capacity of ultra deep mines would normally not be affected to a large extent. Also other parameters, such as the salinity of ground water or the presence of confined aquifers and discontinuities in the regional hydrostatic gradient, can all have similar effects.

The effects of relatively minor changes in the storage capacity of CO_2 is due to the sensitivity of density of CO_2 at conditions close to the saturation line, which may be expected to occur in a typical coal mine reservoir. Density in such conditions is equally sensitive to variations in temperature. This is demonstrated in figures 14 and 15 by modifying the reservoir temperature by respectively varying the surface temperature and the geothermal gradient. It is in both situations again the deep reservoir which is most affected by these variations because its conditions balance that for liquid and gaseous CO_2. A shift of 5°C in the temperature gradient (which corresponds to a 5°C difference of the average surface temperature) can indeed have a significant effect, as well as the slope of the geothermal gradient. These

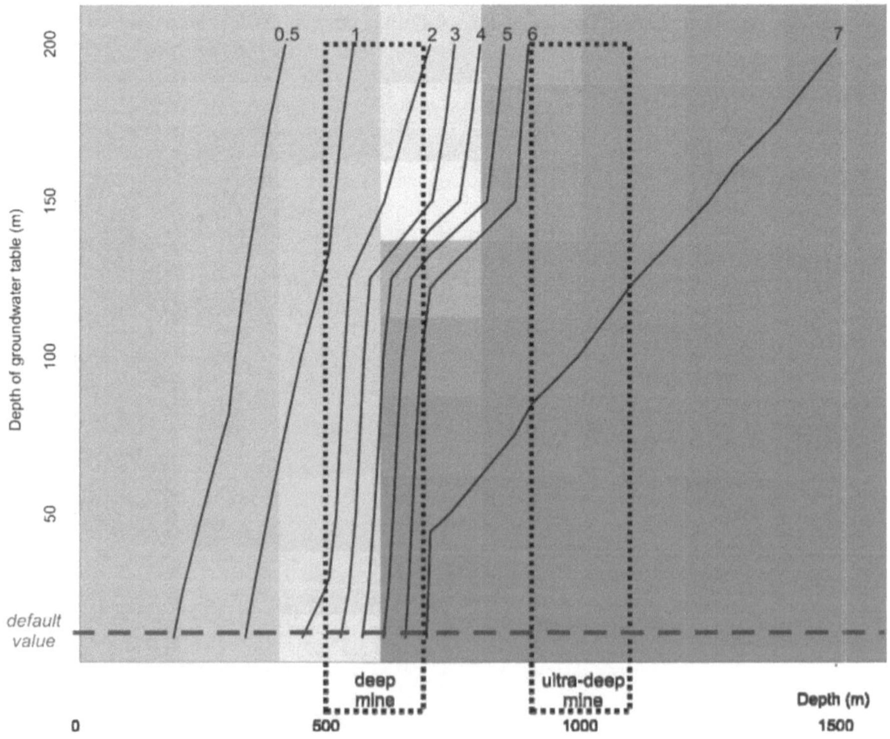

Figure 13. The capacity of a coal mine in function of its depth and the depth of the groundwater table, which also results in differences in the reservoir pressure. This graph demonstrates that the system is sensitive to even minor pressure changes in the reservoir, especially for deep mines, and that pressure gradients should be considered in detail.

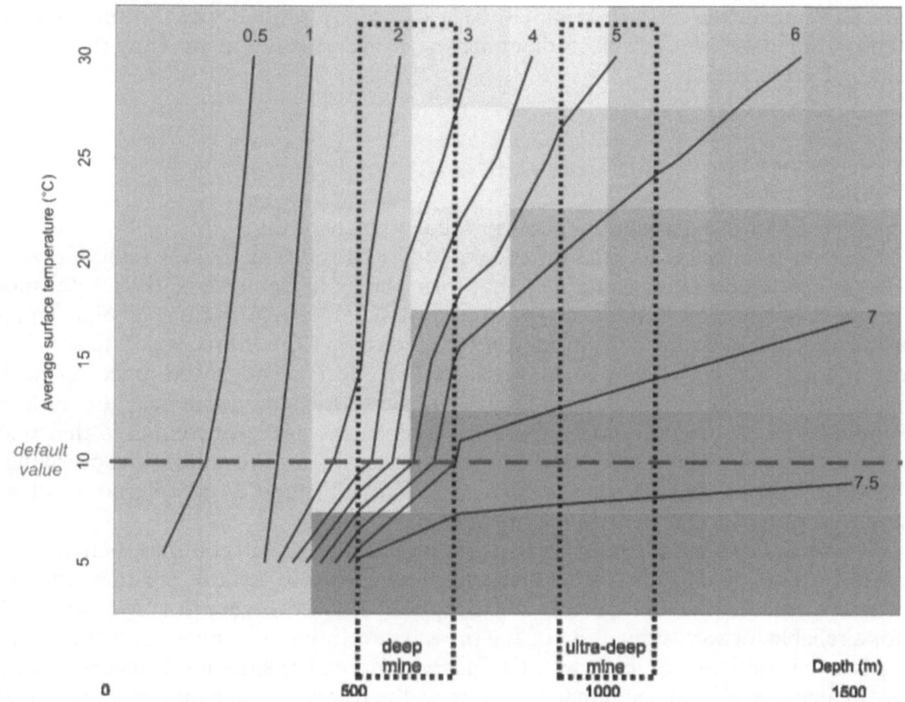

Figure 14. The capacity of a coal mine in function of its depth and the average surface temperature, which is a boundary condition for the calculation of the reservoir temperature. This graph shows that the capacity of deep mines is also sensitive to parameters affecting the reservoir temperature.

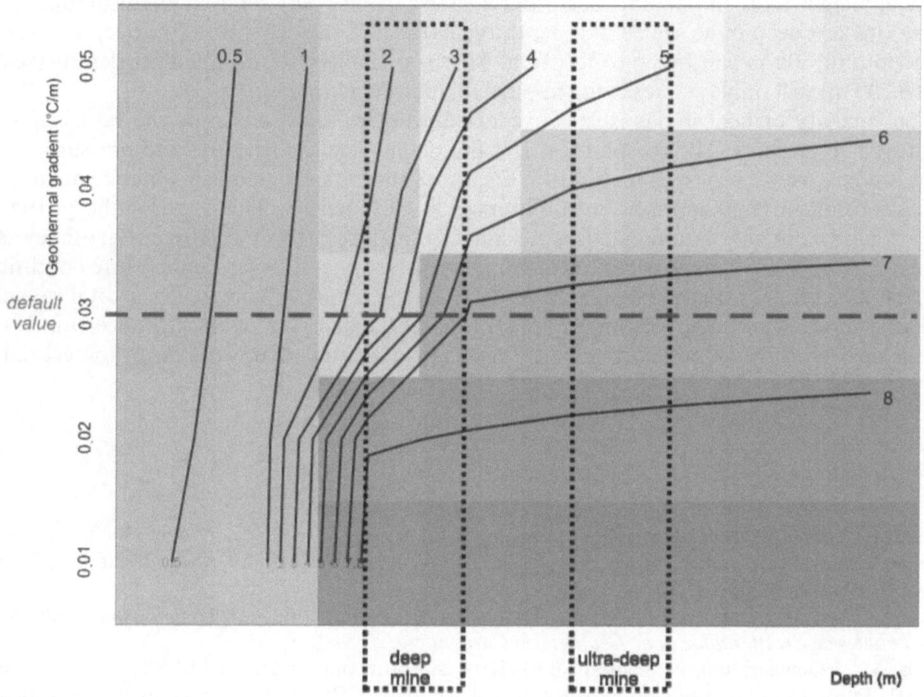

Figure 15. The capacity of a coal mine in function of its depth and the geothermal gradient. Even minor errors in the estimated geothermal gradient can lead to significant errors of the reservoir capacity.

changes also impact the ultra deep mines, be it in a less profound way. An uncertainty of only 0.005°C/m in the geothermal gradient may nevertheless result in errors on the estimated capacities of up to 20%.

8 CONCLUSIONS

Coal mines used for CO_2 storage will show particular behavior that is different from e.g. aquifer storage sites. This behavior can best understood by considering that the reservoir consists of different sub-reservoirs or containers that form the main elements of the jar pot model. In this model, abstraction is made of the complex design of a real-world coal mine, and particular focus is given on the three main storage modes for CO_2: in free space, in solution in formation water, and adsorbed on coal-in-place. The effects of advanced processes, such as the CO_2 pressure gradient and ground-level recovery, on the storage capacity are important and should be taken into account, also in early assessment studies. Essential to the concept of coal mine storage is that the combined top seal, consisting of the natural seal overlying the coal bearing sequence and the sealed shafts, is able to withstand a certain amount of overpressure with respect to the hydrostatic pressure.

This is essential because the pressure in a coal mine reservoir will continue to increase possibly decades after the injection of CO_2 was terminated before a stable pressure situation is reached, in which the pressure exerted on the top seal is always higher than hydrostatic. This calls for a reliable forward simulation of the pressure evolution in a mine reservoir. Monitoring needs to be continued at least until the maximum pressure situation is reached, because the risk of unexpected leakage remains as long as the reservoir pressure increases. It is only after the maximum pressure situation is reached than the mine will start evolving to a final and safe equilibrium situation.

Contrary to what can be expected, it are not the size of the residual space and the amount of coal-in-place that determine in the first place the capacity of a coal mine. The depth of the coal mine is more important. The difference in capacity between two identical mines, but where one is a deep mine sealed at a depth of 500 m and the other an ultra deep one sealed at a depth of 800 m can be up to fourfold. Mines at shallow to intermediate depths (sealed above 500 m) will only be interesting in particular circumstances.

The capacity of coal mines is sensitive to parameters that determine the reservoir temperature and pressure. The reason for this is that geological temperature and pressure conditions are relatively close to the saturation line or its subcritical extension where relative small changes result in relative important differences in CO_2 density. This is particularly true for deep mines where both liquid and gaseous like CO_2 may be present at different depth levels in the same reservoir. Small changes towards higher pressure and lower temperature conditions will result in a higher fraction of high-density CO_2, thus sharply increasing the total capacity, and vice versa. Also deep coal mines are still quite sensitive, and standard uncertainties on geothermal gradients alone correspond e.g. to uncertainties of up to 20% in the estimated storage capacity.

REFERENCES

Barta, Leslie & Daniel J. Bradley. 1985. Extension of the specific interaction model to include gas solubilities in high temperature brines. *Geochimica et Cosmochimica Acta* 49, no. 1 (January): 195–203. doi:10.1016/0016-7037(85)90204-2.

Berding, C.J.A. 1952. Le contrôle des coups d'eau dans les mines de houille. In *Congrès pour l'avancement des études de Stratigraphie et de Géologie de Carbonifère*.

Busch, A., Y. Gensterblum, N. Siemons, B.M. Krooss, Frank van Bergen, H.J.M. Pagnier & P. David. 2003. Investigation of preferential sorption behaviour of CO_2 and CH_4 on coals by high-pressure adsorption/desorption experiments with gas mixtures. In *Proceedings of the 2003 Coalbed Methane Symposium, Tuscalosa*, 14. http://recopol.nitg.tno.nl/downloads/Busch0350.pdf.

Christenson, Niels-Peter & Sam Holloway. 2003. *The GESTCO project: Summary report to the European Commission*. GESTCO report. D:\B04 Publications PSS-CCS\List002\CD_GESTCO\Report\documents\GESTCO1_summary_report.pdf.

Coppens, P.J. 1967. *Synthèse des propriétés chimiques et physiques des houilles. Les houilles belges*. Report of the Institut national de l'industries charbonnière. Inichar.

Devleeschouwer, X., P.-Y. Declercq, M. Dusar & A. Debien. 2008. Contrasting ground movements revealed by radar interferometry over abandoned coal mines (Campine, Belgium). In *Proceedings of Fringe 2007 Workshop*, 12.

Dreisbach F, R. Staudt & J.U. Keller. High Pressure Adsorption Data of Methane, Nitrogen, Carbon Dioxide and their Binary and Ternary Mixtures on Activated Carbon. Text. http://www.ingentaconnect.com/content/klu/adso/1999/00000005/00000003/00230997.

Duan, Zhenhao & Rui Sun. 2003. An improved model calculating CO_2 solubility in pure water and aqueous NaCl solutions from 273 to 533 K and from 0 to 2000 bar. *Chemical Geology* 193, no. 3–4 (February 14): 257–271. doi:10.1016/S0009-2541(02)00263-2.

Enick, Robert M. & Scott M. Klara. 1990. CO_2 Solubility-In-Water-and-Brine-Under-Reservlir-Conditions-PB-Taylor & Francis. *Chemical Engineering Communications*, no. 1: 23. doi:10.1080/00986449008940574.

European Parliament & Council of the European Union. 2009. *Directive 2009/31/EC of the European Parliament and of the Council of 23 April 2009 on the geological storage of carbon dioxide and amending Council Directive 85/337/EEC, European Parliament and Council Directives 2000/60/EC, 2001/80/EC, 2004/35/EC, 2006/12/EC, 2008/1/EC and Regulation (EC) No 1013/2006*. April 23. http://eur-lex.europa.eu/LexUriServ/LexUriServ.do?uri=OJ:L:2009:140:0114:0135:EN:PDF.

Evans, D.J. 2009a. *Underground gas storage: worldwide experiences and future development in the UK and Europe*. Geological Society.

Evans, D.J. 2009b. A review of underground fuel storage events and putting risk into perspective with other areas of the energy supply chain. *Geological Society, London, Special Publications* 313, no. 1 (January 1): 173 –216. doi:10.1144/SP313.12.

Evans, D.J. & R.A. Chadwick. 2009. Underground gas storage: An introduction and UK perspective. *Geological Society, London, Special Publications* 313, no. 1 (January 1): 1–11. doi:10.1144/SP313.1.

Evans, D.J. 2008. *An appraisal of underground gas storage technologies and incidents, for the development of risk assessment methodology*. HSE Research Report. British Geological Survey. http://www.hse.gov.uk/research/rrpdf/rr605.pdf.

Haan, M.F.E. 1998. Enhanced coallbed methane production by carbondioxide injection: theory, experiments and silumations. Master, TU-Delft, faculty of Earth sciences.

Holloway, S. 1996. *The underground disposal of carbon dioxide*. JOULLE II report.

Kunz, E. 2000. Entstehung und Vorkommen von Grubengas im Steinkohlenbergbau. In *Vortrag Obenhausener Grubengas-Tage 2000*.

Labasse, H. 1965. Les pressions de terrains dans les mines de houille; Les movements de terrain. *Extrait des annales des mines de Belgique* 5e, 6e, 7e, and 8e livraison.

Lord, A.S. 2009. *Overview of geologic storage of natural gas with an emphasis on assessing the feasibility of storing hydrogen*. Sandia Report. Sandia National Laboratories. http://prod.sandia.gov/techlib/access-control.cgi/2009/095878.pdf.

Lu, Ming. 2010. Rock engineering problems related to underground hydrocarbon storage. *Journal of Rock Mechanics and Geotechnical Engineering*: 289–297.

Malolepszy, Z. & S. Ostaficzuk. 1999. Geothermal potential of the Upper Silesian Coal Basin, Poland. *Bulletin d'Hydrogéologie* 17.

May, Franz, Michiel Dusar, Jean Barbier, Sam Holloway, Frank van Bergen, George Hatziyannis, Niels-Peter Christenson & Reidulf Boe. 2003. *Study area F: coal mines*. GESTCO report. D:\B04 Publications PSS-CCS\List002\CD_GESTCO\Report\documents\GESTCO17.pdf.

Metz, Bert, Ogunlade Davidson, Heleen de Coninck, Manuela Loos & Leo Meyer. 2005. *Carbon Dioxide Capture and Storage*. IPCC Special Report. Working Group III of the Intergovernmental Panel on Climate Change. D:\B04 Publications PSS-CCS\List004\SRCCS.pdf.

Peele, Robert. 1952. *Mining Engineer's Handbook Vol 1*. John Wiley & Sons.

Peila, D. & S. Pelissa. 1995. Civil reuses of underground mine openings: a summary of international experience. *Tunneling and Underground Space Technology* 10, no. 2: 179–191.

Piessens, Kris & Michiel Dusar. 2002. *Feasibility of CO_2-sequestration in coal mines*. GESTCO report.

Piessens, Kris, and Michiel Dusar. 2003a. Modeling vertical reservoir properties with CO_2-VR. In *2003 International Coalbed Methane Symposium*, 14 p. The University of Alabama, Tuscaloosa: Eds. R. Esposito, C.D. Haynes, J.R. Holland, E. Martin, S. Reeves, K.H. Schultz & R. Tinsley, May 5.

Piessens, Kris, and Michiel Dusar. 2003b. CO_2-geothermics in abandoned coal mines. In *2003 International Coalbed Methane Symposium*, 10 p. The University of Alabama, Tuscaloosa: Eds. R. Esposito, C.D. Haynes, J.R. Holland, E. Martin, S. Reeves, K.H. Schultz & R. Tinsley, May 5.

Piessens, Kris, and Michiel Dusar. 2003c. CO_2-sequestration in abandoned coal mines. In *2003 International Coalbed Methane Symposium*, 11 p. The University of Alabama, Tuscaloosa: Eds. R. Esposito, C.D. Haynes, J.R. Holland, E. Martin, S. Reeves, K.H. Schultz & R. Tinsley, May 5.

Piessens, Kris, and Michiel Dusar. 2004a. Feasibility of CO_2 sequestration in abandoned coal mines in Belgium. *Geologica Belgica* 4, no. 3–4: 165–180.

Piessens, Kris, and Michiel Dusar. 2004b. Integration of CO_2 sequestration and CO_2 geothermics in energy systems for abandoned coal mines. *Geologica Belgica* 7, no. 3–4: 165–180.

Pöttgens, J.J.E. & J.M. van Herk. 2000. Ground movements over abandoned coalmines in relation to rising mine waters in Limburg (The Netherlands). In *French Mining Congress*, 9.

Rumpf, B. & G. Maurer. 1993. An Experimental and Theoretical Investigation on the Solubility of Carbon Dioxide in Aqueous Solutions of Strong Electrolytes. *Berichte der Bunsengesellschaft für physikalische Chemie* 97: 85–97.

Schultz, K. 1998. *Gas storage at the abandoned Leyden Coal Mine near Denver, Colorado*. EPA report. Raven Ridge Resources, Inc. http://www.cmpdi.co.in/cbm/CBM%20papers/Mine%20Studies/stu001.pdf.

Shi, J.Q. & S. Durucan. 2005. CO_2 storage in caverns and mines. *Oil & Gas Science and Technology* 60, no. 3: 569–571.

Span, Roland & Wolfgang Wagner. 1996. A New Equation of State for Carbon Dioxide Covering the Fluid Region from the Triple-Point Temperature to 1100 K at Pressures up to 800 MPa. *Journal of Physical and Chemical Reference Data* 25, no. 6: 1509. doi:10.1063/1.555991.

Spencer, L.K., J. Bradshaw, B.E. Bradshaw, A.L. Lahtinen, and A. Chirinos. 2011. Regional Storage Ca-pacity Estimates: Prospectivity Not Statistics. *Energy Procedia* 4: 4857-4864.

Sroka, A., K. Tajdus & A. Preusse. 2011. Calculation of Subsidence for Room and Pillar and Longwall Panels. In *Underground Coal Operators' Conference*, 83–90. http://ro.uow.edu.au/coal/340.

Takenouchi, Sukune & George C. Kennedy. 1965. The solubility of carbon dioxide in nacl solutions at high temperatures and pressures. *American Journal of Science* 263, no. 5 (May 1): 445–454. doi:10.2475/ajs.263.5.445.

van Tongeren, P. & B. Laenen. 2001. *Residual space volumes in abandoned subsurface coalmines of the Campine Basin (northern Belgium)*. GESTCO report.

Van Der Vaart, Rick, Cindy Huiskes, Hans Bosch & Tom Reith. 2000. Single and Mixed Gas Adsorption Equilibria of Carbon Dioxide/Methane on Activated Carbon.˙ *Adsorption* 6, no. 4 (December 1): 311–323–323.

Van Krevelen, D.W. Geochemistry of coal. *Earth Sciences* 16: 183–247.

Van Krevelen, D.W. 1993. *Coal*. Elsevier Science Publishers.

CO_2 Storage in Carboniferous Formations and Abandoned Coal Mines – He, Ribeiro E Sousa, Elsworth & Vargas Jr. (eds)
© 2012 Taylor & Francis Group, London, ISBN 978-0-415-62079-6

Author index